ANALYSIS OF ORGANIC MICROPOLLUTANTS IN WATER

Commission of the European Communities

ANALYSIS OF ORGANIC MICROPOLLUTANTS IN WATER

Proceedings of the Third European Symposium held in Oslo, Norway, September 19–21, 1983

Edited by

G. ANGELETTI

Directorate-General for Science, Research and Development, Commission of the European Communities, Brussels

and

A. BJØRSETH

Central Institute for Industrial Research, Oslo

D. REIDEL PUBLISHING COMPANY

DORDRECHT / BOSTON / LANCASTER

Library of Congress Cataloging in Publication Data
Main entry under title:
Analysis of organic micropollutants in water.

'Proceedings of the Third European Symposium on 'Analysis of Organic Micropollutants in Water,' held in Oslo (Norway), from 19 to 21 September 1983'-Foreword.
   At head of title: Commission of the European Communities.
   Includes index.
   1. Organic water pollutants-Analysis-Congresses. I. Angeletti, G., 1943- . II. Bjørseth, Alf. III. European Symposium on "Analysis of Organic Micro-pollutants in Water" (3rd: 1983 : Oslo, Norway) IV. Commission of the European Communities.
TD427.O7A528   1984   628.1'68   83-26955
ISBN 90-277-1726-5

The Symposium was jointly organized by
- The Commission of the European Communities, Brussels
- The Central Institute for Industrial Research, Oslo
- The Royal Norwegian Council for Scientific and Industrial Research, Oslo

Publication arrangements by
Commission of the European Communities
Directorate-General Information Market and Innovation, Luxembourg

EUR 8518
© 1984 ECSC, EEC, EAEC, Brussels and Luxembourg

LEGAL NOTICE
Neither the Commission of the European Communities nor any person acting on behalf of the Commission is responsible for the use which might be made of the following information.

Published by D. Reidel Publishing Company
P.O. Box 17, 3300 AA Dordrecht, Holland

Sold and distributed in the U.S.A. and Canada
by Kluwer Academic Publishers,
190 Old Derby Street, Hingham, MA 02043, U.S.A.

In all other countries, sold and distributed
by Kluwer Academic Publishers Group,
P.O. Box 322, 3300 AH Dordrecht, Holland

All Rights Reserved
No part of the material protected by this copyright notice may be reproduced or utilized in any form or by any means, electronic or mechanical, including photocopying, recording or by any information storage and retrieval system, without written permission from the copyright owner.

Printed in The Netherlands

FOREWORD

In this book, the proceedings of the Third European Symposium on "Analysis of Organic Micropollutants in Water", held in Oslo (Norway), from 19 to 21 September 1983, are presented.

The symposium was organized within the framework of the Concerted Action COST 64b bis *, which has the same name and is included in the Third R & D Programme on Environment of the Commission of the European Communities - Indirect and Concerted Actions - 1981 to 1985.

The aim of the symposium was to review the progress and results achieved during the past two years, since the Second symposium, held in Killarney (Ireland) in November 1981.

The programme of the symposium consisted of review papers covering different areas related to the analysis of the organic pollutants in water, including sampling and sample treatment, gas and liquid chromatography, mass spectrometry and specific analytical problems for some types of compounds.

We think that the volume gives a rather complete overview of these activities in Europe. Moreover, the paper presented by D. Hunt reviews the development of the new technique mass spectrometry - mass spectrometry in the United States of America.

Some special sessions concerned the presentation of original contributions in form of poster, the extended versions of which are published in this volume.

The final session was dedicated to an overview of the future environmental problems, especially as far as regards the analytical chemistry, particular field of interest for this symposium and more in general the integration of the research programmes on the environment in the current and future policies of the European Communities.

We believe that the volume contributes to the solution of the problems posed by the organic pollutants in water and that the links between past and present research and the follow-up projects in this field are assured.

Brussels/Oslo, November 1983.

G. ANGELETTI   A. BJØRSETH

* COST 64b bis : Scientific and Technical Cooperation with the Non-Member Countries Norway, Portugal, Spain, Sweden, Switzerland and Yugoslavia in the field of "Analysis of Organic Micropollutants in Water".

CONTENTS

Foreword   v

SESSION I - SAMPLING

Introductory remarks on sampling
    F.J.J. BRINKMANN, National Institute for Water
    Supply, Leidschendam   3

Zur Entnahme von Grundwasserproben
    V. NEUMAYR, Institut für Wasser-, Boden- und
    Lufthygiene des Bundesgesundheitsamtes, Berlin   5

Sampling of surface waters for tracing macro- and micro-pollutants
    W. SCHMITZ, Landesanstalt für Umweltschutz Baden-Württemberg, Karlsruhe   15

Techniques d'échantillonnage pour l'analyse des traces organiques
    A. BRUCHET, L. COGNET, J. MALLEVIALLE, Laboratoire Central-Lyonnaise des Eaux et de l'Eclairage, Le Pecq   27

Extraction of sediments using ultrasonic and steam distillation treatments
    C. O'DONNELL, Water Resources Division, An Foras Forbartha, Dublin   36

SESSION II - ANALYSIS

High resolution gas chromatography in water analysis: 12 years of development
    K. GROB, Institute for Water Resources and Water Pollution Control (EAWAG), Dübendorf   43

Analysis of organics in the environment by functional group using a triple quadrupole mass spectrometer
    D.F. HUNT, J. SHABANOWITZ, T.M. HARVEY, Department of Chemistry, University of Virginia, Charlottesville, Virginia   53

Recent progress in LC/MS
  D.E. GAMES, M.A. McDOWALL, Department of Chemistry,
  University College of Cardiff,
  M. GLENYS FOSTER and O. MERESZ, Ontario Ministry
  of the Environment ... 68

Computerized data handling in gas chromatography/mass
spectrometry
  P. GROLL, Institut für Heisse Chemie, Kernforschungszentrum Karlsruhe ... 77

New developments in selective detection in capillary gas
chromatography
  P. SANDRA, Laboratory of Organic Chemistry, State
  University of Ghent ... 84

Determination of organic water pollutants by the combined
use of high-performance liquid chromatography and high-
resolution gas chromatography
  W. GIGER, M. AHEL, C. SCHAFFNER, Swiss Federal
  Institute for Water Resources and Water Pollution
  Control (EAWAG) and Swiss Federal Institute of
  Technology, Dübendorf ... 91

Analysis of organic micropollutants by HPLC
  J.C. KRAAK, Laboratory for Analytical Chemistry,
  University of Amsterdam ... 110

Identification of non-volatile organics in water using
field desorption mass spectrometry and high performance
liquid chromatography
  C.D. WATTS, B. CRATHORNE, M. FIELDING, C.P. STEEL,
  Water Research Centre, Marlow, Buckinghamshire ... 120

Mass spectrometric identification of surfactants
  K. LEVSEN, E. SCHNEIDER, F.W. ROELLGEN, P. DAEHLING
  Department of Physical Chemistry, The University of Bonn,
  A.J.H. BOERBOOM, P.G. KISTEMAKER, S.A. McLUCKEY, FOM
  Institute for Atomic and Molecular Physics, Amsterdam ... 132

Characterisation of non-volatile organics from water by
pyrolysis-gaschromatography-mass spectrometry
  L. STIEGLITZ, W. ROTH, Nuclear Research Centre,
  Karlsruhe ... 141

Automated GC/MS for analysis of aromatic hydrocarbons in
the marine environment
  S. SPORSTØL, K. URDAL, Central Institute for Industrial
  Research, Oslo ... 147

Analysis of sulphonic acids and other ionic organic compounds
using reversed-phase HPLC
  K.J. CONNOR, A. WAGGOTT, Water Research Centre, Stevenage
  Laboratory ... 153

Closed loop stripping and cryogenic on-column focusing of trihalomethanes and related compounds for capillary gas chromatography
    M. TERMONIA, J. WALRAVENS, P. DOURTE, X. MONSEUR, Institut de Recherches Chimiques, Secrétariat d'Etat à l'Agriculture, Tervuren   162

Observations with cold on-column injection
    P. SANDRA, Laboratory of Organic Chemistry, State University of Ghent   165

Polymethylphenylsilicones in fused silica capillary columns for gas chromatography
    M. VERZELE, F. DAVID, P. SANDRA, Laboratory of Organic Chemistry, State University of Ghent   169

Collaborative study on the mass spectrometric quantitative determination of toluene using isotope dilution technique
    A. BUECHERT, National Food Institute, Søborg   173

Determination of volatile organic substances in water by GC/SIM results of an investigation on the rivers Rhine and Main
    F. KARRENBROCK, K. HABERER, ESWE-Institut für Wasserforschung und Wassertechnologie GmbH, Wiesbaden   179

SESSION III - SPECIFIC ANALYTICAL PROBLEMS

Analytische Bestimmung organischer Halogenverbindungen
    M. SCHNITZLER, W. KUEHN, DVGW-Forschungsstelle am Engler-Bunte-Institut, Universität Karlsruhe   191

Determination of organic sulphur, organic phosphorus and organic nitrogen
    G. VEENENDAAL, KIWA Ltd., The Netherlands Waterworks' Testing and Research Institute Ltd., Rijswijk   205

Results from Round Robin Exercises of phenolic compounds within the COST 64b bis project
    L. RENBERG, National Swedish Environment Protection Board, Special Analytical Laboratory, Stockholm   214

Instrumental analysis of petroleum hydrocarbons
    R.G. LICHTENTHALER, Central Institute for Industrial Research, Oslo   225

Screening of pollution from a former municipal waste dump
at the bank of a Danish inlet
    U. LUND, VKI, Water Quality Institute, Hørsholm and
    Danish Civil Defence, Analytical Chemical Laboratory,
    Copenhagen
    A. KJAER SØRENSEN, Danish Civil Defence, Analytical
    Chemical Laboratory, Copenhagen
    J.A. FARR, VKI, Water Quality Institute, Hørsholm   234

The impact on the ecology of polychlorinated phenols and
other organics dumped at the bank of a small marine inlet
    J. FOLKE, J. BIRKLUND, VKI, Water Quality Institute,
    Hørsholm
    A.KJAER SØRENSEN, Danish Civil Defence, Analytical
    Chemical Laboratory, Copenhagen
    U. LUND, VKI, Water Quality Institute, Hørsholm and
    Danish Civil Defence, Analytical Chemical Laboratory,
    Copenhagen   242

Organic micropollutants from the pulp industry: analysis
of lake Päijänne water
    E. TOIVANEN, I. KUNINGAS, S. LAINE, Helsinki City
    Waterworks, Water Examination Bureau, Helsinki   255

Ground water pollution by organic solvents and their
microbial degradation products
    D. BOTTA, L. CASTELLANI PIRRI, E. MANTICA, Dept.
    of Industrial Chemistry and Chemical Engineering,
    Politecnico di Milano   261

Comparison of three methods for organic halogen determination in industrial effluents
    G.E. CARLBERG, A. KRINGSTAD, Central Institute for
    Industrial Research, Oslo   276

Organic micropollutants in surface waters of the Glatt
Valley, Switzerland
    M. AHEL, W. GIGER, E. MOLNAR-KUBICA, C. SCHAFFNER,
    Swiss Federal Institute for Water Resources and
    Water Pollution Control (EAWAG), Dübendorf   280

Species and persistence of pollutants in the pond water
from an orchard area treated with organophosphorus
pesticides
    V. DREVENKAR, Z. FROEBE, B. ŠTENGL, B. TKALČEVIĆ
    Institute for Medical Research and Occupational
    Health, Zagreb   289

Occurrence and origin of brominated phenols in Barcelona's water supply
    J. RIVERA, Institut de Quimica Bio-Orgànica, Barcelona
    F. VENTURA, Sociedad General de Aguas de Barcelona   294

The relationship between the concentration of organic matter in natural waters and the production of lipophilic volatile organohalogen compounds during their chlorination
    M. PICER, V. HOCENSKI, N. PICER, Center for Marine Research Zagreb, "Rudjer Bošković" Institute, Zagreb   301

SESSION IV - FUTURE ENVIRONMENTAL PROBLEMS

Standard setting principles
    R.F. PACKHAM, Water Research Centre, Marlow, Buckinghamshire   309

Integration of the environment research action programme into the framework programme for community scientific and technical activities 1984-1987
    A. KLOSE, G. ANGELETTI, C. WHITE, Commission of the European Communities, Brussels   320

COST 64b bis project - Now and the future
    A. BJØRSETH, Central Institute for Industrial Research, Oslo
    H. OTT, Commission of the European Communities, Brussels   328

LIST OF PARTICIPANTS   333

INDEX OF AUTHORS   339

SESSION I - SAMPLING

Chairman: F.J. BRINKMANN

Introductory remarks on sampling

Zur Entnahme von Grundwasserproben

Sampling of surface waters for tracing macro- and micro-pollutants

Techniques d'échantillonnage pour l'analyse des traces organiques

Extraction of sediments using ultrasonic and steam distillation treatments

INTRODUCTORY REMARKS ON SAMPLING

F.J.J. BRINKMANN
National Institute for Water Supply
Leidschendam, The Netherlands

Before giving the floor to the speakers of the sampling session I shall use this opportunity to describe the activities and results of working party I. This will also serve as a state-of-the-art report on sampling.

WP I has been concerned with sampling of organic micropollutants in water as well as in matrices related with water. WP I will finish its work at the end of this year but it is clear to the members of the working party – and as I hope also to the participants of this symposium – that sampling will be a permanent item in discussions on environmental water-chemistry.

Highlights of WP I have been:

- the working party meeting in Gothenburg, where sample treatment was the main topic,
- the European Scientific Symposium in Berlin and especially the session devoted to sampling and the animated discussions which followed,
- the European Symposium in Killarney: again a session was devoted to sampling, and was also successful because of the input of Dr. Josefsson,
- the working party meeting in Rome: sampling procedures, installations, apparatus and devices, as welll as choice of sampling points were the main items and the discussions covering surface water, sediments, groundwater and rainwater resulted in a long and valuable report,
- the working party meeting in Luxembourg on sampling of drinking water.

I am sure that this session, with its introduction by very experienced colleagues, can be added to this list of highlights.

In the literature as well as in practice most of the attention on sampling has been paid to surface water. Nevertheless this has not yet resulted in general prescriptions covering all possible situations.

The local situation and the purpose of the sampling affect the precise choice of sampling points, the time of sampling and, as concerns the method of sampling, the choice between composite sampling and grab-sampling. As well as discontinuities in water quality caused by stream patterns, sampling in relation to organic compounds can be hampered by the presence of suspended solids. There is an enormous lack of knowledge concerning sampling of sediments.

Drinking water is in general free from suspended solids (< 1 mg/l) and therefore easier to sample. In special cases however, sampling of drinking water can be hampered by bitumin particles (bitumin-coated iron pipes), by lead carbonate/hydroxide particles (lead pipes), by other hydroxides/ carbonates (iron and calcium), and even by the presence of animals such as worms. Flow patterns in the pipes also play a role, and the problem of stagnant versus flushing water should be mentioned. The results of the Luxembourg meeting on sampling of drinking water have not yet been reported, but in advance of this report it can be said that in the opinion of the working party the sampling programmmes listed in the EC directive and the WHO standards/guidelines are adequate but should be considered as a minimum. Furthermore, the listed parameters will be relevant in normal cases, but in special cases the programme should be extended to the extra parameters expected to be present.

Research on organic components in rainwater is still in its infancy. This is primarily because of the extremely low concentrations shown for PAH's, for lower aromates, and for halogenated hydrocarbons, which together with acids are the most frequently found groups of compounds. The low concentrations in rainwater also imply long collection-times giving rise to losses by adsorption, evaporation and transformation. The problem of wet-only versus dry deposition (or total deposition) is great. On the other hand a well-designed sampling programme on deposition will even give information about the pathway of organic micropollutants in the air-compartment.

When it concerns organics, sampling of groundwater gains extra dimensions. The Rome meeting has resulted in a report containing much information, but to prevent misunderstanding it has to be realized that sampling of groundwater with the aim of measuring organic micropollutants is still a new field of environmental research. Particularly as it concerns the sampling of anaerobic water much less information is available.

ZUR ENTNAHME VON GRUNDWASSERPROBEN

V. NEUMAYR
Institut für Wasser-, Boden- und Lufthygiene des
Bundesgesundheitsamtes
1000 Berlin, BRD

Summary

Depending on questions and aims leading to groundwater-quality-classification there is allways need of a change for sensible sample-techniques. In praxis, however, there is seldom a possibility to use specific sampling equipment for the most different criteria of analysis. As a matter of fact - negotiating some exceptions - the equipment of wells with well polymerized PVC-pipes and well screens leads to considerable results in common praxis. Furthermore a prepared and clean-washed screening area of silicate gravel as well as a sufficient pipe-diameter, allowing to install a strong pump of the minimum size of 4 inches stands the test. By means of these sampling wells there is normally the possibility to guarantee analysis of inorganic chemical, organic chemical, physical and microbiological parameters with respect to the assessment of hydraulic, hydrogeological and health aspects.
A quality survey of organic pollutants in groundwater is possible by means of a direct application of XAD or activated carbon in a sampling-well. The determination of concentration-developements concerning volatile organic substances in groundwater is practicable by contuous sampling in gas-tight glass-syringes and an infusion control-system.

Zusammenfassung

Je nach Fragestellung und Absicht einer Grundwasserqualitätsbewertung ändert sich die Methodik einer sinnvollen Probenahmetechnik von Grundwässern. In der Praxis ist es jedoch kaum möglich, für die unterschiedlichsten Untersuchungskriterien jeweils spezifische Probeentnahmeeinrichtungen vorzusehen. Von wenigen Ausnahmen abgesehen, hat sich die Einrichtung von Grundwassermeßstellen mittels gut polymerisierter PVC-Rohre und -filter, einer gewaschenen Quarzkiesschüttung im Filterbereich und eines Innendurchmessers des Meßstellenrohres, der das Abpumpen mit leistungsfähigen 4 Zoll Tauchmotorpumpen erlaubt, bewährt. In der Regel lassen sich über diese Meßstellen anorganisch-chemische, organisch-chemische, physikalische und mikrobiologische Parameter ermitteln, um sie nach hydraulischen, hydrogeologischen und gesundheitlichen Aspekten bewerten zu können.
Eine qualitative Überwachung von organischen Verunreinigungen in Grundwässern ist durch den direkten Einsatz von XAD oder Aktivkohle in einer Grundwassermeßstelle

möglich. Zur Überwachung von Konzentrationsgängen flüchtiger und leichtsiedender organischer Inhaltsstoffe in Grundwässern kann eine kontinuierliche Probenahme über gasdichte Glasspritzen und ein Infusionssteuergerät durchgeführt werden.

1. Einleitung

Aufgrund der zunehmend intensiven Nutzung und Belastung von Grundwässern kommen Untersuchungen der physikalischen, chemischen und mikrobiologischen Beschaffenheit des Grundwassers eine wachsende Bedeutung zu. Neben direkten und örtlich begrenzten Boden- und Grundwasserschadensfällen sind in steigendem Maße auch diffuse Beeinflussungen durch die Industrialisierung, landwirtschaftliche Nutzung, Verkehr, Abwasser und Abfall erkennbar. Diesbezügliche Veränderungen der Wasserqualität werden durch den Vergleich von Wasseranalysen offenkund. Da somit der Vergleichbarkeit von Analysenergebnissen eine besondere Bedeutung zukommt, ist es wichtig, die kritischen Untersuchungsschritte - nämlich die Probenahme und die analytische Methode - zu optimieren und zu standardisieren. Hinsichtlich einer geeigneten Probenahme von Grundwässern gewährt nur eine nach Ausbau, Material und Örtlichkeit sinnvoll angelegte Probenahmestelle verbunden mit einer fachgerechten Probenahmetechnik eine repräsentative Wasserprobe eines spezifischen Grundwasserleiters.

2. Probeentnahmestellen

Die Bewertung von Grundwasserqualitäten erfolgt in der Regel aus der Sicht
- der Hydraulik und Hydrologie,
- der Hydrogeologie und Geochemie,
- der Trophologie und Toxikologie und
- der Stoffgenese (geogen - anthropogen).

Somit sollte unter diesen Gesichtspunkten die Einflußgröße einer Entnahmestelle nach Art und Ausbau so gering wie möglich sein. Da ein Wasser nach Verlassen eines Grundwasserleiters unter Umständen sehr rasch seine Beschaffenheit verändert, muß die eigentliche Probenahme zeitlich und räumlich unmittelbar erfolgen. Darüberhinaus ist darauf zu achten, daß die Einrichtung der Probenahmestelle keinen entscheidenden Eingriff in das natürliche System eines Grundwasserleiters darstellt.

Grundwasserentnahmen sind möglich aus:
- Förderbrunnen,
- Grundwassermeßstellen,
- Quellen,
- wasserführenden Stollen und
- Grundwasserblänken.

Während Grundwasserblänken durch ihre limnologische Überprägung kaum repräsentative Proben des entsprechenden Grundwasserleiters liefern können und Stollenwasser durch den Kontakt mit der Umgebungsluft schnell seine physikalischen, chemischen und mikrobiologischen Eigenschaften ändern kann, bleiben repräsentative Probenahmen im wesentlichen nur den Förderbrunnen, Grundwassermeßstellen und Quellen vorbehalten.

2.1 Quellen

Geeignete Probenahmen sind bei frei schüttenden Quellen in der Regel ohne größeren Aufwand durchführbar. Meist genügt eine Probenahme durch Schöpfen direkt an der Stelle, wo das Grundwasser das Gestein oder Sediment verläßt. Es muß jedoch sichergestellt sein, daß keinerlei Oberflächen-, Hang- oder Niederschlagswasser in das Grundwasser gelangen kann.

2.2 Förderbrunnen

Grundwassergewinnung und künstliche Absenkung des natürlichen Grundwasserspiegels stellen einen Eingriff in den Grundwasserleiter dar, so daß eine entsprechende Bewertung von analytischen Untersuchungsparametern immer unter dem Aspekt der jeweiligen Fördermenge vorgenommen werden muß. Da Förderbrunnen meist über elektrische Tauchpumpen große Förderleistungen aufbieten, stellt sich nur eine geringe Aufenthalts- und Kontaktzeit des Grundwassers mit dem Brunnenfilter und dem Pumprohrleitungsmaterial ein. Somit wird die Probe meist kaum oder nur geringfügig durch den Ausbaustoff des Brunnens beeinträchtigt. Bei entsprechenden Förderleistungen ist deshalb durchaus eine Schwermetallanalyse des Grundwassers möglich, obwohl Brunnenrohr und -filter aus Metall gefertigt sind, bzw. eine Analyse auf organische Inhaltsstoffe, obwohl Kunststoffverrohrungen verwendet worden sind.

2.3 Grundwassermeßstellen

Grundwassermeßstellen dienen in der Regel der Erfassung von Grundwasserständen und werden meist nur sporadisch abgepumpt. Lange Standzeiten können jedoch den Umgebungsbereich der Brunnenfilter beeinträchtigen (Versandung, Verokerung, Aufbau eines mikrobiologischen Rasens etc.) und verändern somit auch das im Brunnen befindliche Wasser. Überdies ist eine Beeinflussung des Wassers durch das Material des Brunnenfilters sowie der Verrohrung möglich. Metallrohre und -filter verändern die Metall- und Schwermetallgehalte des Brunnenwassers; Kunststoffrohre können organische Substanzen an das Wasser abgeben oder aber auch im Grundwasser vorhandene, organische Inhaltsstoffe adsorptiv binden. Im Normalfall ist es jedoch trotz der Materialprobleme kaum vertretbar, Grundwassermeßstellen analog anstehender analytischer Fragestellungen in vielfältigen Ausbauvarianten einzurichten. Vielmehr ist es sinnvoll, ein Material zu finden, daß auch unter Einbezug von Kostenüberlegungen, eine geringst mögliche Beeinträchtigung des Wassers gewährleistet. Zahlreiche Untersuchungen haben gezeigt, daß neben dem sehr teuern Einsatz von Edelstahl, gut polymerisiertes PVC eine zufriedenstellende Kompromißlösung darstellt.

Für die zur Einrichtung einer Grundwassermeßstelle notwendigen Bohrarbeiten stehen vereinfacht Trocken- und Spülbohrverfahren zur Verfügung. Da bei Spühlbohrverfahren der Eintrag von Fremdstoffen in den Grundwasserleiter (Spühlwasser,

Bohröle etc.) unumgänglich ist, sollte das Trockenbohrverfahren bevorzugt werden. Während der Bereich des zu beprobenden Grundwasserleiters zwischen Bohrlochwand und Verrohrung mit einer Schüttung gewaschenen Quarzkieses (ähnlich der Korngrößenverteilung des umgebenden Gesteins) gefüllt wird, sollten die hangenden und liegenden Schichten mit dichtendem Material abgeschlossen werden. Bei unterschiedlichen Grundwasserstockwerken bietet sich eine getrennte Verfilterung zur individuellen Probenahme an.

In den Grundwassermeßstellen kann die eigentliche Probenahme erst nach eingehender Reinigung des Brunnenfilters erfolgen. Dies ist in der Regel nur durch Abpumpen des Brunnens und gleichzeitiger Spühlung des Filters möglich. Eine effektive Reinigung stellt sich nur bei entsprechend erhöhten Durchflußmengen des Grundwassers durch den Brunnenfilter ein. Die dazu notwendigen Pumpleistungen werden durch Pumpen ab etwa 4 Zoll Durchmesser aufgebracht. Aus diesem Grunde ist darauf zu achten, daß die Rohrduchmesser der Grundwassermeßstellen das Einbringen solcher Pumpen gewährleisten.

3. Probenahme

Üblicherweise können Grundwasserprobenahmen mit nachstehenden Probenahmegeräten durchgeführt werden:
- Schöpfgeräte,
- Kolbenprober,
- Saugpumpe,
- Tauchschwingkolbenpumpe,
- Hubkolbenpumpe,
- Tiefsauger,
- Impulspumpe,
- Mammutpumpe.

Außer in Quellen und in wasserführenden Stollen sollten Schöpfgeräte nur in Ausnahmefällen benutzt werden. Man unterscheidet einfache Schöpfgeräte, die nur in der Lage sind, Wasser von der Wasseroberfläche aufzunehmen und Schöpfer, wie der "RUTTNER-Schöpfer", der Wasserproben aus bestimmten Tiefen entnehmen kann. Für große Tiefen (ca. 500 - 1000 m) sind spezielle Geräte entwickelt worden (1,2).

Bis in eine Wassertiefe von etwa 7 bis 8 m können der Kolbenprober, eine mit der Hand bedienbare Kolbenpumpe meist aus Glas (Förderleistung: ca. 0,5 l/min), und die Saugpumpe, die als Kreiselpumpe durch Elektro- oder Benzinmotoren betrieben wird (Förderleistung: 1 bis 2 l/sec) eingesetzt werden.

Die Tauchschwingkolbenpumpe ist eine Unterwasserpumpe mit geringer Förderleistung und kleinem Durchmesser. Diese elektrisch betriebene Pumpe eignet sich vor allem für Grundwassermeßstellen mit geringem Rohrdurchmesser (ab 5 cm) und Förderhöhen bis etwa 50 m. Die gängigste Unterwasserpumpe zur Gewährleistung großer Fördermengen und Bewältigung großer Wassertiefen ist die Tauchmotorpumpe. Zum Antrieb dieser Pumpe ist ein Stromerzeuger erforderlich.

Wenig verbreitete Pumpen sind die Hubkolbenpumpe, die über Elektro- oder Benzinmotor angetrieben, tiefe Wasserstände bewältigt und selbst in engen Rohren eingesetzt werden kann, der Tiefsauger, der nach dem Wasserstrahlprinzip arbeitet, und bis etwa 30 m Wassertiefe eingesetzt werden kann, die Impulspumpe, die über Druckluftimpulse Wasser bis aus einer Tiefe von ca. 50 m fördern kann und die Mammutpumpe, die nach dem Luftheberverfahren arbeitet, jedoch bei einer Analyse auf gelöste gasförmige Stoffe oder leichtflüchtige Substanzen nicht eingesetzt werden kann.

Neben der Beprobung von Grundwässern über die gesamte Filterstrecke eines Brunnens, besteht auch die Möglichkeit, Wasserproben aus bestimmten Filterrohrabschnitten zu entnehmen. Hierzu muß oberhalb und unterhalb des gewünschten Entnahmepunktes das Brunnenrohr abgepackert werden (3).

Um eine repräsentative Wasserprobe eines bestimmten Grundwasserleiters zu gewinnen, bedarf es einer Vorreinigung des Brunnens, insbesondere des Brunnenfilters. Dies ist in der Regel durch Abpumpen mit erhöhter Förderleistung (0,5 bis 3 l/s) über eine hinreichende Zeit erreichbar. Als minimal notwendige Fördermenge ist hierbei das 2 1/2-fache Austauschen der Wassersäule des Brunnenvolumens, einschließlich des Filtervolumens anzusehen. Repräsentative Grundwasserproben können beispielsweise nach folgender methodischen Vorgehensweise gezogen werden:
- die zu beprobende Grundwassermeßstelle wird am Vortag der eigentlichen Probenahme je nach Ausbaugröße des Brunnens 2 bis 4 Stunden (Förderleistung: 0,5 bis 3 l/sec) abgepumpt (Reinigungsphase);
- etwa 24 Stunden nach der Reinigungsphase wird der Brunnen wiederholt abgepumpt (0,5 bis 3 l/sec) bis das Wasservolumen des Brunnens, einschließlich Filtervolumen, 2 1/2-fach ausgetauscht ist. Erst dann erfolgt die Probenahme mit stark gedrosselter Förderleistung, um eine blasenfreie Überführung des Wassers in die Probenahmeflaschen zu gewährleisten (Probenahmephase).

Die Art des Probenahmegefäßes, Maßnahmen zur Konservierung und der Probentransport orientieren sich an der Zielsetzung der jeweiligen Analysen. Es muß jedoch darauf hingewiesen werden, daß eine Reihe physikalischer und chemischer Parameter bereits am Probenahmeort bestimmt werden müssen. Dazu gehören beispielsweise die Temperaturmessungen von Wasser und Luft, der pH-Wert, die Leitfähigkeit, Basenkapazität, der Sauerstoffgehalt, das Redoxpotential und sämtliche Sinnesprüfungen.

4. Qualitative Überwachung organischer Wasserinhaltsstoffe im Grundwasser durch direkten Einsatz von Aktivkohle und XAD

Eine sehr einfache Methode zur qualitativen Überwachung der Grundwasserqualität ist durch den direkten Einsatz von Aktivkohle und XAD-Harz möglich. An einer Metallschnur befestigt, werden Aktivkohle oder XAD-Harz in einem Glasfrittengefäß (siehe Abb. 1) mit guter Wasserleitfähigkeit in den Filterbereich von Grundwassermeßstellen eingehängt. Hier kommt es über die Dauer von etwa einer Woche bis zu einem Monat zu einer Anreicherung organischer Wasserinhaltsstoffe. Die Aktivkohleprobe wird über Strip-Verfahren nach GROB behandelt, gaschromatographisch-massenspektrometrisch untersucht, wobei eine Reihe organischer Substanzen bis zum Siedepunkt 350°C und bis zu einem Molekulargewicht von 800 erfaßt werden. Die XAD-Probe wird unter angelegtem Stickstoffdruck mit Äther extrahiert, eingeengt, zum Teil derivatisiert und ebenfalls gaschromatographisch-massenspektrometrisch untersucht.

In der Tabelle 1 sind die Ergebnisse unterschiedlicher Probenahmetechniken - anhand einer Stichprobe (10 Liter Wasser über XAD-Säule und mit Äther extrahiert) nach Abpumpen, direkter Einsatz von Aktivkohle und direkter Einsatz von XAD aufgeführt. Die hierbei beprobte Grundwassermeßstelle liegt unmittelbar im Grundwasserunterstrom eines Regenüberlaufes im Abwasserkanalsystem einer Stadt. Das zum Versickern gelangende Wasser enthält häusliches und industrielles Abwasser, Drainwasser aus landwirtschaftlich genutzten Gebieten und Niederschlagswasser. Wie die Ergebnisse der Tabelle 1 zeigen, ist der Vorteil des kontinuierlichen Anreicherns an Aktivkohle oder XAD im Brunnen selbst gegenüber einer einmaligen Stichprobe, im zusätzlichen Erfassen von Substanzen zu sehen, die nur episodisch oder unter starken Konzentrationsschwankungen in das Grundwasser gelangen. Während der Einsatz der XAD-Anreicherung durch die mögliche Derivatisierung das größte Spektrum erfaßbarer organischer Inhaltsstoffe liefert, eignet sich die Aktivkohleanreicherung hervorragend zum Nachweis flüchtiger und leichtsiedender organischer Substanzen.

5. Methode einer kontinuierlichen Probenahme zur Erfassung flüchtiger, leichtsiedender organischer Wasserinhaltsstoffe

Die Probenahme von Wässern im Hinblick auf die Analyse flüchtiger, leichtsiedender Substanzen stellt generell ein Problem dar. Die Durchführung kontinuierlicher Probenahmen und das Erstellen von repräsentativen Mischproben ist mit herkömmlichen Probenahmegeräten diesbezüglich kaum durchzuführen. Insbesondere Grenzwertfestlegungen, Richtlinien oder beispielsweise die Richtwertfestlegung des Bundesgesundheitsamtes (BRD), die für die Summe chlorierter Lösemittel im Trinkwasser ein Jahresmittel von 25 µg/l vorsieht (4), lassen jedoch solche Probenahmetechniken notwendig werden.

Gasdichte kontinuierliche Probenahmetechniken zur Ermittlung von 1- und 2-Stunden-Mischproben oder Tages- bzw-Wochenmischproben sind relativ einfach über Glasspritzen in Verbindung mit einem Infusionssteuergerät durchzuführen. Ein Infusionssteuergerät, das durch Umpolung des regelbaren Antriebmotors eine Reihe von gasdichten Glasspritzen kontinuierlich befüllt, liefert unbeeinflußte Mischproben. Bei dem, im Bild 1 dargestellten Gerät, wird das Wasser über eine Edelstahlkapillare angesaugt und gleichmäßig in 6 einzelne Glasspritzen (50 ml) verteilt. Die Füllgeschwindigkeit des Steuergerätes ist zwischen 7,5 µl/h bis 300 ml/h regelbar, so daß Mischproben von 10 min bis in eine Größenordnung von einigen Wochen gezogen werden können.

Als praktisches Anwendungsbeispiel soll die kontinuierliche Überwachung eines mit chlorierten Lösemitteln belasteter Förderbrunnen und des abgegebenen Trinkwassers des entsprechenden Wasserwerkes vorgestellt werden. In der Abbildung 2 ist der Konzentrationsverlauf von Tetrachlorethylen und Trichlorethylen in einem Förderbrunnen über 70 Tage dargestellt. Die Probenahme erfolgte im Wechsel von 3- bzw 4-Tagesmischproben. Es ist deutlich zu erkennen, daß die Konzentrationsentwicklung nach Einschalten des Brunnens zunächst über zwei Wochen anstieg, um dann allmählich abzusinken. In der Abbildung 3 ist das

Einschalten des belasteten Förderbrunnens durch den deutlichen Konzentrationsanstieg von etwa 5 µg/l Summe chlorierter Lösemittel auf nahezu 40 µg/l im abgegebenen Trinkwasser des Wasserwerkes erkennbar. Die rapide Rückentwicklung der Konzentrationen in den Folgetagen ist auf die Anwendung eines Aufbereitungsschrittes durch intensives Belüften zurückzuführen. Der Einsatz einer lückenlosen kontinuierlichen Probenahme überwacht in diesem Falle mit Erfolg die Belastungssituation des kontaminierten Brunnens, sowie die Abgabe eines einwandfreien Trinkwassers an die Bevölkerung.

6. Literatur

(1) REPSOLD,H. & FRIEDRICH,H.: Über eine elektromechanische Sonde zur Wasserprobeentnahme in Bohrlöchern.-
    Geol. Jb.,E 9: 35 - 39, Hannover (1976)
(2) HOFREITER,G.: Ein Gerät zur Entnahme von Grundwasserproben.
    gwf - Wasser/Abwasser, 118: 384 - 385, München (1977)
(3) OBERMANN, P.: Möglichkeiten der Anwendung des Doppelpackers in Beobachtungsbrunnen bei der Grundwassererkundung.-
    Brunnenbau, Bau v. Wasserwerken, Rohrleitungsbau: 27: 93 - 96, Köln (1976)
(4) BUNDESGESUNDHEITSAMT: Empfehlungen zum Vorkommen von flüchtigen Halogenkohlenwasserstoffen im Grund- und Trinkwasser. Bundesgesundheitsblatt 25, Nr. 3, März (1982)

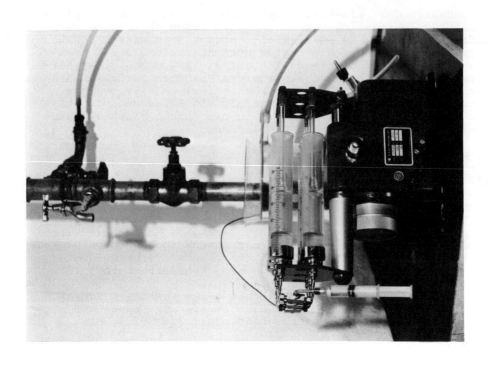

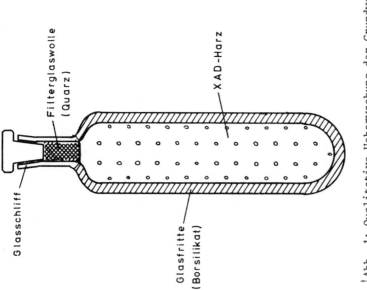

Abb. 1: Qualitative Ueberwachung der Grundwasserqualität mittels XAD

Bild 1: Probenahmegerät zur kontinuierlichen Probenahme und zur Erstellung von Mischproben

Tab. 1: Grundwasseruntersuchungen eines Brunnens in der Nähe eines Regenüberlaufes mit Abwasseranteilen

Stichprobe nach Pumpen XAD - GC-MS	Aktivkohle in Brunnen GC-MS	XAD in Brunnen GC-MS
Trichlorethylen	Trichlorethylen	Trichlorethylen
Tetrachlorethylen	Tetrachlorethylen	Tetrachlorethylen
-	1,1,1-Trichlorethan	-
Toluol	Toluol.	Toluol
Xylol (3 Verbindungen)	Xylol (2 Verbindungen)	Xylol (3 Verbindungen)
-	Dichlorbenzol	Dichlorbenzol
-	-	Bromkresol
-	-	Dimethylphthalat
Dibutylphthalat	-	Dibutylphthalat
Laurinsäuremethyl	Laurinsäuremethyl	Laurinsäuremethyl
-	-	Myristinsäuremethyl
Palmitinsäuremethyl	Palmitinsäuremethyl	Palmitinsäuremethyl
Stearinsäuremethyl	-	Stearinsäuremethyl
Terpene (1 Verbindung)	Terpene (3 Verbindungen)	Terpene (2 Verbindungen)
-	-	Sulfonattenside
-	-	Methylnaphthalin
-	Hexachlorbenzol	Hexachlorbenzol
-	γ Hexachlorcyclohexan	γ Hexachlorcyclohexan

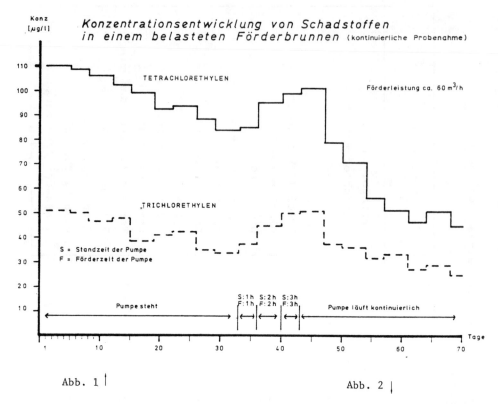

Abb. 1

Abb. 2

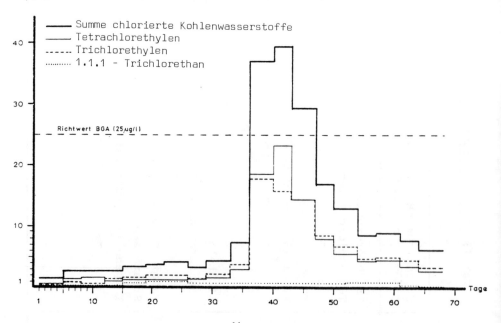

SAMPLING OF SURFACE WATERS FOR TRACING MACRO- AND MICROPOLLUTANTS

WOLFGANG SCHMITZ
Landesanstalt für Umweltschutz Baden-Württemberg, Karlsruhe

Summary

The paper reports on 15 years experience in water quality monitoring of surface waters in the State of Baden-Württemberg, which is based on techniques of representative continuous sampling and multiple parameter analytical procedures in the central water laboratory. The technique of sampling is in close relation to handling many problems as sample preservation, transport organisation, emergency measures etc. The procedures have been adopted by several environmental agencies in the F.R. Germany and other states of the European Communities. The paper describes the procedures, their advantages, applications, restrictions and problems not yet solved.

The appearance of specific concentrations of chemical constituents in natural waters is irregular with regard to location and time. For technical reasons studying the occurance of micropollutants as also the majority of other substances in natural waters is necessarily bound to steps of discrete sampling of water, followed by analytical procedures in the laboratory rather than on continuous on-line operations.

To obtain useful information on spatial and temporal distribution of chemical concentration data in natural waters, a planful concept has to be established. which aims to the achievement of an optimum of information at the expenses of reasonable efforts mainly arising with respect to chemical analytical work.

This philosophy was applied to the set-up for quality supervision of surface waters in the State of Baden-Württemberg in Germany. This system was established 1972 after preparatory work and has grown meantime to about 50 permanent sampling stations mainly placed along the rivers Rhine and Neckar, which are the most important waters in terms of usage in our country.

Perment sampling can be easily performed, but analytical work must be based on discrete samples mostly. The appropriate solution of this discrepancy is collecting of water from a continuous or quasi-continuous water flow into a sampling device, which composes a proportional mixed sample during a certain period of time.

Our working hypothesis was that a sampling period of two weeks was suitable to obtain a time-proportional composed sample, which would be fairly representative for concentration levels and fluctuations characteristic for the particular situation of the river under investigation. Further on this appeared to be proper also for calculations of freights of constituents from biweekly averaged concentration and runoff data.

The samplers generally used were submergence samplers as designed by AURAND and developped into a reliable instrument in close cooperation between his and our institutes (Fig. 1). The sampler consists simply of a 2 liter glass bottle with a head piece, which regulates a constant flow of water over the sampling period into the bottle by means of a capillary valve placed in the pathway of the outflowing air. It is submerged into the water in a definite depth, in our case .5 meter. So filling is achieved by the hydrostatic pressure and no electrical power supply is necessary.

The samplers are placed into a protection pipe, which is fastened either on floating devices or on concrete walls of the river bank at those locations, where normally no movements of water level occur, as on most of our observation stations at the river Neckar, where the samplers have been installed near the inlet of the water into power plants (Fig. 2). The samplers provide 2 liters of time-proportionally sampled water within 24 hours (filling time can be changed by exchanging capillary valves of appropriate diameter). After this time the bottle is taken off by an attendent and the sample is stored generally by deep freezing. Before deep freezing an aliquote part is taken to procure a biweekly composite sample by mixing with aliquote parts of consequent samples, which again is preserved by deep freezing. Normally analytical determinations are made on the biweekly conposite samples. Exceptionally as in cases of hazards or suspected violations determinations are made on the samples of a single day.

The scope of the analytical program and the sequence of procedures involved between sampling and obtaining analytical data is given in figure 3.

In the course of our operations we were compelled to equipe observation stations, where no daily personal attendance was possible, with automatic sampling devices, which could collect and preserve samples over a period of 1-2 weeks. A sampler, which fulfilles these requirements was constructed for us by the BÜHLER-COMPANY, Tübingen. This instrument samples water into one or several bottles placed into a deep freezed container according to variable programs.

In connection with the installation of the sampling system we considered and tested several procedures of sample preservation. Realizing that no single method could solve all the problems of preservation, we found the deep freezing technique as the one, which works fairly well with regard to most of the parameters, we were interested in to detect. We noticed restrictions to be made in case of preserving orthophosphate, hydrogene carbonate and calcium at higher levels of water hardness. I also have to mention that phase distribution of soluted and particulate constituents in the water is a problem connected with sampling and preserving water, which is not yet sufficiently solved.

Sometimes the selection of an appropriate alternative parameter is a way out of methodical difficulties. So the determination of total phosphorus instead of orthophosphate may deliver sufficient information for the problem under investigation. Another example: Since we found that there is no proper way of correctly estimating BOD_5 in samples after freezing, except they were taken from effluents of sewage works, we found BOD_{30}, which can be determined reliable in frozen samples, being a parameter, which could replace the BOD_5 properly.

Another problem well known presents the material of sampling or storing equipment. Our submergence samplers collect into glass bottles, the material being proper for storing water without loss or constamination of constituents, of course not for all of them. Glass is hardly suitable for deep freezing preservation. Here polyethylen bottles are appropriate in most cases, because contact of bottle walls and the liquid phase of water is only of short duration, if deep freezing technique is applied. In special situation (highly volatile organics) aluminium bottles were used by us succesfully.

One should keep well in mind these methodological restrictions, but one should not overstress this point of view and should be ready to compromise if nessessary and should reject scruples, if their consequence would mean the rennunciation of any information at all. Our set-up and supervision program in Baden-Württemberg is much based on such compromising and nevertheless has confirmed our expectations and produced data, which resulted in exellent information and insights into problems of water pollution and river limnology. I want to demonstrate this briefly in a few examples of results of our technique of continuously sampling.

Figure 4 is a plotter graph of biweekly average concentrations of ammonia occurring in the river Neckar during the year 1979 at the locations Poppenweiler, a short distance below the discharge of the city of Stuttgart, and at the location Feudenheim near the confluence of the river into the river Rhine. For the purpose of comparison and interpretation the tolerance limits as proposed by the IAWR for raw water supply A (normal requirements) and B (required for advanced purification technique) are inserted into the graphs. The graphs indicate that at Poppenweiler levels of A are exceeded all the time, levels of B with a frequency of 22/26, whilst at Feudenheim B is passed over only with a frequency of 1/26. The sanitation program Neckar aims to chemical concentrations below the limits of B at the confluence with the river Rhine. This status has not yet been achieved.

Figure 5 expresses a synoptical view of the concentration ranges of a constituent along the river Neckar arranged in percentiles of frequency of occurance. Frequencies of exceedence of given tolerance limits can be taken from these graphs easily. The frequency distributions refere to biweekly concentration averages and not to short time top or minimum concentrations, which may occur occasionally. According to our experiences such extraordinary data are generally not very significant in pollution supervision. Exemptions I shall describe later.

Figure 6 demonstrates the local changes of concentrations of ammonia and nitrate in the course of the river Neckar. The data are given as lines of annual average concentrations. From this presentation we can conclude that ammonia oxydation and transformation into nitrate and a corresponding loss of oxygen is an important process of biological self-purification in the river Neckar. The transformation of ammonia into nitrate occurs nearly stocheometrically and has increased by the time.

Very successful were evaluations of the data regarding the relations between freight of constituants and runoff. For as we found, this kind of consideration permits longtime trend analyses and offers the possibility to make clear statements of quality changes in a river in ranges of time.

As figure 7 shows, the freights of constituants plotted versus runoff in many cases form a characteristic line. Comparisons between characteristic freight lines obtained for different years give a clear indication of reduction or rise of loads of pollutants in the rivers. To produce the characteristic freight lines of our examples we did not use all the freight data of our data-sets, but omitted data of higher runoff above a certain level and restricted our analysis to data only obtained within water temperature levels of 15-20 ^{o}C, in other words to quasi-isothermic conditions. Since the water temperature has a large influence on destruction and transformation of substances in the water, the inclusion of data of all temperature levels creates a data-set of concentrations which does not give any clear insight.

The methods of obtaining information of continuous sampling proved to be successful also in the field of supervision of sewage plant effluents. Figure 8 is an example of our studies of the fluctuations of effluent concentrations of sewage plants along the river Neckar. Here we determined daily average concentrations of BOD, DOC, ammonia, nitrate, phosphorus and chloride over a period of several months. The data were used to establish a river model relating loads of organics to oxygen situations in the stream.

One may argue that in all these cases the method of composite sampling will not hit the very highest concentrations, but only averages below these. My opinion is that such extreme data are relevant only exceptionally in water pollution problems. Extremes of concentrations induced by sudden short time discharge of pollutants are flattened readily by effects of dispersion in turbulent flowing streams. The last 2 examples, which I shall present here are a good proof of the validity of this hypothesis.

These examples refer to special situations of sudden pollution of a river by hazardous substances. The one occured on October 8th 1976 as an alarm was released by Swizz Water Authorities reporting that approximately two tons of tri-chlorobenzene had been drained into the river Rhine from a sewage outlet of the CIBA-GEIGY-COMPANY at Basel in consequence of a transport accident. Considering the toxicity of this substance and regarding the information on the amount of the substance, the Water Authorities in Germany came to the conclusion that no direct harm could be caused by this accident. But nevertheless we thought it to be wise to observe the passage of the substance in the river Rhine and to obtain an independant confirmation of the reported data. We made use of our daily samples stored during the critical period before, during and following the accident and analysed samples of the stations Weisweiler, Maxau and Mannheim. The results given in figure 9 showed that we had a basic level of tri-chlor-benzene in the water of the river Rhine already before and also after the time, when the accident had occured. Adding up the freights, calculated from concentrations and runoffs at Weisweiler, 80 km below Basel, we found indeed an additional amount of 2300 kg of the substance having passed the river Rhine at this place, approximately the amount, which had been reported as released. On the other hand further downstream at Maxau we found only a slightly raised level and further downstream at Mannheim no significant increase of the level of the substance at all. A suggestion for the disappearance of the substance in the course of the river Rhine is either loss into the air, favoured by water falling over wheirs, or adsorption on suspended matter. In our submergence samplers suspended matter is extensively separated from the liquid phase and will not appear in the analytical result therefore.

Another accident occured last year at the river Main. Dutch Water Work Authorities detected an occurance of higher concentrations of 1,2,4-tri-chloro-nitrophenol in the river Rhine and suspected the source of this pollution being in Germany. The Dutch proclaimed that an amount of 500 kg/day had appeared at the position of their water uptake. They sent a ship to take samples at different positions in the river Rhine and tributaries. One single sample taken from the river Main contained 140 mg $TCNP/m^3$. From this it was suggested that a chemical factory near Frankfurt was the source of pollution and a prosecution of the case was started by the German legal authorities. The case would have been left uncleared without proof samples of the critical time interval. 24-hours average composite samples were at hand from Koblenz, rather far below the river Main. These samples, analysed by the institute WABOLU, Berlin, lead to the sound conclusion, that indeed an amount of approximately 500 kg/day had passed this location, deriving obviously from somewhere above. The analysis of daily samples preserved at the locations river Rhine and Neckar at Mannheim showed that no significant amounts of this substance had occured here and therefore the source should be further downstream. Unfortunately there were no regular daily samples taken and preserved between Mannheim and Koblenz by State Agencies. But some information could be obtained from samples of a station, which had been set up by the WABOLU at the lower part of the river Main for the purpose of a research project. The evaluation of the daily taken and preserved 24-hours average samples resulted in the following facts. The substance had a regular concentration level of 3 mg/m^3, but during a few days per week higher concentrations up to the tenfold used to occur. During the week of the suspected accident the levels in the 24-hours average samples reached 14 mg/m^3 in the maximum and lay at 8 mg/m^3 in the average. In this period a daily freight of 20-30 kg of the substance can be derived from the concentration and runoff data. So it did not seem very likely that the river Main would be the source of this case of pollution. Also the finding of 140 mg/m^3 in a random sample must not be conclusive, if a 24-hours average is 14 mg/m^3 and the pollutant comes from nearby. We are lacking the final evidence of this case from a very simple reason: During the critical period there was no sample taken on Sunday, because no attended could be found to do the work on a holiday. From a analysis of runoff and flowtime relations it seems reasonable to assume that the pollutant had been discharged into the river Main on Sunday and could be observed dispersed over a longer period in downstream positions after a corresponding gap of time. Consequent sampling procedures on places between Mannheim and Koblenz could have cleared the case easily instead of having left it to suggestions.

From these few given examples, which could be expanded easily it seems to be obvious that this simple strategy and method of permanent and time-proportional sampling of water, resulting in mixed samples has proven to be a very successful mean of supervision of river and waste waters, and its further use should be encouraged. Meantime apart from Baden-Württemberg the German Commission and also the International Commission for the Protection of the River Rhine have changed their program from the schedule of obtaining single samples at fixed days during a year into a system of taking biweekly composite samples and preserving these deep frozen. I am convinced that this method will be suitable for many more applications than hitherto, in particular in the still neglected field of supervision of emissions of waste waters, which easily would become proof powerful.

It was a pleasure to me to have had the opportunity to report on our experiences during this symposium, and I am looking forward for your opinions in the forthcoming discussion.

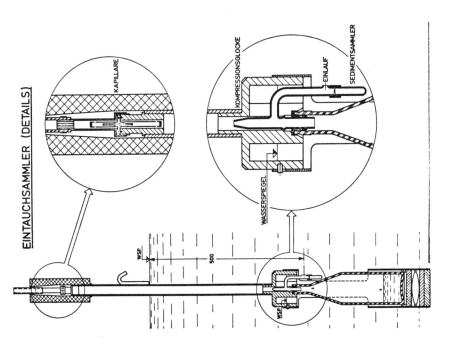

Fig. 1 Submergence Sampler

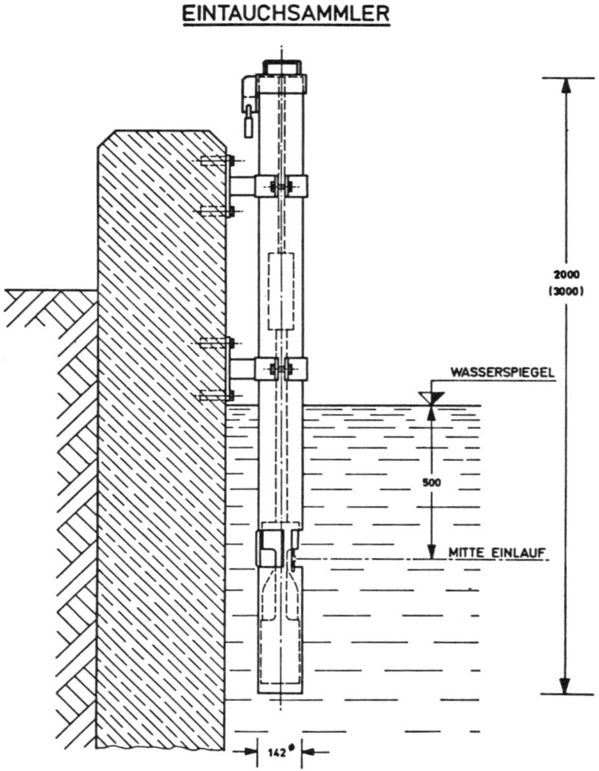

Fig. 2 Submergence Sampler; Placing into the river

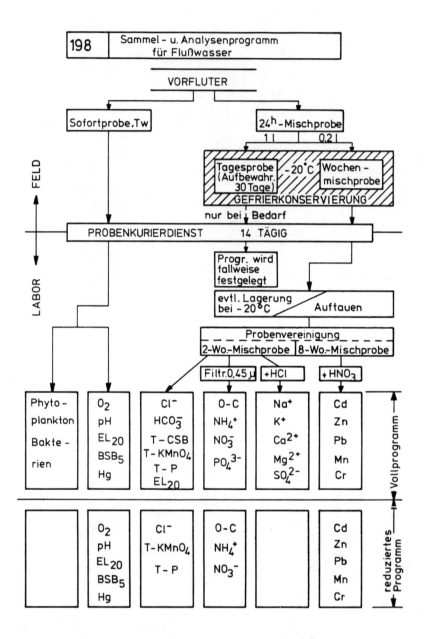

Fig. 3 Standard program surface water quality supervision Baden-Württemberg

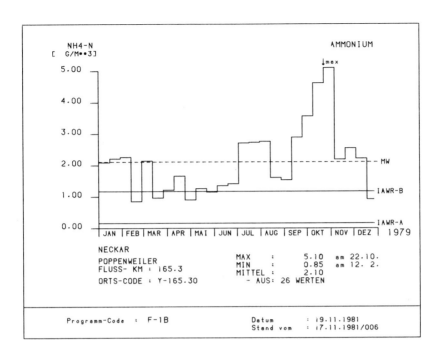

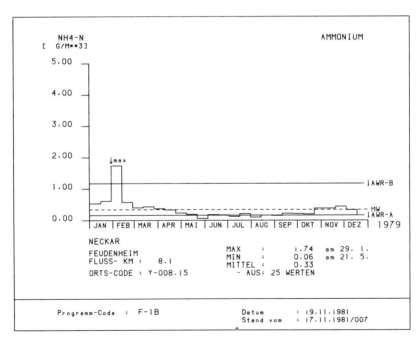

Fig. 4 Ammonia in the Neckar river (computer plots)

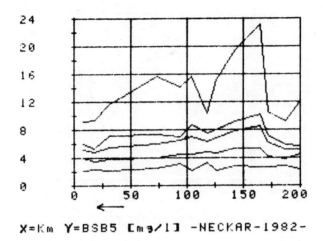

Fig. 5 BOD$_5$ in the river Neckar (computer plot: frequency of occurrence of concentrations)

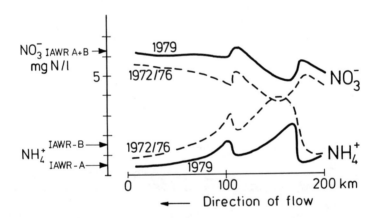

Fig. 6 Nitrate and ammonia in the river Neckar. Average concentrations 1972-76, 1979. Direction of flow from right to left (km 200 to km 0)

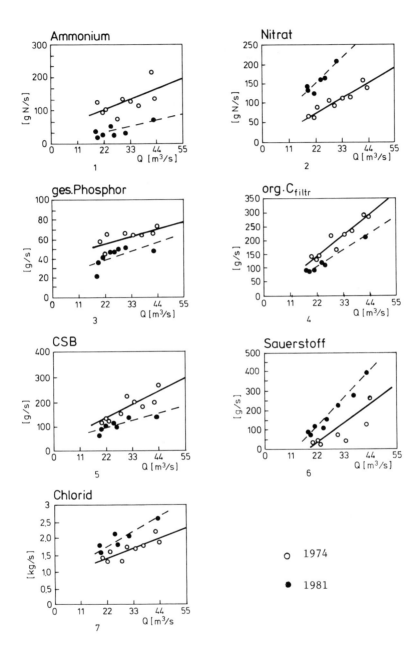

Fig. 7 Characteristic freight lines of constituents in the water of the river Neckar at Poppenweiler

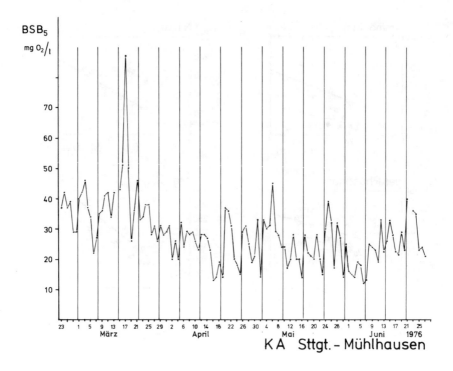

Fig. 8 BOD$_5$ in effluents of the sewage treatment plant of the city of Stuttgart

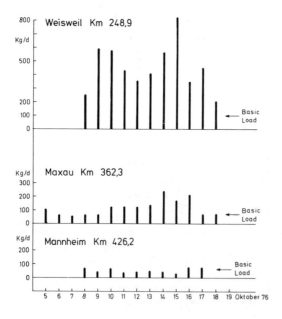

Fig. 9 Tri-chloro-benzene in the river Rhine October 1976

TECHNIQUES D'ECHANTILLONNAGE POUR L'ANALYSE DES TRACES ORGANIQUES

A. BRUCHET, L. COGNET et J. MALLEVIALLE
Laboratoire Central-Lyonnaise des Eaux et de l'Eclairage
38 rue du Président Wilson - Le Pecq - France

Summary

Due to the large number and variability, both in type and concentration, of trace organics in water supplies, a concentration step is required for their analysis. For these reasons, complementary continuous sampling techniques are used by the Laboratory of S.L.E.E. Three types of samplers are used : a composite liquid sampler for the analysis of substances by closed loop stripping, head space..., a continuous liquid-liquid extraction cell, and a contactor on macroporous resins or other adsorbents for subsequent elution followed by GC-MS analysis or biological assays. These devices enable composite samples of average concentration to be obtained over one or several days.

Résumé

L'analyse des traces organiques présentes dans une eau brute ou de consommation nécessite une étape de concentration du fait de leur grand nombre et de leur variabilité en concentration et en nature. A cet effet des appareillages complémentaires de prélèvement en continu ont été mis au point par le Laboratoire de la S.L.E.E. Trois types d'appareil sont décrits ici :
Un échantillonneur d'eau pour l'analyse des substances volatiles par des techniques du type "stripping" en boucle fermée, espace de tête, etc...
Un système continu d'adsorption sur résine macroporeuse ou tout autre support solide et un extracteur liquide-liquide en continu permettant l'analyse par CG-SM ou la réalisation de tests biologiques.
Ces appareils permettent d'obtenir un prélèvement composite de concentration moyenne sur un ou plusieurs jours.

1. INTRODUCTION

L'analyse des traces organiques dans les eaux de surface ou eaux souterraines destinées à la production d'eau potable reste un domaine vaste et complexe. En effet bien que ces eaux ne contiennent que de faibles teneurs en carbone organique (1 à 10 mg/l de carbone), les traces organiques englobent plusieurs milliers de composés spécifiques d'origine naturelle (acides humiques protéines) ou provenant de l'activité humaine (pesticides, hydrocarbures etc...). Ces différents composés peuvent être classés, en première approximation, suivant de grandes gammes de concentration avec des bornes à 1 mg/l, 1 ug/l et 1 ng/l (1).

Composés dans la gamme de concentration supérieure à 1 mg/l : deux grandes classes de composés se trouvent dans cette catégorie : les protéines et les acides humiques ou fulviques. Il s'agit en fait de familles de substances de poids molaire élevé assez peu ou mal caractérisées et qu'il est difficile de doser autrement que par des indices globaux. Il est généralement admis que ces composés représentent environ 80% de l'ensemble des matières organiques.

Composés dans la gamme de concentration comprise entre 1 µg/l et 1 mg/l. On retrouve des produits naturels, acides aminés libres, sucres, pigments du type chlorophylle aussi bien que des polluants synthétiques solvants chlorés, détergents, dans certains cas hydrocarbures et phtalates.

Composés dans la gamme de concentration comprise entre 1 ng/l et 1 µg/l. Pratiquement toutes les familles de la chimie organique se trouvent représentées dans cette fourchette de concentration. Leur dosage est en général basé sur l'utilisation de techniques sophistiquées (chromatographie en phase gazeuse et liquide haute pression avec divers détecteurs) qui dans la plupart des cas nécessitent le recours à une phase de concentration. Le propos de cet article est de décrire différentes techniques automatiques de préparation d'échantillons.

2. **POSITION DU PROBLEME**

Un certain nombre de remarques importantes doivent être faites avant le choix d'une ou deplusieurs techniques de préparation d'échantillons :

2.1 - Des volumes importants (plusieurs dizaines de litres) sont souvent nécessaires pour une identification en spectrométrie de masse de composés à l'échelle du nanogramme/litre. Le transport d'importants échantillons d'eau du point de prélèvement à un laboratoire central suffisamment équipé pose un certain nombre de problèmes, risques de pollution, stabilité de l'échantillon, stockage etc... Ceci conduit tout naturellement au choix d'une technique automatique utilisée sur le site.

2.2 - A de tels niveaux de concentration, la qualité des eaux de surface varie souvent de façon considérable d'un jour àl'autre de la semaine et bien entendu en fonction des conditions climatiques (2). Dans le suivi de certaines étapes de traitement, comme la filtration sur charbon en grains des effets de compétition et de chromatographie rendent difficile la comparaison entre l'amont et l'aval du traitement (3). En outre la plupart des usines de traitement d'eau fonctionnent à des débits variables dans le temps qui sont fonctions de la demande des consommateurs, des horaires à tarif préférentiel pour l'électricité, etc... Tout ceci conduit à une grande variabilité de concentration dans le temps qui rend peu représentatifs les prélèvements instantannés y compris s'ils concernent de grands volumes d'eau. Cet inconvénient est encore aggravé par le fait qu'en raison de leur prix élevé, ces analyses sont souvent effectuées à des fréquences insuffisantes.

L'idéal serait donc de disposer de capteurs permettant la mesure continue de différents paramètres. De tels appareils sont disponibles sur le marché pour des paramètres globaux (carbone organique dissous, absorption à 254 nm, etc...) ou des indices représentatifs de familles chimiques (phénols par colorimétrie, hydrocarbures par infra-rouge...) mais ces appareillages posent des problèmes aux faibles concentrations et ne correspondent pas réellement à des analyses spécifiques. Il est cependant vraisemblable que d'ici quelques années seront disponibles sur le marché des appareillages de process combinant les deux phases de concentration et de mesure : adsorption et électrodes enzymatiques (4) pour certains pesticides, stripping et chromatographie pour les solvants chlorés, etc...

Devant l'impossibilité de faire des mesures en continu, il parait opportun d'avoir recours à des techniques d'échantillonnage permettant l'intégration des résultats sur des périodes de temps allant de un à plusieurs jours.

2.3 - Dans la mesure où de très nombreuses familles chimiques sont présentes dans les eaux, le recours à différentes techniques complémentaires de préparation d'échantillon permets d'obtenir une préséparation grossière des substances organiques et ainsi faciliter leur analyse ultérieure (par exemple séparer les composés volatils des non volatils).

2.4 - En raison de l'extrême sensibilité des techniques utiliséesil est préférable d'éviter au maximum d'utiliser des appareillages susceptibles d'entrainer une pollution des échantillons, ce qui est le cas avec de nombreux matériaux plastiques. C'est pourquoi les appareillages décrits ci-dessous ne comportent que du verre, du téflon et dans certains cas de l'acier inox.

3. DESCRIPTION DES APPAREILLAGES

Trois appareillages automatiques fonctionnant sur le site selon les critères définis ci-dessus ont été mis au point :

3.1 Echantillonneur pour analyse de substances volatiles

Le dosage de solvants du type chloroforme ou trichlorethylène se fait en règle générale par des techniques d'espace de tête. La technique de "closed loop stripping analysis" décrite par K. Grob (5) s'utilise de plus en plus, notamment dans la recherche de substances responsables de goûts et d'odeurs (6).

Le système schématisé sur la figure 1 est un appareil dérivé de celui proposé par J.J. Westrick (7). Il permet d'obtenir des prélèvements d'eau intégrés dans le temps (de 1 heure à plusieurs jours si nécessaire) avec un minimum de contact entre l'atmosphère et l'échantillon pour éviter les pertes. Dans la version présentée ici le volume prélevé est au maximum de 3 litres ce qui correspond au volume utilisé dans la technique CLSA.

Il se compose d'une électrovanne à 3 voies montée sur une horloge élecTrique qui permet de régler la fréquence et la durée du prélèvement. Un flotteur en téflon ou acier inox augmente la capacité du récipient au fur et à mesure que le volume de l'échantillon augmente. Le robinet téflon permettant la récupération de l'échantillon peut être remplacé par une électrovanne couplée à une horloge électrique dans le cas où l'appareillage est utilisé comme témoin de pollution.

Il est bien évident que ce système, prévu au départ pour les composés volatils, peut être utilisé pour l'analyse de tout autre paramètre global ou spécifique. Son faible encombrement permet en outre de le placer, si nécessaire, dans une enceinte réfrigérée.

3.2 Système d'adsorption sur support solide (résines macroporeuses)

La concentration de traces organiques sur divers adsorbants a été abondamment décrite ces dernières années, charbon actif, phases inverses et surtout résines macroréticulaires du type Amberlite XAD (8) (9) (10). La plupart du temps ces phases d'adsorption sont faites au Laboratoire sur des prélèvements instantannés pouvant aller de 1 litre à plusieurs mètres cubes pour des volumes de résine de 1 à plusieurs centaines de millilitres, ce qui, dans certains cas, parait défier les lois de l'adsorption.

En accord avec I.H. Suffet (11), l'appareil mis au point à la S.L.E.E. (fig. 2) et actuellement commercialisé par la SERES (Aix en Provence- France) permet l'utilisation de plusieurs centaines de millilitres de résines pour des volumes d'eau allant de 5 à 200 litres maximum. Deux réactifs peuvent être ajoutés simulanément dans l'échantillon, une solution tampon qui permet d'ajuster le pH et une solution de neutralisation d'oxydants résiduels.

En fait le système est constitué de deux colonnes en série permettant l'utilisation de deux adsorbants différents, par exemple, une résine XAD8 qui fixe les matières humiques ce qui conduit à une meilleure fixation des composés non polaires sur le second adsorbant (XAD2).

Le temps de contact est un paramètre très important, c'est pourquoi il est préférable de se situer à des vitesses de filtration de l'ordre de 10 volumes/volume/heure. Le pH d'utilisation retenu est de 2, des essais d'adsorption à des pH basiques se sont révélés jusqu'à présent peu fructueux.

La préparation des résines est évidemment soumise à un protocole expérimental précis qui doit conduire à des blancs convenables. Il s'agit en général de plusieurs lavages successifs par des solvants différents (exemple dichlorométhane et méthanol).

L'élution des résines peut se faire en laboratoire par simple démontage des colonnes téflon qui peuvent ensuite être séparées en deux parties si l'on souhaite réaliser des élutions distinctes des deux adsorbants.

Suivant le but recherché le solvant d'élution peut être du dichlorométhane un alcool, du DMSO ou un mélange de solvants. En routine le laboratoire de la S.L.E.E. utilise des élutions successives par le dichlorométhane, le méthanol et une solution de soude. Le premier solvant se révèle contenir la grande majorité des composés chromatographiables en phase gazeuse (un exemple en est donné sur le tableau I). Dans la phase méthanol n'ont pu être identifés par GC-MS que des acides gras. Cette phase beaucoup plus colorée se révèle cependant contenir un grand nombre de substances organiques comme le montrent les chromatogrammes obtenus par HPLC de la figure 3. L'élution à la soude conduit à l'obTention d'acides humiques et fulviques (12).

Cet appareillage fonctionne pendant une semaine sans surveillance et un système d'électrodes couplées à un compteur à impulsion donne avec précision le volume d'eau passé sur la résine. Il est en outre équipé de sécurités qui évitent des débordements ou la mise à sec du lit de résines.

La présence de matières en suspension en quantité notable dans l'eau à analyser nécessite une filtration préalable si le volume de l'échantillon doit dépasser quelques dizaines de litres.

Il est bon de souligner toutefois les inconvénients de cette technique qui sont principalement liés aux lois de l'adsorption. Pour une eau donnée il convient de s'assurer que le volume d'eau passé ne conduit pas à dépasser les capacités d'adsorption de la résine. Pour un composé déterminé l'efficacité de récupération dépendra des autres composés dissous dans l'eau. Cet appareillage sera donc particulièrement approprié à la comparaison d'eaux de qualités voisines, par exemple dans le cas d'études de comparaison de chaines de traitement.

3.3 **Extracteur liquide-liquide**

Les inconvénients posés par le système précédent ont conduit le laboratoire de la S.L.E.E. à développer une cellule d'extraction liquide-liquide en continu. I.H. Suffet (13) a montré qu'un appareillage de ce type permet d'obtenir d'excellents rendements d'extraction pour un grand nombre de composés chromatographiables en phase gazeuse et notamment les substances organiques citées dans les normes européennes de l'eau potable (pesticides, polycycles aromatiques etc...).

L'appareillage schématisé sur la figure 4 (SERES, Aix en Provence, France) comprend trois phases successives :
- extraction par un solvant des composés organiques dissous dans l'eau par mélange à co-courant au moyen d'une bande tournante hélicoîdale ;
- séparation des deux phases par simple décantation dans une cellule conçue pour diminuer au maximum les problèmes d'émulsion ;
- recyclage du solvant (dichlorométhane en général) par régénération dans une unité d'évaporation-recondensation, alors que l'eau (phase supérieure) est évacuée par un trop plein.

Ce système est régulé par des électrodes placées dans le séparateur de phase.

L'appareillage permet l'addition automatique d'un acide ou d'une base destinés à modifier le pH de l'eau. Le rapport solvant/eau est réglable autour de la valeur de 0,1. Les débits d'eau peuvent aller de 0,5 à 2 l/h. Des systèmes de sécurité permettent le fonctionnement en automatique sur le site durant plusieurs jours. En fin d'opération il suffit de récupérer l'extrait organique, opération qui peut d'ailleurs être également automatisée.

4. **DOMAINES D'APPLICATION PARTICULIERS**

Ces appareils peuvent être utilisés dans le cadre d'un contrôle de fabrication ou d'études visant à mettre au point des chaines de traitement mais ils sont également susceptibles de répondre à des besoins différents.

Dans le cas de pollutions accidentelles d'eaux de surface les appareillages 1 et 3 peuvent être utilisés comme "mouchards", qui fonctionnent en continu et rejettent automatiquement à l'égout les prélèvements inintéressants. Ils peuvent également être couplés à des systèmes d'alarme comme un ichtyomètre qui ne déclencherait le prélèvement ou l'extraction qu'en cas de réponse positive du test biologique. Un système de ce type sera installé dans la station d'alerte de l'usine du Mont Valérien (Compagnie des Eaux de Banlieue) en aval de Paris à la fin de l'année 1984.

Les deux derniers systèmes peuvent être couplés à un système de mesure comme une électode enzymatique ou un détecteur ultraviolet et être ainsi utilisés directement comme systèmes d'alarme. Des études sont actuellement en cours pour évaluer la faisabilité d'un tel système.

5. **CONCLUSIONS**

L'analyse des substances organiques à l'état de traces dans les eaux passe par un certain nombre de contraintes (larges volumes d'eau représentativité des échantillons...) qui sont en partie résolues par l'utilisation des trois systèmes décrits dans cet article. Ils permettent en effet d'effectuer sur le site des préparations d'échantillons intégrés dans le temps. Il suffit ensuite d'envoyer des cartouches ou des extraits de faible volume vers un laboratoire central disposant du matériel approprié à l'analyse (CG-SM, etc...).

Ces appareillages peuvent être utilisés dans le contrôle et le suivi de la qualité d'une eau, l'étude de l'efficacité d'une chaine de traitement en combinant analyses chimiques spécifiques et tests biologiques. Ils peuvent également être couplés à des appareillages d'alarme du type Ichtyotest pour les problèmes de pollutions accidentelles.

Il serait cependant intéressant de développer des systèmes permettant la mesure continue de paramètres organiques spécifiques susceptibles de poser des problèmes au niveau de l'eau potable, solvants chlorés, pesticides, etc...

REFERENCES

(1) MALLEVIALLE, J., E. SCHMITT, A. BRUCHET. "Composés Organiques Azotés dans les Eaux : Inventaire et Evolution dans Différentes Filières Industrielles de Production d'Eau Potable", présenté aux "Journées Information Eau de Poitiers, France, le 30 septembre et 1er octobre 1982.
(2) SUFFET, I.H, L. BRENNER, P.R. CAIRO. "GC/MS Identification of Trace Organics in Philadelphia Drinking Water During a 2 Year Period", Water Research 14, 853 (1980).
(3) YOHE, T.L., I.H. SUFFET and P.R. CAIRO. "Specific Organic Removals of Granular Activated Carbon", J.A.W.W.A. 13, 402 (1981).
(4) MANEM, J., J. MALLEVIALLE, P. DURAT and E. CHABERT. "La Détection des Pesticides Organophosphorés et des Carbamates à l'Aide d'une Electrode Enzymatique à Butylcholinestérase", l'Eau, l'Industrie, les Nuisances 74, 31 (1983).
(5) GROB, K. "Organic Substances in Potable Water and in its Precursor. Part 1- Methods for their Determination by Gas-liquid Chromatography, J. of chromato. 84, 255 (1973).
(6) Mc GUIRE, M.J., S.W. KRASNER, C.J. HWANG and G. IZAGUIRRE. "Closed-loop Stripping Analysis as a Tool for Solving Taste and Odor Problems", J.A.W.W.A. oct, 530 (1981).
(7) WESTRICK, J.J and M.D. CUMMINS. "Collection of Automatic Composite Samples Without Atmospheric Exposure" JWPCF 51 (12), 2948 (1979).
(8) SUFFET, I.H., L. BRENNER, J.T. COYLE and P.R. CAIRO. "Evaluation of the Capability of Granular Activated Carbon and XAD 2 Resin to Remove Trace Organics from Treated Drinking Water", E.S.T. 12, (12), 1315 (1978).
(9) JUNK, G.A. and Coll. "Use of Macroreticular Resins in the Analysis of Water for Trace Organic Contaminants", J of Chromato. 99, 745 (1974).
(10) THURMAN, E.M., R.L. MALCOLM and G.R. AIKEN. "Prediction of Capacity Factors for Aqueous Organic Solutes Adsorbed on a Porous Acrylic Resin", Anal. Chem. 50 (6), 775 (1978).
(11) SUFFET, I.H. Communication personnelle.
(12) MANTOURA, R.F.C; and J.P. RILEY. "The Analytical Concentration of Humic Substances from Natural Water", Analytica Chim. Acta. 76, 97 (1975).
(13) YOHE, T.L. and R.S. GROCHOWSKY. "Development of a Teflon Helix Continuous Liquid-liquid Extractor Apparatus and its Application for the Analysis of Organic Pollutants in Drinking Water, Measurement of Organic Pollutants in Water and Waste Water", A.S.T.M., S.T.P. 686, CE, Ed American Society for Testing and Materials, pp 47-67 (1979).

PRODUITS	CONCENTRATIONS EN ng/l		
	EB	EFI	EFP
Halogénés			
Dichloroacétone	< 10	10-20	< 10
Trichloroacétone	< 10	10-20	< 10
Tétrachloracétone	< 10	< 10	< 10
CHBrCl2	< 10	< 10	< 10
Bromocyclohexane	< 10	< 10	< 10
Atrazine	10-100	< 10	< 10
Acides carboxyliques			
Acide benzoïque	< 10	< 1	1000-10000
Acide dodecanoique	10-100	10-50	10-100
Acide pentadecanoique	< 10	< 10	1-10
Acide hexadecanoique	< 10	< 10	10-100
Azotés			
Methylethylmaleimide			
Butylbenzenesulfonamide	10-100	1000-10000	10-100
Aldehydes			
Nonanal	< 10	< 1	1-10
Benzaldehyde	< 10	10-50	10-100
Ethylaldehyde	< 10	< 10	10-100
Dimethylaldehyde	< 10	100-1000	10-100
Cetones			
Acetophenone	< 10	Masqué par les dérivés alcanes	10-20
Dérivés alcanes			
Somme	< 10	10000-100000	< 10
Alcane C9	< 10	100-1000	100-1000
Alcane C10	< 10	1000-10000	100-1000

TABLEAU I- Analyse par GC-MS de l'éluat de résines par le dichlorométhane (EB : eau brute de l'usine de Moulle à Dunkerque, EFI : eau flottée après une préchloration, EFP : eau flottée après une préozonation)

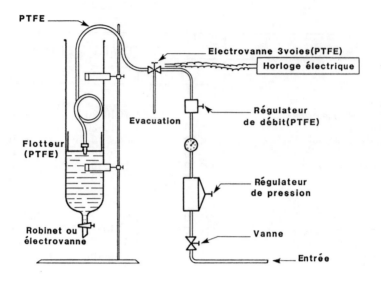

Figure 1 - Echantillonneur d'eau pour l'analyse des substances volatiles

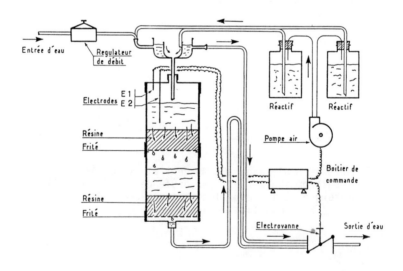

Figure 2 - Système d'adsorption sur support solide (résines macroporeuses)

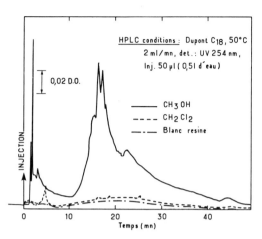

Figure 3 - Comparaison des élutions CH$_2$Cl$_2$ et MeOH par chromatographie HPLC

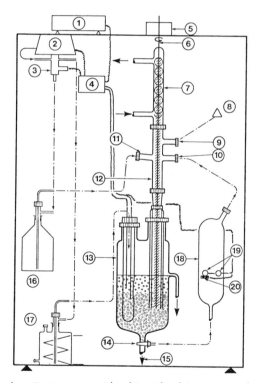

Figure 4 - Extracteur liquide-liquide en continu

1 et 4 : boitiers de commande - 2 : pompe à air - 3 : électrovanne -
5 : moteur - 6 : bande tournante - 7 : réfrigérant - 8 : régulation de
l'arrivée d'eau - 9,10,11 : tubulures d'admission d'eau, de solvant et de
solution tampon - 12 : réacteurs - 13 : décanteur - 14 : vanne 3 voies -
15 : soutirage de l'extrait - 16 : solution tampon - 17 : solvant -
18 : distillateur - 19 : bougie chauffante - 20 : thermocouple.

EXTRACTION OF SEDIMENTS

USING ULTRASONIC AND STEAM DISTILLATION TREATMENTS

C. O'DONNELL

Water Resources Division, An Foras Forbartha,
St. Martin's House, Waterloo Road, Dublin 4, Ireland

Summary

A sediment sample was spiked with organochlorine pesticides, polychlorinated biphenyls (PCB) and polynuclear aromatic hydrocarbons (PAH), and extracted by methods involving shaking with solvents or ultrasonic agitation, and steam distillation. The ultrasonic agitation gave similar results to shaking but showed considerable savings in time. Steam distillation gave cleaner extracts but with lower recoveries and it did not remove the heavier PAH from the sediment.

INTRODUCTION

The importance of sediments as collectors for many organic micropollutants is well known, particularly for compounds which are relatively non-polar and practically insoluble in water. Such compounds tend to be adsorbed onto suspended particulate matter and to settle out into the sediments. This is a major mechanism in the removal of micropollutants from the water column, but such compounds may still be made available to the food chain through benthic organisms, so that high pollutant levels in sediment are still a cause for concern. Since the precipitation process is continuous, sediments are a good source of integrated samples. Micropollutant levels are generally several thousand times higher than in the water which should make for their easier analysis. However, to take advantage of this the extraction conditions must be optimised.

A wide range of methods is available for the extraction of organics from sediment. Sub-samples may be extracted continuously as in Soxhlet extraction or by successive batch treatments with solvents, the latter method being more versatile allowing rapid changes of solvent. This is particularly useful where the water content of the samples is likely to vary, so that a polar solvent may be used to mix the wet sample, followed by a non-polar solvent for extraction of the micropollutants of interest. The combination of steam distillation with solvent extraction described by Godefroot et al (1) concentrated the PCB from a water sample and therefore shows promise as a sediment treatment procedure for relatively involatile compounds.

The choice of extraction method depends on the target compounds to be analysed and on the physical problems posed by the sediment, in particular the organic and moisture content. Moisture content, for example, has most effect on Soxhlet extraction and no effect on steam distillation.

The purpose of the present work was to find a method which would be applicable to a range of target compounds in different sediment types - from muddy to sandy - which would be rapid and convenient, give good recoveries and a clean extract for chromatographic analysis.

PROCEDURE

The acetone/hexane extraction described by Goerlitz and Law (2) was used as a reference for the other methods. Ultrasonic agitation was then introduced instead of shaking with a view to making savings in time. A standard laboratory cleaning bath was used and test tubes containing the samples were simply immersed in the liquid for one minute agitations. As the standard organochlorine mixture was easily separated by GC the extracts were cleaned up without fractionation by passing the hexane solution through a florisil "Sep-Pak" and eluting with 20 per cent diethyl ether in petroleum ether (10 ml) before concentration to 1 ml for analysis.

Steam distillation/extraction (S.D.E.) as described by Godefroot et al gave good recoveries of pesticide and PCB from solution and was included in the tests with mixed results as described below.

For the comparison of recovery rates from the different extraction methods a spiked sample was prepared. The mechanism of incorporation of organic micropollutants into sediment particles over a long period of time cannot be easily reproduced in the laboratory, so an attempt was made to distribute the added compounds as evenly as possible onto the sediment particles in order that they would be adsorbed on active sites rather than existing as microscopic oily droplets which would be easily extracted. The compounds chosen as spikes covered a range of volatilities and were detectable by selective detectors for easy quantitation.

EXPERIMENTAL

Materials: Methanol acetone and hexane were distilled in glass before use. Diethyl ether and 2,2,4-trimethylpentane were "Analar" grade (BDH Chemicals). Florisil "Sep-Paks" (Waters Associates) and sodium sulphate ("Analar"; BDH Chemicals) were used for clean-up.
Standards: BHC (10 ppm mixed isomers in benzene), Polychlorobiphenyl 1260 (10 ppm in benzene), benzo(b)fluoranthene, benzo(k)fluoranthene and indeno(1,2,3-c,d)pyrene (10 ppm in toluene) were "Nanogen" standards. Aldrin, dieldrin and endrin were 20 ppm in isooctane from Chrompack. Fluoranthene, benzo(a)pyrene and benzo(g,h,i)perylene (Koch-Light laboratories) were dissolved in chloroform and diluted to 180 ppm.
Apparatus: The ultrasonic bath was a Model FS200 (Decon Ultrasonics Ltd.). The HPLC system comprised a Waters 6000A pump, Rheodyne valve Model 7125, Waters RCM-100 radial compression system with PAH cartridge, and Perkin-Elmer LC1000 fluorescence detector with excitation at 360 ± 15 nm and emission >430 nm. The mobile phase was acetonitrile/water 9:1 at 1 ml/min. GC analysis was performed on a Hewlett-Packard 5794A Gas chromatograph with on-column injection and ECD (Ni-63) detection. The column was fused silica (10 m x 0.25 mm I.D.), coated with Cp-Sil 19 CB bonded phase (Chrompack). The carrier gas was hydrogen at 32 kPa and the ECD purge gas was argon/methane (5%) at 100 ml/min. Programme: 70 °C rising at 10 °C/min to 250 °C, holding for 5 mins.
Preparation of sediment sample: A dried sediment sample was sieved through a 150 μm screen and used as the substrate. The characteristics are given in Table 1. A portion of this sediment (200 g) was covered with 200 ml of a diethyl ether solution containing 20 μg of PCB 1260 and 10 μg each of BHC isomers, aldrin, dieldrin, endrin and the six PAH. The mixture was agitated and evaporated slowly to dryness on a rotary evaporator (bath temperature 40 °C). As the final weight was 200 g, each gram contained 0.1 μg of PCB and 0.05 μg of the other standards.

Table 1. Physical Characteristics of Sediment Substrate

Loss on drying	4.0% (3 hr at 105 °C)
Organic Matter	8.8% (24 hr at 550 °C)
Fine Sand 63-150 µm	27%
Coarse Silt 20-60 µm	32%
Clay and Fine Silt <20 µm	41%

Extraction: Spiked sediment (5 g) was extracted with acetone/hexane according to Goerlitz and Law. This involved shaking for a total of 90 minutes. The hexane extract was passed through a florisil "Sep-Pak" which was then washed with 10 ml of 20% ether in petroleum ether. The combined solutions were concentrated to 1 ml. The same procedure was then repeated with ultrasonic agitation for one minute in place of each shaking, a total of 6 minutes. Both extracts were analysed for the target organochlorines by GC/ECD. On-column injection was used to minimise the effects of solvent composition. The results (Table 2) show recoveries of 73 - 91% for the shaking method, with slightly higher values for the ultrasonic treatment which is considerably shorter. The ultrasonic method can easily be extended to multiple samples of 1 - 5 g in test tubes.

Steam distillation/Extraction: Sediment samples (5 g) were mixed with pure water (50 ml) and extracted by steam distillation/extraction. The final extract in pentane was evaporated to 1 ml after the addition of 2,2,4-trimethylpentane (1 ml) and analysed by GC. For PAH analysis by reversed phase HPLC, 1 ml of methanol was used in place of 2,2,4-trimethylpentane.

The recovery of organochlorine compounds was lower than for the batch extractions, but the extracts were cleaner even without the florisil treatment. Aldrin however was still not recovered satisfactorily and this requires further investigation.

The recovery of PAH varied much more than the organochlorines as shown in Table 3. The fluoranthene peak was obscured by co-extracted material and the other five compounds ranged from 41% for benzo(b)fluoranthene to trace recoveries for indeno(1,2,3-c,d)pyrene.

The recovery for the PAH from aqueous solution was tested by adding 200 ml of standard solution (0.1 g/ml) to 50 ml of water and extracting as before. The results in Table 3 show full recovery of fluoranthene falling to about 50% for the benzo(g,h,i)perylene. This confirms that the steam distillation step is responsible for only a fraction of the losses in sediment extraction.

In order to check the ability of the solvent extraction to recover PAH from the sediment, a 5 g sample was extracted as before using the ultrasonic method. The resulting extract was subjected to steam distillation/extraction and taken up in 1 ml of methanol. Analysis by HPLC showed (Table 3) that this treatment still would not recover the heavier PAH completely.

CONCLUSION

The wide variation in water content of sediment samples is easily accommodated using batch extraction with polar followed by non-polar solvents such as acetone and hexane, or by steam distillation. The recovery of a selection of organochlorine compounds from a prepared sample indicates that ultrasonic agitation in a standard laboratory cleaning bath can be used effectively for the extraction with considerable savings in time over shaking methods.

Steam distillation/extraction is easily applied to sediments of high moisture content and give a cleaner extract than solvent extraction but the

Table 2. RECOVERY OF ORGANOCHLORINES

Comparison of Shaking, Ultrasonic Agitation and Steam Distillation/Extraction

Compound	No.	Shaking 5 g Sediment	Ultrasonic Agitation 5 g Sediment	S.D.E. 5 g Sediment	Standards
BHC	1	87%, 76%	114%, 126%	66%	125%
BHC	2	73%, 77%	87%, 146%	56%	113%
Aldrin	3	− −	− −	44%	117%
Dieldrin	4	88%, 86%	97%, 100%	78%	123%
Endrin	5	89%, 76%	100%, 97%	66%	124%
PCB 1260	6	91%, 80%	100%, 86%	36%	105%

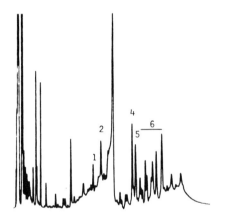

Shaking, 5 g Sediment

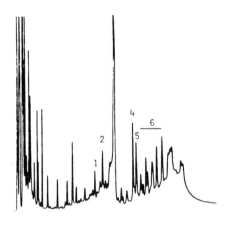

Ultrasonic Agitation, 5 g Sediment

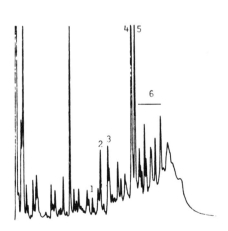

S.D.E., 5 g Sediment

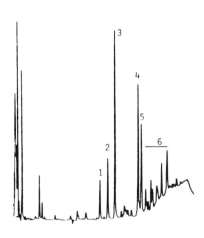

S.D.E., Recovery of Standards

For GC conditions see text

recovery of organochlorines was lower, in the range 40 - 80%. The PAH were more difficult to recover, with only traces of the two heaviest compounds being recovered.

Table 3. RECOVERY OF PAH

Method Sample	No.	S.D.E. 5 g Sediment	S.D.E. Aqueous Standards	Solvent/S.D.E. 5 g Sediment
Fluoranthene	1	672%	101%	230%
Benzo(b)fluoranthene	2	58%	93%	155%
Benzo(k)fluoranthene	3	30%	99%	88%
Benzo(a)pyrene	4	28%	86.5%	92%
Benzo(g,h,i)perylene	5	-	44.5%	21%
Indeno(1,2,3-c,d)pyrene	6	4%	55%	20%

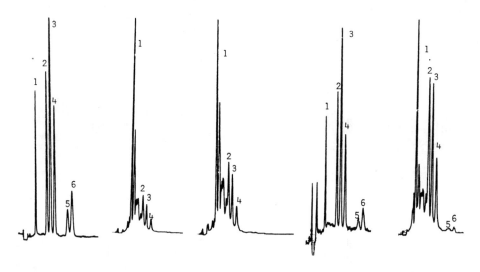

Standards 0.1 ppm   Sediment Blank 5g   Sediment 5g   Aqueous Standards   Solvent/S.D.E. 5g Sediment

References

1. Godefroot M., Stechele, M., Sandra, P. and Versele, M. "A new method for the quantitative analysis of organochlorine pesticides and polychlorinated biphenyls "Proceedings of the Second European Symposium on "Analysis of Organic Micropollutants" held in Killarney (Ireland) November 1981, A. Bjorseth and G. Angeletti (eds.) D. Reidel Publishing Co., Dordrecht, Holland, 1982, pp 16-23.
2. Goerlitz, Donald F. and Law, Leroy M., Determination of Chlorinated insecticides in suspended sediment and bottom material, J.A.O.A.C. 57 (1) 176-181 (1974).

SESSION II - ANALYSIS

Chairmen: E. MANTICA, R. FERRAND, H. KNOEPPEL and M. FIELDING

High resolution gas chromatography in water analysis: 12 years of development

Analysis of organics in the environment by functional group using a triple quadrupole mass spectrometer

Recent progress in LC/MS

Computerized data handling in gas chromatography/mass spectrometry

New developments in selective detection in capillary gas chromatography

Determination of organic water pollutants by the combined use of high-performance liquid chromatography and high-resolution gas chromatography

Analysis of organic micropollutants by HPLC

Identification of non-volatile organics in water using field desorption mass spectrometry and high performance liquid chromatography

Mass spectrometric identification of surfactants

Characterisation of non-volatile organics from water by pyrolysis-gas-chromatography-mass spectrometry

Automated GC/MS for analysis of aromatic hydrocarbons in the marine environment

Analysis of sulphonic acids and other ionic organic compounds using reversed-phase HPLC

Closed loop stripping and cryogenic on-column focusing of trihalomethanes and related compounds for capillary gas chromatography

Observations with cold on-column injection

Polymethylphenylsilicones in fused silica capillary columns for gas chromatography

Collaborative study on the mass spectrometric quantitative determination of toluene using isotope dilution technique

Determination of volatile organic substances in water by GC/SIM results of an investigation on the rivers Rhine and Main

HIGH RESOLUTION GAS CHROMATOGRAPHY IN
WATER ANALYSIS : 12 YEARS OF DEVELOPMENT*

K. GROB

EAWAG, Dübendorf, CH - 8600 Switzerland

Abstract

The development of glass capillary gas chromatography is reviewed with
special emphasis on application to water analysis. Column technology is
discussed in detail, in particular with respect to deactivation of
columns for direct injection of aqueous solution. In addition, the various
injection techniques, their advantages and problems, are summarized.
Finally, the relations between commercial and laboratory made columns
are discussed.

1. INTRODUCTION

The purpose of this paper is to review the development of glass capillary gas chromatography (GC) since the initiation of the COST 64b project in 1972. In particular, this will be related to the GC analysis of organic micropollutants in water. A short summary of the development of glass capillary columns is shown in Table 1.
Prof. Desty invented the glass capillary columns in 1959. He had for one year been successfully working with steel capillary columns. However, he felt that the stainless steel was quite expensive, and development started based on economy, and nothing else. The continuation was similar. Professor Desty had his first capillary GC separation in the fall of '59. However, then in spring of '60 he gave up. This first separation of a very light mineral oil fraction occurred on a glass column with temperature limit of 80°C, and this column lasted one day. The group of Desty continued to try this, and half a year later they gave up. They felt it useless to continue work with such a strange and hopeless thing.

* Prepared from a tape recording of Professor K. Grob's talk
  by A. B.

In Switzerland, we were at that point convinced that cigarette smoke analysis on steel capillary columns, which was our main activity at that time, could not be continued because of loss of essential substances due to the activity of the column. So we looked for glass because of its inertness. We had heard about Desty's group and the construction of a drawing machine, but we never heard about the fact that the group gave up. If we had heard this, we would never have started our activity in glass capillary columns.

In 1961 we got the same problems as Professor Desty did, and so we had our 3 years from 1961 to 1965 almost without any results. However, in 1965 the results improved. We had in 1965 again from tobacco smoke, the first GC/MS coupling which had ever been done with a capillary column, as far as I know. As seen from Figure 1, it was a heavy loss of separation due to the interface. One year later we had a Ryhage interface and it was not much difference between the pure GC and GC/MS traces (Figure 2).

2. GLASS CAPILLARY GAS CHROMATOGRAPHY IN THE COST PROJECT

In 1971, my wife and myself wanted to try more applications on relevant samples and not just struggle with columns. And so we wanted to see what we could do with water samples (with no connections to water research). Through a winding road we came on the close loop stripping analysis which then in 1972 produced such results as shown in Figure 3. This is an extract from contaminated ground water and we tried here to identify the source by comparing this extract with pure materials. The water research people in Switzerland was informed about this method, and soon we were invited to come to EAWAG where we learned about COST 64b. We had scientists for training in Switzerland and we also met them elsewhere and we found where the analysis had their shortages. They had to do too much work to extract samples, to clean up samples, to prepare them, so that finally they could inject the samples on to glass capillary columns. And so we felt what we should try to do was making columns, suitable for direct water injection.

We had no water resistant column at that time. In 1974 we had what now is called the barium carbonate column. The motivation to make it was to make a water resistant column. And when we coated it with a Carbowax, it was really fully water resistant. Shortly after this the disappointment came. When we injected water samples onto such a column, particularly by splitless injection, what came out could hardly be called a chromatogram. We could not even distinguish what were peaks and what was a strange base line. Later we recognized that with stream splitting it did work. Stream splitting had to be used on these columns also to analyze liquors, wines, and beverages of any kind. Later, exactly I should say only in 1981, we learned why it is impossible to do real organic separations by direct water injection on to a water resistant column with splitless injection.

To illustrate this, I prepared a few days ago the example shown in Figure 4. It is heavily simplified, but nevertheless I hope it may be helpful. The column is a short apolar column of a little bit more than standard film thickness and the column temperature is 70°C, i.e. 30 degrees below the boiling point of water (which is the solvent). We have 3 solutes, all dissolved directly in water. The solutes are butane diol, a very easily water soluble substances, primary hexanol, as an example of a substance which is to some extent soluble in water, but not very easily soluble in water, and lastly a hydrocarbon which is very poorly soluble in water. This mixture was first injected with stream splitting, which is the simplest way, and we got 3 regular peaks from a 12 meter column.

One problem should be noted. When you inject a high amount of an aqueous solution or a G.C. with FID, then immediately after the injection the flame is extinguished. After 20 seconds or so it can be reignited. I have tried many other ways, but that is how it has to be done.

Let us try to understand the comparison between those two simplified chromatograms shown in Figure 4. 2 µl of water were injected directly on column at 70°C oven temperature. Liquid water fully retains the water soluble substances. The butane diol which appears as a normal peak is fully trapped. Slowly, the water dries out. The organic material is concentrated in the last short trace of liquid, and when this liquid finally is evaporated, the organic material starts migrating through the column. Compared to stream splitting, there is only a delay, and this delay is identical to the time this material was dissolved in water and therefore could not migrate.

The primary hexanol is less soluble in water. It is partially soluble and that is the bad thing. Part of this material was trapped in the water. However, another part was constantly evaporated by the flowing carrier gas from the condensed water layer. Because of this, the corresponding peak starts in the same moment as the first trace of this material goes to the detector. However, we have a long delay for the last portion, because the last portion of this material disappeared only when the last trace of water is evaporated. And during all this time the FID receives hexanol and records it. The hexanol peak might have strange shapes. It can be one very large peak, or it can be split into 10 different peaks. The most frequent is a very flat peak with a sharp edge just before the last portion of material enters into the detector.

The third compound, the hydrocarbon, is practically not water soluble. Consequently, this material was never trapped in the liquid in the inlet portion of our column, but was constantly transported over it. We inject into an open cavity and here we have heavy mixing with surrounding hydrogen as carrier gas. So, this heavily diluted mixture is now blown during a long time over the condensed water in the column inlet. And that is why the peak is broadened.

In Figure 4 we also have an example of on-column injection, again with 2 µl injected. The first peak is again perfect. This is the same full trapping effect, the same focusing effect as we had already in splitless injection. We also have the same delay and we have the same broadening, which is due to partial trapping. The most interesting point is why we have a sharp peak after the broad and distorted peak. The reason is a non-trapping under on-column conditions. It occurred in the very narrow capillary where there is virtually no dilution by cold on-column injection. After injection, the column was immediately heated to 70 degrees. At this temperature hexanol is almost immediately vapourized from the liquid and goes on the way as a sharp, almost ideal band and so gives an almost ideal peak. Hence, with on-column injection we can have nice peaks with practically non water soluble materials using direct water injection.

What we can analyze with this technique suffers from two heavy limitations. It has to be practically non water soluble as all mineral oil constituents and many organic materials. And the second thing - it has to be done on column. And this is a heavy limitation again, because in many cases these water samples are relatively dirty, and we have no interest at all to inject all the dirt, all the things of no interest into the column, because this obviously reduces column lifetime.

First, the influence on columns. In Table 2, the particularities of direct aqueous injection are summarized. We had for a long time only two types of water resistant columns the polyglycol type (for instance carbowaxes) and particularly on a barium carbonate surface, which is a very suitable surface and the apolar inert columns with immobilized phases. It was a surprise when we had these immobilized phases which so long before were found to be absolutely water sensitive, now became very highly water resistant. Onto many of these columns, we have injected 400 to 500 water samples on-column. There was little change in the column properties. The column can also be repaired by resilylation if there is a harmful effect, which is certainly due to some hydrolysis of the stationary face.

Then we have the influence on the detector. It has already been mentioned that FID gets extinguished. This is not quite convenient, but we have to live with it. In Table 2, you also see mentioning of ECD. It was previously stated that direct water injection suffers from limitations. The heaviest limitation certainly is sensitivity. Water samples very rarely contain sufficient organic materials to produce a reasonable chromatogram for instance with a FID. So, what can be done? There are two ways to solve this problem. One way is to heavily increase sample size and this is something which is under study now, at groups in Scandinavia and elsewhere.

For the moment, however, what we can do, is to use very selective detection and here obviously ECD is the best choice. Using ECD with direct water injection was a kind of surprise to us. For many years we knew that water was an enemy of the ECD because it completely quenches the standing current. It picks away all the electrons in the ECD chamber and we knew that, as long as water is passing through an ECD, the ECD will behave strangely.

Nevertheless we have tried and it does work. However, there have to be precautions. It works only when all the water passes the ECD practically immediately. It has really to go through at once. This means three things : the injection part must not retain water. This means again that i.a. splitless injection cannot be used because a splitless injector has so many angles and small volumes into which the water will diffuse and from which it will travel onto the column with heavy delay, that it also will reach the ECD with heavy delay. So, for direct water injection and ECD, on-column injection has to be used. Furthermore, the column must not retain water. We need absolutely inert columns which really have the water amount pass immediately.

Another thing is of course the construction of the ECD itself. It has to have a geometry in which no water is trapped. And there I must say that not automatically, the most modern ECDs are the best. According to our experience, the older ECDs with a simpler cylindric geometry are the best ones.

Finally, table 2 lists the influence of the solute band, this means the influence of the water vapour. It should be repeated that only the practically not soluble substances and the very easily soluble ones can be expected to appear with nice peaks in the chromatogram. The latter ones are rare and usually not of high importance in water research.

Figure 5 shows an example of such direct water injection including all the technical details. First, a few comments about the base line. When you inject 2 microlitres of water on column, you produce a heavy change of flow rate. This is because of water vapour flows through your column with a totally different flow resistance as compared to hydrogen. This flow rate will produce strange changes in the shape of the base line. After the injection (arrow) a sharp drop because of resistance is seen. Then the base line recovers somewhat, and you have peaks 1, 2, 3. This GC has a new ECD with a complex geometry, so we have some water retardation which means that the water signal is broadened. As shown in Figure 5, methylene chloride already suffers from reduced sensitivity in the ECD. That is why it is so important to have the best geometry of the detector.

3. CONCLUSIONS

In conclusion I would like to make some personal remarks. In retrospect, the questions always arise : what about column design, column development. Some of my reflections are summarized in Table 3. For a long time, we had only the packed columns, and they were almost exclusively developed by their users. Those who used it, also made it. Many packed columns have been purchased, but I do not think that the companies really making commercial packed columns, did an essential contribution to the development of the columns. That was the job of the users. And what the users did, was a tremendous number of very sophisticated optimizations for different purposes. I think this is the strong point which allowed packed columns to become so accepted and to do so good a job over such a long time.

For capillary columns, the same parameters as for packed columns can be varied. However, they can all be varied in a far broader range.

For instance, we can change film thickness on capillary columns within a range of 1 to 400. We can change other parameters, we can select, we can optimize far more.

What about the present situation? The bulk of capillary GC today is done on a few column types, with a few stationary phases and a few internal diameters. It is all standardized. This means that practically no optimization for a given application is done, regardless of the larger possibilities of optimization we do have now. We have a drastic shift from individual to industrial column preparation.

The stationary phase is in most cases not optimized. Most users apply columns described in the literature (which again was copied from a previous paper) or columns which were on stock.

In addition and quite generally, far too long columns are used. Some times I try to demonstrate to the users how much they could gain with a shorter column. This includes time, economy, far better sensitivity, far better analysis of delicate substances because of reduced residence time and less contact with the column surface. However, no analyst today is going to cut a purchased expensive column to see whether part of it will do the job. Consequently, optimation is not done.

It is the same with column dimensions, they are hardly optimized and it is the same with the film thickness. Here we have a tremendous potential which is generally unknown. The same with the support surface. It appears as the support surface and all technical details have now become secrets. You cannot expect from your producer that they will tell you how they really did it. And this prevents you from exactly predicting what you can do with a certain column and what you cannot do. One should know how the column is made, otherwise you are working with a black box.

Table 3 also lists some disadvantages. The primary point is that only a fraction of the potential is really used.

The only way out of the problems discussed above is improved column technology in the laboratory. This was earlier done so nicely with packed columns a bit less with capillary column but is now almost on the point to vanish. My advice is : do not just depend on commercial sources.

In Switzerland, there are today 23 independent groups making their own columns. They know basic column technology and how to select optimized columns. These are important things many of you cannot do.

This is an encouragement not just to have this slow degradation of GC techniques. That would be a pity to the field of organic analysis, may be particularly to the water research field.

TABLE I - Development of Glass Capillary Columns

1956	Martin: Narrow bore column
1957	Golay: Wall-coated capillary column
1959, spring	Desty's group constructs capillary drawing machine
1959, fall	Desty's group runs first analysis (gasoline) on a glass capillary column
1960, spring	Desty discontinues developing glass capillary columns because of trouble with coating
1961	First trials on glass capillary in Zürich
1964	First useful glass capillary column in Zürich (Emulphor on carbonized surface)
1965	First gc/ms coupling with glass capillary column (ms Hitachi, fritted interface)
1967	Tesarik etches glass with HCl gas (1968 with HF)
1967	Organic derivatization of KOH-etched glass in Zürich
1972	Alexander and Rutten detect NaCl crystals
1975	Crystallization of $BaCO_3$ in Zürich
1977	Welsch detects effect of high temperature silylation (non-treated glass, 300°)
1978	Persilylation in Zürich (leached glass, 400°)
1979	Dandeneau produces first fused silica column
1980	Immobilization of apolar silicone coatings with peroxide in Zürich

TABLE II - Particularities of direct aqueous injection

Influence on column

In most columns support surface or stat. phase hydrolyzed.

Resisting column types:
Polyglycols (e.g. Carbowaxes) on BaCO$_3$ surface.
    Attention: water very strongly retained
              possible problems with missing immobilization
Inert columns with immobilized silicone coatings (apolar).
    Almost no retention for water

Influence on detector

FID: flame extinguished
ECD: ground current effectively quenched by water
     consequence: no response for organics as long as water is
                  eluted
     rule: all potential causes of retarded water elution to be
                                                        avoided
           (primary causes: injector, column, detector)

Influence on solute bands

Primary characteristic of water: poor solvent for most organics
                                 analyzed by gc.
Consequences:
Injection with splitting: no problems
On-column injection: no problems with strongly water-soluble
                              (fully trapped) solutes
                     little or negligible problems with water-insoluble (non-trapped)
                              solutes
                     problems with solutes of intermediate solubility (partially trap.)
Splitless injection: working only with strongly soluble solutes

TABLE III - Commercial versus laboratory-made columns

Typical features

Packed columns: developed primarily by users. Enormous number of exact
                optimizations for special applications.
Capillary columns: broader range of variable parameters for optimization
                   e.g. length, int. width, film thickness.

Present situation

Bulk of work done on a few column types. Practically no optimization for given application.
Drastic shift from individual to industrial preparation.
Column characteristics disappearing from analytical literature.

Optimization

feasible only through comparison of various columns which have to be on stock
or are quickly prepared. In contrast, the situation is characterized by
- stat. phase not optimized, but copied from literature.
- generally too long columns used (with reduced sensitivity, excessive
  side effects, reduced life, time wasted) Who cuts a purchased column
  to see whether a shorter column does a better job?
- internal width and sampling technique not adapted on each other.
- film thickness not optimized.
- support surface and immobilization technique unknown, thus important
  details of column behaviour not predictable.

Disadvantages of dependence on commercial col.

- only a fraction of the potential of capillary gc utilized, basically
  due to missing optimization.
- the user knows his tool insufficiently, thus he is unable of properly
  knowing how to use or not to use it.
- the working conditions are excessively limited by the principle of
  avoiding everything that could damage the column.
- characteristics of commercial columns can hardly be constant over a
  long period, due to continuous incorporation of improvements or simplifications of the manufacturing procedures (essential in case of severly standardized analyses).
- the further column development by manufacturers can hardly replace the
  development by the users.

Trennung von Gasphasenlösung auf dem Gaschromatograph
2,0 µl Lösung, Stromteilung 1:20; Wasserstoff 1,0 at. entsprechend
5,6 ml/min bei 25°; 8 min 0°, Programm 3°/min bis 95°; FID

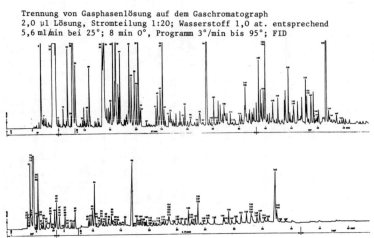

Trennung von Gasphasenlösung auf der kombinierten Apparatur
4,0 µl Lösung, Stromteilung 1:30; Wasserstoff 4 ml/min; 7 min 28°,
Programm 2,1°/min, bis 100°; Totalionenstrom

FIGURE I

4,0 µl Rauchlösung ohne Stromteilung injiziert auf 1 m Kapillarstück
mit Abtrennung des Lösungsmittels. Analyse auf 55 m/0,35 mm Glaskapillare,
belegt mit Emulphor O. Anzeige für GC: Flammenionisation; für GC-MS:
Totalionenstrom

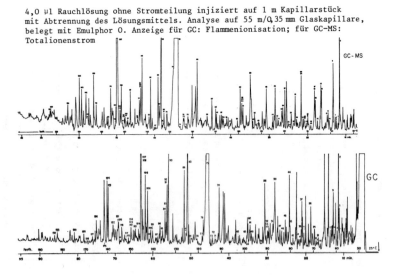

FIGURE II

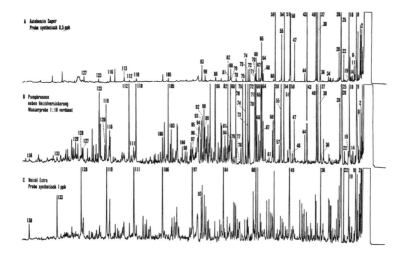

FIGURE III

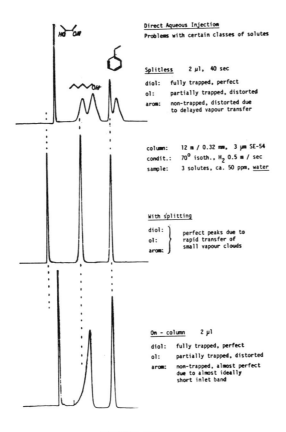

FIGURE IV

– 51 –

Direct Water Analysis with ECD,

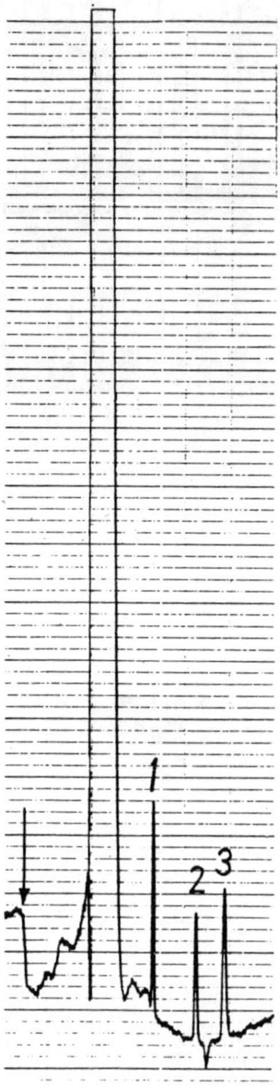

a practical example

Column: 3om / 0.33mm, 5 μm PS-255
both ends 1.5 m non-coated

Carrier: H_2, 0,4 bar (o.55 m/sec)

Gas chromatograph: C. Erba, Mod. 4160
on-column injector
oven $104°$ isotherm.

ECD: Carlo Erba, Mod. 25
const. frequency mode (CP, CF)
4o V, 0.5 μs, 1.1 nA

Recorder: 1 mV
chart speed 10 mm / min
Attenuation x 512

Injection: 2 μl on-column
syringe Hamilton, 75 SN
needle GA 32, 75 mm

Sample: 1: dichloro methane, 10 ppb
2: trichloro methane, 1 ppb
3: 1,1,1-trichloro ethane, 1 ppb
in deionized water
(spiked with methanol solution,
5 μl per 100 ml water)

FIGURE V

ANALYSIS OF ORGANICS IN THE ENVIRONMENT BY FUNCTIONAL GROUP USING A TRIPLE QUADRUPOLE MASS SPECTROMETER

D.F. HUNT, J. SHABANOWITZ, AND T.M. HARVEY
Department of Chemistry, University of Virginia,
Charlottesville, Virginia 22901, U.S.A.

Summary

Part of a comprehensive scheme for the direct analysis of organics in the environment is described. Liquid and solid chemical wastes and residues from lyophilized aqueous solutions are volatilized directly into the ion source of the triple quadrupole instrument. All or most wet chemical and chromatographic separation steps are eliminated. Analysis of organics by functional group and molecular weight is accomplished using the technique of collision activated dissociation and a series of 0.5 s neutral loss and parent ion scans under data system control on the triple quadrupole instrument. Results from the analysis of influent to a waste water treatment plant obtained by both GC/MS and triple quadrupole mass spectrometry are compared. In the triple quadrupole method, both knowns and unknowns are characterized, detection limits are at the 10-100 ppb level, and the total analysis time per sample is typically only 25-30 min.

1. INTRODUCTION

Analysis of water samples for the ca. 114 toxic organic chemicals on the U.S. Environmental Protection Agency (EPA) priority pollutant list presently involves separation of the organics from the matrix using a combination of wet chemical extractions steps, sample concentration and clean-up by various types of chromatography, and final analysis by a gas chromatograph-mass spectrometer-data system (1). This approach has proved to be highly reliable and applicable to a variety of different matrices. Efforts to extend the methodology and develop a master analytical scheme for the analysis of all organics in water that can be made to pass through a gas chromatograph are presently in progress (2).

Disadvantages of the above approach include: (a) the inability to detect highly polar compounds too involatile or thermally labile to pass through the gas chromatograph, (b) the high labor costs dictated by the need for extensive sample clean-up and preseparation prior to analysis, and (c) the large amount of time required to perform even a single analysis on a relatively expensive gas chromatograph-mass spectrometer-data system.

Here we describe part of an alternate comprehensive analytical scheme that overcomes many of the limitations inherent in the GC/MS methodology. The new method employs the technique of collision activated dissociation (CAD) on a triple quadrupole mass spectrometer and facilitates direct, rapid, qualitative-semiquantitative analysis of organics in both liquid and solid environmental matrices. Elimination of most, if not all, wet chemical and chromatographic separation steps, detection of both knowns and unknowns by molecular weight and functional group at the 0.01 to 1.0 ppm level, and a total analysis time per sample of under 30 min are additional features of the triple quadrupole method.

Preliminary results obtained on analysis of industrial sludge using the above approach were reported in an earlier paper (3). Here we compare results obtained on the same influent to a waste water treatment plant by conventional gas-chromatography/mass spectrometry and triple quadrupole mass spectrometry methods, respectively. Additional examples of the utility of tandem mass spectrometry for mixture analysis can be found in three, recent, excellent reviews (4-6).

2. EXPERIMENTAL

All experiments were performed on a Finnigan, Model 4500 triple quadrupole mass spectrometer-data system. This instrument consists of a conventional electron ionization/chemical ionization ion source, three quadrupole filters, Q1, Q2, and Q3, and a conversion dynode electron multiplier detector (Figure 1) (7). Mixture analysis on this instrumentation can be accomplished with or without prior chromatographic separation of the components.

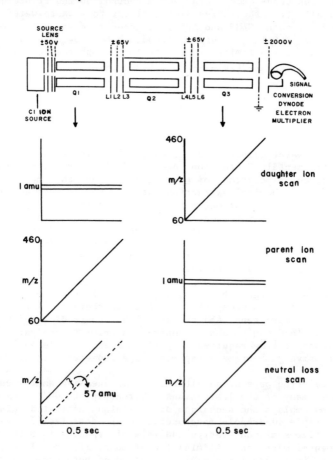

Figure 1. Modes of operation of the Finnigan triple quadrupole mass spectrometer.

Operation of the Triple Quadrupole as a Conventional Single Analyzer Instrument

When a quadrupole mass filter operates with only low radio frequency (rf) on the rods, it functions as an ion-focusing device and transmits ions of all m/z values. Accordingly, if Q1 and Q2 function in the rf only mode, and Q3 is operated with both rf and dc potentials on the rods, the triple quadrupole mass spectrometer behaves like a conventional single analyzer instrument. All ions exiting the ion source are transmitted through Q1 and Q2 and are mass analyzed in Q3. Standard electron ionization or chemical ionization mass spectra of each component eluting from the gas chromatograph result.

Operation of the Triple Quadrupole Mass Spectrometer in the Daughter Ion Scan Mode

Daughter ion scans (Figure 1) are particularly useful when a complex matrix is to be analyzed for the presence or absence of a small number of known compounds. To obtain collision-activated dissociation mass spectra, matrix components are volatilized into the ion source and Q2 is filled with argon gas to a pressure of 1-4 mtorr and operated in the rf only mode. The rf and dc potentials on Q1 are then set to pass ions with a particular m/z value (Figure 1). When the selected ions enter Q2, they suffer collisions with argon atoms, become vibrationally excited, and dissociate to produce ions characteristic of their structure. In the presence of a weak rf field, all of these fragment ions are transmitted to Q3 where they undergo mass analysis. A mass spectrum of the fragment ions derived from each ion entering Q2 results. Other matrix components will usually contribute to the total signal at the particular m/z value selected by Q1 and the resulting collision-activated dissociation mass spectrum will contain fragment ions derived from several components. All ions in the collision activated-dissociation mass spectrum of the pure analyte should appear in the mixture spectrum if the known compound is present in the matrix.

Previous reports from this laboratory have described the use of this approach for the detection of nitrophenols and phthalates in industrial sludge at the 100 ppb level (8) and for sequence analysis of polypeptides in mixtures (9).

Operation of the Triple Quadrupole Mass Spectrometer in the Parent Ion Mode

If Q3 is set under data system control at a particular m/z value, Q2 is operated as a collision cell, and Q1 is scanned over the desired mass range, the resulting spectra contain all of the parent ions that afford a particular fragment in the collision-activated dissociation process. This is referred to as a parent ion scan and is particularly useful for the analysis of either a homologous series of compounds or a class of compounds containing the same functional group. In the present work, phthalates are analyzed using this scan mode.

Operation of the Triple Quadrupole Mass Spectrometer in the Neutral Loss Scan Mode

Neutral loss scans are employed for the rapid analysis of complex mixtures for members of a class of compounds that undergo the same type of fragmentation, loss of the same neutral moiety, under collision-activated dissociation conditions. Q2 is employed as a collision chamber and both Q1 and Q3 are set under data system control to scan repetitively at a fixed mass separation over the desired mass range (Figure 1). Mixture components are volatilized into the ion source and converted to ions characteristic of sample molecular weight under either positive or negative chemical ionization conditions. At any point in the scan of Q1, all ions of a particular m/z value will be transmitted to Q2. Only those ions that lose a particular neutral of specified mass will have the right m/z ratio for transmission through Q3. Nitroaromatic compounds, for example, form abundant (M+H)+ ions under methane, positive-ion, chemical-ionization conditions and then suffer loss of 17 amu (OH·) on collision-activated dissociation in Q2. Since loss of 17 amu is highly charateristic of nitroaromatic compounds, the neutral loss scan with Q1 and Q3 separated by 17 amu facilitates detection of these compounds and is transparent to most other matrix components. Fragment ions produced by loss of neutrals having masses other than 17 amu fail to pass through Q3 and are never detected. Spectra recorded in the neutral loss scan mode yield both the molecular weight and relative abundance of all members of a particular class of compounds in the matrix.

In the past this approach has been used in our laboratory for the analysis of heterocyclic organosulfur compounds (10), carboxylic acids in urine (11), nitrated polycyclic aromatic hydrocarbons in diesel particulates (12) and fuels treated with gaseous NO_2 (13). Here, neutral loss scans are employed for the analysis of chlorocarbons, polycyclic aromatic hydrocarbons, and phenols.

Chemicals

Priority-pollutant analytical standards used in this study were obtained from Supelco Inc., Bellefonte, PA. Bis(2-chloroisopropyl) ether and the deuterated compounds employed as internal standards for GC/MS were purchased from Radian Corp., Austin, TX. All other reagents were purchased from Aldrich Chemical Co., Milwaukee, WI. Methane (99.9%), nitrous oxide (98.5%), argon (99.8%), and dinitrogen tetroxide (98%) were purchased from Matheson Gas Products, Inc., East Rutherford, NJ.

Instrumental Parameters

The mass spectrometer was operated under chemical ionization conditions with a mixture of methane and nitrous oxide as the reagent gas. Electron bombardment of methane at 0.5 torr generates the reactant ions, CH_5^+ and $C_2H_5^+$. Interaction of nitrous oxide with thermal electrons in the ion source produces the strong, Bronsted-base, reactant ion, OH-, by a two step process involving dissociative electron capture to produce N_2 and $O^{\underline{-}}$ and subsequent reaction of the oxygen radical anion with methane to generate OH- and CH_3·. Optimum reactant ion abundance is obtained by metering nitrous oxide into the methane gas at 0.5 torr until the intensity of the OH- signal maximizes. This usually occurs when the ratio of methane to nitrous oxide is ca. 10/1.

The ion source was maintained at a temperature of $150°$ C. Argon was employed as the collision gas at a pressure of 1.5 mtorr in all collision-activated dissociation experiments. The energy of the ions entering the collision cell was 10 eV.

Analysis of Phthalates, Aromatic Hydrocarbons, Chlorocarbons and Phenols.

Solid matrices (5-10 mg) and residues from lyophilized aqueous samples are placed on or between glass-wool plugs in a 4.5 cm piece of 4 mm x 2.5 mm i.d. glass tubing. If necessary, 10-20 µl of dichloromethane can be added to the glass tube to facilitate extraction of the organics out of the matrix and onto the glass wool. After 5 min the dichloromethane is removed by purging the glass tube with a stream of nitrogen for 10-15 s at a flow rate of 20 ml/min.

Phenols in the matrix are converted to carbamates by saturating the sample with 10 µl of ethyl acetate-methyl isocyanate (4:1) containing 100 ppm of triethylamine and allowing the reaction to proceed for 5 min at room temperature. (Methyl isocyanate is freshly distilled daily with a Wheaton micro-distillation kit.) Excess reagents are removed from the reaction mixture by purging the sample tube with a stream of nitrogen for 10-15 s. Polycyclic aromatic hydrocarbons in the sample are then converted to nitroaromatic compounds by treating the matrix with 1 µl of trifluoroacetic acid followed by a stream of gaseous dinitrogen tetroxide for 3-5 s (14). A stream of nitrogen is again employed to remove excess reagent and the sample is then placed on the end of a specially constructed heated solids probe and inserted into the removable ion-source volume of the Finnigan triple quadrupole mass spectrometer.

A mixture of methane and nitrous oxide sufficient to maintain a pressure of 0.5 torr in the ion source is employed as both the carrier gas and to generate the chemical ionization reactant ions, CH_5^+ and OH^-. Gas flows through the sample tube into the ion source at all times. Slow thermal volatilization of organics from the matrix is facilitated by heating the sample from 25 to 360°C over a period of 5-9 min. During this time the instrument, under data system control, is cycled repetitively through a series of eight 0.5 s experiments, one parent ion scan at m/z 149 for phthalates and seven neutral loss scans for the other three classes of compounds studied. At the end of the heating period all data acquired from each type of scan is summed together and printed out in a conventional bar graph format. The result is a plot of molecular weight vs. relative abundance for all mixture components containing the particular functional group being detected. Total time for sample derivatization, data acquisition and data processing in the above procedure is typically 25 min.

2. RESULTS AND DISCUSSION

Success of the functional group analysis approach depends on the assumption that ions characteristic of the molecular weights of all members of a particular class of organic compounds will suffer collision activated dissociation with high efficiency and either lose a highly characteristic neutral or form a highly characteristic charged fragment. Unfortunately this is not the case for several types of organic compounds under the low energy conditions employed in the collision cell of the triple quadrupole instrument (10-20 eV). Those compounds that fail to meet the above criteria must be derivatized in order to promote the desired behavior in the collision-activated dissociation process. Ideally the derivatization reaction should (a) employ volatile reagents that can be removed easily from the matrix, (b) proceed in high yield under mild conditions, and (c) introduce a functional group that is sufficiently basic or acidic to localize proton addition or abstraction at the site of the newly introduced substituent.

Phthalates

Collision activated dissociation of (M+H)+ ions from all phthalates except dimethyl phthalate affords the protonated anhydride, m/z 149, in high yield (eq 1). Since this fragment is highly characteristic of phthalate

$$\underset{m/z\ 223}{\underset{H^+}{\text{COOEt}}\text{COOEt}} \longrightarrow \underset{m/z\ 149}{\text{[phthalic anhydride]}^{+OH}} + C_2H_4 + EtOH \quad (1)$$

esters, members of this class of compounds can be monitored in mixtures using parent ion scans on the triple quadrupole instrument (Table I).

Table I. Diagnostic Ions Observed for Phthalates in the 149 amu Parent Ion Scan (PIS).

compound	mol wt	PIS 149 m/z (RA)[a]
diethyl phthalate	222	223 (10)
		177 (100)
di-n-butyl phthalate	278	205 (100)
		279 (50)
n-butyl benzyl phthalate	312	205 (100)
		313 (35)
di-2-ethylhexyl phthalate	390	391 (100)
		261 (4)
di-n-octyl phthalate	390	391 (100)
		261 (30)

[a] % relative abundance values from the spectrum generated by summing together NLS 17, 30, 46, 47, 57, 35, 36 and PIS 149 spectra.

Results from the analysis of a sample from an influent to a waste water treatment plant are shown in Figure 2, (top). Signals at m/z 223, 279, 313 and 391 correspond to (M+H)+ ions from diethyl-, dibutyl-, butylbenzyl-, and either di-2-ethylhexyl- or di-n-octyl phthalate, respectively. Fragment ions in the main-beam, methane, chemical-ionization mass spectrum of these four phthalates at m/z 167, 177, 205, and 261 are also observed in the parent ion scan because they too undergo further dissociation in the collision chamber to produce m/z 149. The signal at m/z 149 represents that fraction of the protonated anhydride species generated in the ion source that passes through the collision chamber intact. Comparison of spectra recorded on spiked (Figure 2, bottom) and unspiked samples indicates that the phthalates in this particular waste water sample are present at the 30-500 ppb level.

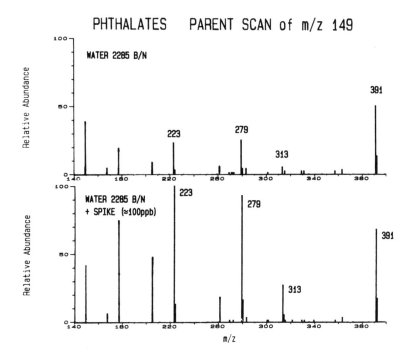

Figure 2. (Top) Results of a m/z 149 parent ion scan for detection of phthalates in a sample from an influent to a waste water treatment plant. (Bottom) Results from the same water sample spiked at the 100 ppb level with diethyl-, di-n-butyl-, butyl benzyl-, and di-n-octyl phthalates.

Polynuclear Aromatic Hydrocarbons

Collision activated dissociation of either M+ or (M+H)+ ions from polynuclear aromatic hydrocarbons occurs with low efficiency during collision experiments conducted with ion energies in the range of 10-20 eV. Derivatization of this class of compounds for analysis on the triple quadrupole is accomplished by exposing the sample matrix to gaseous N_2O_4 for 3-5 s. Under these conditions mono- and/or dinitro derivatives are formed in good yield from all polynuclear aromatic hydrocarbons on the EPA priority pollutant list. Carbamate derivatives of phenols on the same list are not nitrated in the above procedure.

Nitroaromatic compounds make excellent derivatives for analysis by the tandem mass spectrometry approach since the nitro group is highly basic and easily protonated when methane is used as the chemical ionization reagent gas. The resulting (M+H)+ ions dissociate readily in the collision cell with loss of one or more of the neutrals, OH, NO, NO_2, and HNO_2 (Figure 3). Losses of OH (17 amu), NO (30 amu), NO_2 (46 amu) and HNO_2 (47 amu) are highly characteristic of the nitro functional group and one or more of these pathways is observed for all of the polynuclear aromatic hydrocarbons on the priority pollutant list. Analysis of this class of compounds can be carried out, therefore, using neutral loss scans for one of the above moieties. If it is necessary to distinguish between nitrated polynuclear aromatic hydrocarbons already in the environmental matrix and those produced in the derivatization step, a second sample of the untreated matrix must be examined directly.

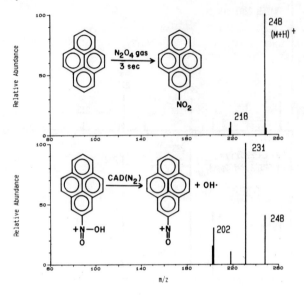

Figure 3. (Top) Main beam positive ion CI methane spectrum of nitropyrene. (Bottom) Collision activated dissociation mass spectrum of the (M+H)+ ion from nitropyrene.

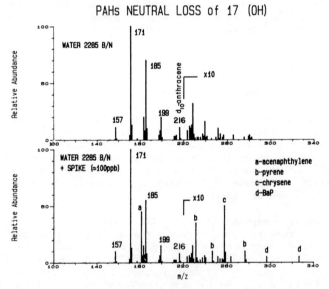

Figure 4. (Top) Results of a 17 amu neutral loss scan for detection of aromatic hydrocarbons in a sample from an influent to a waste water treatment plant. (Bottom) Results on the same sample spiked at the 100 ppb level with four polynuclear aromatic hydrocarbon standards.

Results of 17 amu neutral loss scans on a sample from an influent to a waste water treatment plant and the same sample spiked with a standard mixture of four polynuclear aromatic hydrocarbons are presented in Figure 4. Most of the ions in the spectra occur at m/z values 16 amu (M+H-NO)+ below the molecular weight of the nitro derivative and, therefore, 29 amu above the molecular weight of the parent aromatic hydrocarbon. Signals at even m/z values contain an even number of nitrogen atoms and are formed by loss of OH from either (M+H)+ ions or (M+H-NO)+ fragment ions in the main beam

Table II. Diagnostic Ions Observed for Hazardous Chemicals in the 17, 30, 46 and 47 amu Neutral Loss Scan (NLS) Modes.

compound	mol wt	NLS 17[a] m/z	(RA)[b]	NLS 30[a] m/z	(RA)[b]	NLS 46[a] m/z	(RA)[b]	NLS 47[a] m/z	(RA)[b]
benzene	77	-	-	94	(25)	78	(12)	77	(100)
toluene	92	-	-	108	(10)	92	(32)	91	(100)
phenol	94	123	(93)	-	-	94	(3)	93	(9)
dimethyl phthalate	194	-	-	135	(100)	-	-	-	-
2-nitrophenol	139	-	-	139	(100)	93	(35)	92	(3)
4-nitrophenol	139	123	(100)	139	(100)	94	(6)	93	(18)
ethyl benzene	106	135	(6)	121	(1)	106	(43)	105	(100)
xylenes	106	135	(4)	122	(5)	106	(66)	105	(100)
2,4-dimethylphenol	122	-	-	-	-	122	(80)	121	(30)
chlorobenzene	112	141	(100)	128	(7)	112	(10)	111	(22)
naphthalene	128	157	(43)	144	(12)	128	(44)	127	(100)
2,4-dinitrotoluene	182	166	(75)	-	-	137	(35)	136	(100)
2,6-dinitrotoluene	182	166	(21)	-	-	137	(11)	136	(100)
2,4-dinitrophenol	184	168	(100)	-	-	109	(10)	138	(3)
benzidine[c]	184	168	(21)	-	-	-	-	-	-
acenaphthylene	152	181	(100)	167	(10)	152	(15)	151	(7)
2-methyl-4,6-dinitrophenol	198	182	(100)	-	-	123	(18)	122	(3)
2-chloronaphthalene	162	191	(100)	178	(9)	162	(52)	161	(46)
fluorene	166	195	(100)	-	-	166	(39)	165	(50)
4-bromophenyl phenyl ether	248	199	(100)	185	(15)	170	(10)	169	(2)
anthracene	178	207	(100)	193	(7)	178	(5)	177	(5)
phenanthrene	178	207	(72)	194	(8)	178	(85)	177	(100)
acenaphthene	154	183	(9)	-	-	199	(100)	198	(18)
fluoranthene	202	231	(83)	217	(7)	202	(100)	201	(39)
4-chlorophenyl phenyl ether	204	233	(100)	219	(3)	204	(26)	203	(5)
chrysene	228	257	(100)	244	(8)	228	(60)	227	(33)
benz(a)anthracene	228	257	(40)	244	(7)	273	(100)	272	(19)
pyrene	202	276	(100)	-	-	202	(11)	246	(10)
N-nitrosodiphenylamine[c]	198	288	(100)	-	-	169	(100)	-	-
benzo(k)fluoranthene	252	296	(100)	283	(5)	267	(28)	266	(20)
benzo(b)fluoranthene	252	296	(100)	313	(15)	267	(95)	296	(100)
benzo(a)pyrene	252	326	(100)	-	-	-	-	-	-
dibenz(a,h)anthracene	278	322	(11)	339	(100)	293	(16)	322	(16)
benz(ghi)perylene	276	350	(89)	-	-	336	(100)	335	(21)
o-phenylenepyrene	276	350	(100)	-	-	321	(55)	320	(23)

[a]For isotope clusters, the reported ion current is for the ion of lowest mass. [b]% Relative abundance values obtained from the spectrum generated by summing together NLS 17, 30, 46, 47, 57, 35, 36 and PIS 149 spectra. [c]Derivatized with methyl isocyanate only.

chemical ionization spectrum of the corresponding dinitro derivatives. Dinitration is observed for most polynuclear aromatic hydrocarbons. Diagnostic ions in the neutral loss scan modes used for nitrated polyaromatic hydrocarbons are given in Table II.

The internal standard of d_{10}-anthracene (m/z 216) at 50 ppb as well as the standard addition of four polyaromatic hydrocarbons (a,b,c and d), at 100 ppb, are readily detected (Figure 4, bottom). Comparison of spectra obtained from spiked and unspiked samples (Figure 4) indicates that naphthalene (m/z 157), Cl- (m/z 171), C2- (m/z 185), and C3- (m/z 199) naphthalenes, plus acenaphthene-biphenyl (m/z 183), fluorene (m/z 195), and phenanthrene-anthracene (m/z 207) are all present in this particular sample matrix at the 3-300 ppb level.

Phenols

Like the situation with polynuclear aromatic hydrocarbons, M+ and (M+H)+ ions from phenols also dissociate inefficiently under the low energy conditions employed in the triple quadrupole instrument. Accordingly, members of this class of compounds are also derivatized prior to analysis. Treatment of the sample matrix with methyl isocyanate converts phenols into carbamates. This derivative is highly basic and readily forms (M+H)+ ions that in turn suffer facile dissociation with loss of methyl isocyanate (57 amu) in the collision cell (eq 2). Analysis of phenols in mixtures is accomplished, therefore, using 57 amu neutral loss scans on the carbamate derivatives. Diagnostic ions for phenols using this scan mode are shown in Table III.

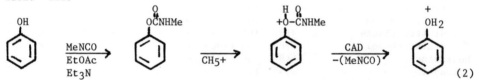

(2)

Table III. Diagnostic Ions Observed for Hazardous Chemicals in the 57 amu Neutral Loss Scan Mode

compound	mol wt	NLS 57[a] m/z (RA)[b]
phenol	94	95 (100)
2,4-dimethylphenol	122	123 (100)
2-chlorophenol	128	129 (100)
2-nitrophenol	139	140 (11)
4-nitrophenol	139	140 (19)
4-chloro-3-methylphenol	142	143 (100)
2,4-dichlorophenol	162	163 (100)
benzidine[c]	184	185 (100)
2,4,6-trichlorophenol	196	197 (100)
pentachlorophenol	264	265 (100)

[a]For isotope clusters, the reported ion current is for the ion of lowest mass. [b]% relative abundance values obtained from the spectrum generated by summing together NLS 17, 30, 46, 47, 57, 35, 36 and PIS 149 spectra. [c]Derivatized with methyl isocyanate only.

Results obtained on a waste water sample and the same sample spiked at the 100 ppb level with four phenols from the EPA priority pollutant list are shown in Figure 5. The signals observed correspond to (M+H)+ ions of the parent phenols since these are the ions transmitted through the second mass analyzer. Ions at m/z 85, 140, 143, and 265 (Figure 8, bottom) are assigned to phenol, nitrophenol, chloromethylphenol, and pentachlorophenol, respectively, and are the result of the 100 ppb standard addition. The signal at m/z 100 occurring in both the top and bottom spectra is the result of d_5-phenol, which was added as an internal standard at the 50 ppb level. The signals at m/z 95, 109, 123, and 171 (Figure 5, top) correspond to phenol, methylphenol and dimethylphenol, and are detected in this water sample at the 5-50 ppb level.

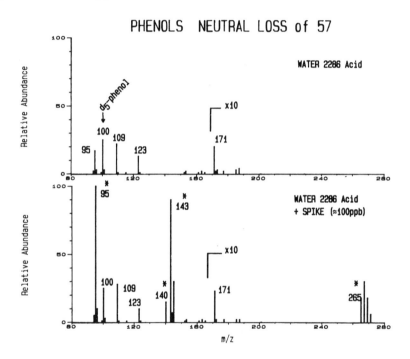

Figure 5. (Top) Results of a 57 amu neutral loss scan for detection of phenols in a sample from an influent to a waste water treatment plant. (Bottom) Results from the same sample spiked at the 100 ppb level with phenol, 4-nitrophenol, chloromethylphenol, and pentachlorophenol.

Chlorocarbons

The EPA priority pollutant list contains ca. 60 chlorocarbons and/or preparations. We find that these can be analyzed directly under positive ion chemical ionization conditions using neutral loss scans that monitor for the loss of either Cl (35 amu) or HCl (36 amu). Diagnostic ions for a large number of chlorocarbons are shown in Table IV.

Results obtained using 35 and 36 amu neutral loss scans for the analysis of chlorocarbons in a sample consisting of influent to a waste water treatment plant are shown in Figures 6 and 7, respectively. Data on the same sample spiked with two chlorocarbons standards, 1,4-dichlorobenzene and 1,2,4-trichlorobenzene, at the 100 ppb level are also presented. In the spiked sample, signals for these two standards appear at m/z 112 and

Table IV. Diagnostic Ions Observed for Hazardous Chemicals in the 35 and 36 amu Neutral Loss Scan Mode

compound	mol mw	NLS 35[a] m/z (RA)[b]		NLS 36[a] m/z (RA)[b]	
1,1-dichloroethane	98	-	-	27	(100)
1,2-dichloroethane	98	-	-	27	(100)
2-chloroethyl vinyl ether	106	-	-	27	(100)
cis- and trans-1,3-dichloropropene	110	-	-	39	(100)
1,2-dichloropropane	112	-	-	41	(100)
chloroform	118	-	-	47	(100)
dichloromethane	84	49	(50)	49	(100)
1,1-dichloroethylene	96	-	-	61	(100)
t-1,2-dichloroethylene	96	-	-	61	(100)
1,1,1-trichloroethane	132	-	-	61	(100)
1,1,2-trichloroethane	132	-	-	61	(100)
Chlordane	406	66	(10)	65	(100)
chlorobenzene	112	-	-	77	(74)
trichlorofluoromethane	136	82	(100)	81	(15)
carbon tetrachloride	152	82	(100)	81	(5)
Heptachlor epoxide	386	82	(66)	81	(100)
trichloroethylene	130	-	-	95	(100)
1,1,2,2-tetrachloroethane	166	96	(4)	95	(100)
alpha-BHC	288	-	-	111	(100)
beta-BHC	288	-	-	111	(100)
delta-BHC	288	-	-	111	(100)
1,2-dichlorobenzene	146	-	-	111	(100)
1,3-dichlorobenzene	146	-	-	111	(100)
1,4-dichlorobenzene	146	-	-	111	(100)
Deldrin	378	-	-	111	(100)
tetrachloroethylene	164	130	(14)	129	(100)
1,2,4-trichlorobenzene	180	146	(30)	145	(100)
Arochlor 1221		154	(30)	153	(100)
Arochlor 1232		154	(28)	153	(100)
4,4'- DDD	318	172	(8)	171	(100)
gamma-BHC (Lindane)	288	-	-	181	(100)
hexchlorobutadiene	258	188	(100)	187	(15)
Arochlor 1016		188	(70)	187	(100)
Arochlor 1242		188	(70)	187	(100)
4,4'-DDT	352	206	(85)	205	(100)
3,3'-dichlorobenzidine[c]	252	218	(100)	217	(84)
Endosulfan sulfate	420	218	(30)	217	(50)
Arochlor 1248		222	(100)	221	(90)
pentachlorophenol	264	230	(19)	229	(7)
hexachlorocyclopentadiene	270	235	(100)	235	(60)
Toxaphene[d]	410	-	-	235	(100)

Table IV. (Con't) Diagnostic Ions Observed for Hazardous Chemicals in the 35 and 36 amu Neutral Loss Scan Mode

compound	mol mw	NLS 35[a] m/z (RA)[b]	NLS 36[a] m/z (RA)[b]
alpha-Endosulfan	404	240 (30)	239 (100)
beta-Endosulfan	404	240 (25)	239 (100)
Endrin	378	244 (50)	243 (100)
hexachlorobenzene	282	248 (100)	247 (10)
Aldrin	362	256 (50)	235 (100)
4,4'-DDE	316	- -	281 (100)
Arochlor 1254		290 (100)	289 (70)
Heptachlor	370	300 (100)	299 (100)
Arochlor 1260		324 (100)	323 (30)

[a]In the case of isotope clusters, the reported ion current is for the ion of lowest mass. [b]% Relative abundance values from the spectrum generated by summing together NLS 17, 30, 46, 47, 57, 35, 36, and PIS 149 spectra. [c]Derivatized with methyl isocyanate only. [d]The most abundant peak in the isotope cluster is reported.

146 in the 35 amu neutral loss scan and at 111 and 145 in the 36 neutral loss scan. Signals at the same m/z values in the unspiked sample indicate that dichloro- and trichlorobenzenes are only present at the 3-10 ppb level in the influent to the waste water treatment plant (note the difference in relative abundance scales shown on the spectra recorded for unspiked and spiked samples in Figures 6 and 7). Other ions observed in the 36 amu neutral loss scan are probably due to the loss of two water molecules from polyhydroxylated compounds in the sample matrix.

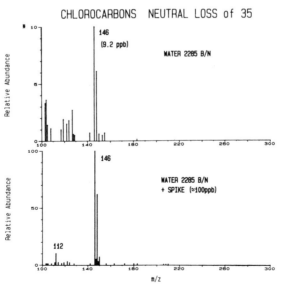

Figure 6. (Top) Results from a 35 neutral loss scan for chlorocarbons in a sample from an influent to a waste water treatment plant. (Bottom) Results from the same sample spiked with 1,4-dichlorobenzene and 1,2,4-trichlorobenzene at the 100 ppb level.

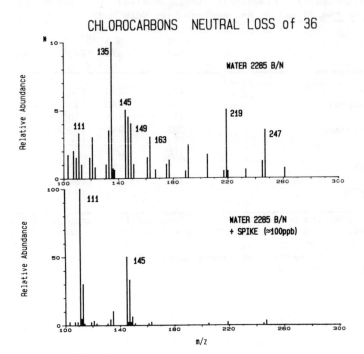

Figure 7. (Top) Results from a 36 amu neutral loss scan for chlorocarbons in a sample from an influent to a waste water treatment plant. (Bottom) Results from the same sample spiked with 1,4-dichlorobenzene and 1,2,4-trichlorobenzene at the 100 ppb level.

Comparison of Results Obtained by GC-Mass Spectrometry and Triple Quadrupole Mass Spectrometry

Results from a double blind analysis of the same sample of influent to a waste water treatment plant by both GC-mass spectrometry and triple quadrupole mass spectrometry are shown in Table V. Quantitation by GC/MS was accomplished with the aid of internal standards, d_{10}-anthracene and d_5-phenol. Since relative response data on priority pollutants under collision activated dissociation conditions are not available, quantitation in the triple quadrupole mode was performed by recording spectra before and after addition of known amounts of the identified compounds.

Both methods of analysis detected the same priority pollutants in the waste water influent. Levels of the identified compounds reported by the two methods are also in excellent agreement. Note that the GC-mass spectrometry method provides data on isomeric compounds separated on the chromatographic column. The triple quadrupole approach usually provides only data corresponding to the total quantity of all isomers contributing to a particular signal. Conventional GC-mass spectrometry uses a reverse library search routine to look for a list of known compounds in an environmental matrix. The triple quadrupole approach detects both known and unknowns (hydroxybiphenyl and the alkylnaphthalenes) with no additional time or effort. Each analysis costs less than 1/10 of that performed by GC/MS.

Table V. Quantitative Analysis of Hazardous Chemicals in a Sample from the Influent to a Waste Water Treatment Plant: Comparison of Results Obtained by GC-Mass Spectrometry and Triple Quadrupole Mass Spectrometry.

compound	scan mode		GC/MS level (ppb)		TQ/MS level (ppb)
diethyl phthalate	PIS	149	36.8		42
dibutyl phthalate	"	"	43.6		43
butyl benzyl phthalate	"	"	25.0		31
di-2-ethylhexyl phthalate	"	"	537		483
naphthalene	NLS	17	28.0		27
acenaphthene mw 154	"	"	0.8		
biphenyl mw 154	"	"	51.0	51.8	56
fluorene	"	"	1.2		3
anthracene/phenanthrene	"	"	1.9		3
c1-naphthalene	"	"	nd^a		90
c2-naphthalene	"	"	nd^a		129
c3-naphthalene	"	"	nd^a		69
phenol	NLS	57	36.0		36
dimethylphenol	"	"	22.0		24
dichlorophenol	"	"	1.2		3
methylphenol	"	"	nd^a		44
hydroxybiphenyl	"	"	nd^a		4
1,2-dichlorobenzene	NLS	35	0.4		
1,3-dichlorobenzene	"	"	1.4		
1,4-dichlorobenzene	"	"	0.4	2.2	3
1,2,4-trichlorobenzene	"	"	6.8		
1,2,3-trichlorobenzene	"	"	2.4	9.2	9

^{a}Not detected.

REFERENCES
1. Medz, R.B. Fed. Reg. 1979, 44, 69464-69551.
2. Ryan, J.F; et. al., In "Advances in the Identification of Organic Pollutants in Water"; Keith, L., Ed.; Ann Arbor Science: Ann Arbor, MI, 1981
3. Hunt, D.F.; Shabanowitz, J.; Harvey, T.M.; and Coates, M.L. J. Chromatogr. 1983, 271, 93-105.
4. McLafferty, F.W. (Editor), "Tandem Mass Spectrometry", Wiley-Interscience, New York, NY (in press)
5. McLafferty, F.W. Science 1981, 214, 280-286.
6. Yost, R.A.; Fetterolf, D.D. Mass Spectr. Reviews, 1983, 2, 1-45.
7. Slayback, J.R.B.; Story, M.S. Ind. Res. Dev. 1981, 129-139.
8. Hunt, D.F.; Shabanowitz, J.; Giordani, A.B. Anal. Chem. 1980, 52, 386-390.
9. Hunt, D.F.; Buko, A.M.; Ballard, J.M.; Shabanowitz, J.; Giordani, A.B. Biomed. Mass Spectrom. 1981, 8, 397-408.
10. Hunt, D.F.; Shabanowitz, J. Anal. Chem. 1982, 54, 574-578.
11. Hunt, D.F.; Giordani, A.B.; Rhodes, G.; Herold, D.A. Clin. Chem. 1982, 28, 2387-2392.
12. Schuetzle, D.; Riley, T.L.; Prater, T.J.; Harvey, T.M.; Hunt, D.F. Anal. Chem., 1982, 54, 265-271.
13. Henderson, T.R.; Royer, R.E.; Clark, R.C.; Harvey, T.M.; Hunt, D.F. J. App. Toxicology, 1982, 2, 231-237.
14. Pitts, J.N., Jr.; et. al., Science 202, 515-519.

RECENT PROGRESS IN LC/MS

D.E. GAMES and M.A. McDOWALL
Department of Chemistry, University College Cardiff.
M. GLENYS FOSTER and O. MERESZ
Ontario Ministry of the Environment.

Summary

Recent developments in the coupling of a liquid chromatograph with a mass spectrometer are reviewed with particular emphasis on commercially available systems. A new interface of the moving belt type enables molecular weight to be obtained by electron impact or chemical ionization from a wider range of compounds than was previously possible with interfaces of this type. Use of fast atom bombardment mass spectrometry with a moving belt interfaces shows promise for further extension of the compounds amenable to LC/MS study. Improvements in the methods of feeding the sample onto the belt, by use of a nebulizer or microbore liquid chromatography have resulted in improved sample detection limits and easier handling of mobile phases with a high water content. The range of compounds amenable to study by interfaces of the direct liquid introduction type have also been extended. Thermospray ionization systems show considerable potential for handling more difficult molecules and in providing improved detection limits. Advances in the use of atmospheric pressure ionization in this area are also described. Finally the utility of LC/MS in the environmental area is reviewed and analysis of testwell samples from a landfill site using the technique is described.

1. INTRODUCTION

Combined gas chromatography/mass spectrometry (GC/MS) has been extensively used for the analysis of organic micropollutants in water. However a large proportion of organic molecules are not amenable to GC because of their low volatility and/or thermal instability. In recent years liquid chromatography (LC) has become the method of choice for the analysis of compounds of this type and the technique is also extensively used for the analysis of compounds which are amenable to gas chromatographic study, but for which LC offers an easier analytical method. In view of the extensive use of LC, development of combined liquid chromatographic/mass spectrometric systems is a desirable objective, since it enables the types of study currently undertaken by GC/MS to be extended to a wider range of compounds. Additionally it would allow compounds to be detected which might be overlooked with the use of a selective wavelength ultraviolet liquid chromatographic detector.
    In this paper we review recent developments in the combination of a liquid chromatograph with a mass spectrometer. This is followed by a review of areas where the technique has been applied to studies relevant to the analysis of micropollutants in water. Finally the utility of the technique is illustrated by a brief description of studies we have been undertaking on the analysis of testwell samples from landfill sites.

2. COMBINED LIQUID CHROMATOGRAPHY/MASS SPECTROMETRY (LC/MS)

Combination of a liquid chromatograph with a mass spectrometer involves the marriage of two basically incompatible techniques. Since one operates in the liquid phase and the other in the gas phase. Three major problems have to be overcome.
a) Development of methods by which the high gas volumes which would be generated by vaporization of the mobile phase from the liquid chromatograph can be tolerated by the mass spectrometer.
b) Vaporization of the solute so that it does not thermally decompose and can thus be ionized by the conventional mass spectral ionization techniques of electron impact or chemical ionization. Alternatively systems involving other mass spectral ionization methods have to be developed.
c) Retention of chromatographic performance.

A variety of different methods have been developed to overcome these problems and the various approaches have been extensively reviewed (1-8). In this section we will concentrate our attention on recent developments in what currently appear to be the most effective methods for carrying out LC/MS.

2.1 Atmospheric pressure ionization

Use of atmospheric pressure ionization for LC/MS has the advantage that all of the eluant from the liquid chromatograph can be introduced into the ion source. Ionization of the solvent molecules is effected by a discharge or radioactive source and these ions form solute ions which are mass analyzed in a quadrupole mass analyzer. Early systems (9) suffered from, difficulties in vaporizing even moderately volatile compounds, production of simple spectra with little structural information apart from molecular weight and because of the presence of cluster ions from the solvent difficulty in obtaining total ion current traces. Introduction of samples via a nebulizer enables lower volatility compounds to be handled (10,11) and use of liquid ion evaporation techniques is an excellent methods for producing spectra from ionic compounds (12,13). Variation of the potential at the pinhole leading to the mass analyzer (14,15) is one method of obtaining more structural information and another is use of collision induced dissociation in a triple quadrupole mass spectrometer (12,13,16). A commercial system for LC/MS which uses a nebulizer for sample introduction into the ion source of triple quadrupole mass spectrometer is available from Sciex.

2.2 Enrichment of solute relative to solvent

Use of a gas nebulizing system in combination with a jet separator for introduction of a portion of the eluent from a liquid chromatograph shows some promise for LC/MS. The most recently described system contains a small water cooling jacket and bubble saturator for the nebulizing gas (17). Data presented to date do not show that this approach has any particular advantages over other systems.

Of more interest is the interface based on the jet principle which has been developed for magnetic mass spectrometers by Kratos Analytical Instruments (18). The liquid flow from the chromatograph passes through a narrow bore capillary which is heated at its tip and provides rapid vaporization of the eluent. This results in production of a well collimated and fast moving beam of solvated sample molecules and solvent vapour. A large proportion of the vapour flow is removed by an auxillary rotary pump in the separator stage and the sample beam then passes into the ion source via a counter orifice where solvent induced chemical ionization (CI) spectra are produced. The system can handle solvent flow rates up to 150 µl min^{-1}, however at present too little data is available to fully evaluate its potential.

As a result of studies of LC/MS a new ionization technique called thermospray ionization has been developed and shows considerable promise for LC/MS of lower volatility and ionic compounds (19). The system can handle mobile phases including aqueous buffers at flow rates between 0.5 and 2 ml min^{-1}. It consists of a stainless steel capillary which is brazed at one end onto an electrically heated copper block. A supersonic jet of vapour containing solute ions is produced, which traverses the ion source of a mass spectrometer and enters into a 1 cm diameter pumping line which is connected to a 300 ml min^{-1} mechanical vacuum pump. A conical exit aperture is sited at right angles to the jet and ions are sampled through it into a quadrupole mass analyzer. Addition of ammonium salt buffers result in formation of organic ions by chemical ionization reaction with the salt ions during the evaporation process. Because the sample is not subjected to energetic bombardment or has significant high temperature exposure to hot metal surfaces, decomposition is minimized. The source is readily fitted to quadrupole mass spectrometers equipped for CI operation and is commercially available from Vestec and Finnigan MAT(20). Impressive LC/MS data have been obtained with the system from a range of mass spectrometrically difficult to handle compounds. The major problems with the system at its current stage of development are difficulties in the day-to-day reproducibility of ionization efficiency and fragmentation patterns, a severe dependance of the ion intensities on the liquid flow and the fact that different compounds require different temperatures for efficient ionization.

2.3 Direct liquid introduction

This is one of the most widely used types of system for LC/MS. A portion of the eluant from the liquid chromatograph is fed into the ion source of a mass spectrometer configured for CI where solvent induced CI spectra are produced from the solute. Commercial systems, available from Hewlett-Packard and Nermag, have a water cooled probe with a diaphragm incorporating a small diameter hole, which enables nebulization of the eluent from the liquid chromatograph into the mass spectrometer ion source (21,22). Incorporation of a desolvation chamber to assist in the desolvation and ionization of solute molecules is necessary for the handling of low volatility compounds (21, 23-25). One advantage of this approach is that it is relatively easy to construct simple systems and although they have a limited range of compounds which can be handled they have been effectively used for many types of study (26-31). A major problem is that on-column detection limits are limited if conventional liquid chromatography is used, because only a small portion of the eluent from the liquid chromatograph is fed into the mass spectrometer ion source. Use of cryogenic pumping can assist with the handling of higher flow rates of some solvents (21). However the most dramatic improvements in detection limits are obtained by use of microbore LC, since if 0.5-1 mm i.d. columns are used with flow rates of 20 µl min^{-1} all the eluent from the liquid chromatograph can be fed into the ion source of the mass spectrometer (32). A further criticism which can be levelled at this approach is that the spectra obtained are simple and lack structurally useful fragment ions. Recently it has been shown that use of collision induced dissociation with this type of LC/MS can assist in the provision of more structural information (33).

There is no doubt that this is an extremely effective system for LC/MS. It is probably of more utility when known compounds have to be identified than when unknown compounds have to be characterized. However this depends on compound types. For quantitative studies the system has advantages over systems of the moving belt type since thermal decomposition of thermally labile compounds is less likely at low levels.

2.4 Continuous sample preconcentration and direct liquid introduction

An interface which preconcentrates the eluent from the liquid chromatograph by allowing it to flow down a resistance-heated stationary wire has been developed. Some of the residual liquid is drawn into the mass spectrometer through a metal capillary tube which has a needle valve at the ion source end and allows the liquid to spray into the ion source of a quadrupole mass spectrometer (34). Although the system worked satisfactorily with normal-phase LC, difficulties were experienced with aqueous solvents which did not spray well. Ultrasonic vibration of the probe has been used to overcome this problem (35). The system is commercially available from Extranuclear Laboratories and has been used to validate assays for aliphatic acids in shale oil process water, however it is difficult to assess its full potential since as yet studies have been limited to reasonably volatile compounds and there are no details of the system's performance with low volatility thermally labile compounds.

2.5 Removal of solvent using a moving belt

This approach to LC/MS has recently been reviewed (36). Eluent from the liquid chromatograph is fed onto a continuously moving belt usually made of Kapton. Solvent is removed by use of an infrared heater and in two vacuum locks and the solute is flash vaporized into the ion source of the mass spectrometer where EI or CI mass spectra can be obtained. Commercial systems are available from Finnigan (37) and VG Analytical (38). Both these systems are limited in the range of compounds that can be handled. A new commercial system from Finnigan MAT (39) extends the range of compounds amenable to study by this approach. We have recently compared the performances of all three systems with a wide range of compounds (40). The first two systems are equivalent to a good direct insertion probe, whereas the latter system extends the range to compounds which require desorption chemical ionization for provision of molecular weight data.

A number of methods for improving the performances of interfaces of this type have been described. Use of spray deposition of the eluent from the liquid chromatograph has been particularly successful, enabling high percentage aqueous mobile phases to be handled and removing the necessity of using the infrared heater and a splitting device (41,42). Use of microbore LC also results in easier handling of high percentage aqueous mobile phases and since a splitting device is not required results in improved on-column detection limits (43-45).

Current systems which are designed for EI and CI studies are limited in the range of thermally labile/low volatility compounds for which they can provide molecular weight information. They also suffer from thermal decomposition of thermally labile compounds at low levels. An answer to the former problem may be provided by the use of surface ionization techniques. The use of secondary ion mass spectrometry with transport interfaces has been demonstrated (46) as has the potential use of laser desorption mass spectrometry (47). In addition a discontinuous system using ^{252}Cf plasma desorption has been described (48) and a semi-automated laser desorption system (49). Potentially the most useful approach is the use of fast atom bombardment mass spectrometry. Off-line data obtained using a moving belt system have been reported (39,50) and recently the first on-line data has been shown (51).

Moving belt systems have proved to be reliable and relatively easy to use. The provision of both EI and CI data is particularly useful in qualitative studies when the types of compound present are not known. Recent developments described above have improved detection limits and enabled a wider range of compounds to be studied and surface ionization methods show promise of further extension of the technique.

3. APPLICATION OF LIQUID CHROMATOGRAPHY/MASS SPECTROMETRY

Initially studies of LC/MS were confined to examination of mixtures of standard compounds. The technique is now being increasingly applied to the solution of real problems. Most applications have been confined to qualitative studies where known compounds are identified or new compounds located for further structural study. The combination of liquid chromatographic retention time and mass spectral data providing a powerful combination. Although the potential of the technique for quantitative studies has been shown, few examples of real applications in this area are in the literature. A recent comprehensive bibliography of publications on LC/MS has appeared (52). This section will be confined to reviewing areas where LC/MS has been applied or has potential application for the analysis of water samples.

Detailed studies of the LC/MS behaviour of a group of 19 carbamate pesticides using a moving belt interface have been reported (53). These studies were conducted on a prototype interface and newer interfaces would give better data. It was shown that these compounds could be detected and quantified down to the low ng. level. Other studies have confirmed these findings and shown that the technique can be used for the analysis of carbamate and urea pesticides in crop residues at the ppm level (54). Aldicarb, aldicarb sulphoxide and aldicarb sulphone have been analyzed in well water samples using this technique and the limits of detection were below 1 ppb (55). Other classes of potential environmental contaminants which have been studied with moving belt systems include perchloro cage pesticides (56), chloropropham and its metabolites (57), chlorophenols (58) and polychlorinated biphenyls and their metabolites (59). The behaviour of organophosphorous pesticides (60), triazine (61) and phenylurea herbicides (62) using DLI LC/MS has been studied and the technique has been used to identify members of the latter class of herbicide in river water samples (62).

The application described above can be described as being of the target analysis type, where a specific compound or groups of compound are being sought. This is probably the most common type of study performed on environmental samples and means that often samples can be contaminated with potentially dangerous compounds which are overlooked either because the analytical technique is not capable of detecting them or because the analysis is designed only to locate a particular set of compounds. Combined LC/MS has the potential to considerably extend the range of organic compounds identified in water samples. Studies of tannery effluents using a moving belt LC/MS system (63) resulted in the identification of twenty different compounds and one of the major components was found to be binaphthyl sulphone which was not found in gas chromatographic studies of the same samples.

We have recently commenced studies of extracts of samples from test wells near landfill sites using a range of mass spectral techniques inclu-

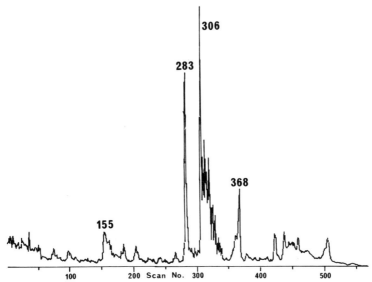

Figure 1 Computer reconstructed total ion current trace obtained using ammonia CI LC/MS of a basic fraction from an extract of a test well sample from a landfill site. A Whatman 250 x 1mm column packed with ODS was used. Mobile phase flow rate was 50 μl min^{-1}. For the first two minutes the mobile phase was methanol + water (40 : 60) for the next fifteen minutes the mobile phase was linear programmed, slope 6, to methanol + water (80 : 20) which was the mobile phase for the remainder of the LC. The solvent systems contained 0.1% acetic acid.

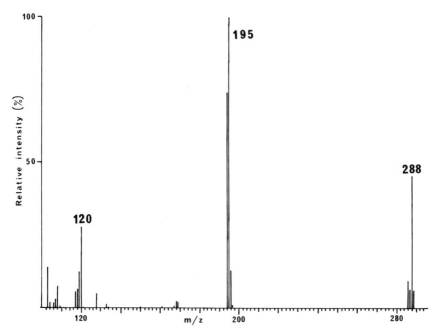

Figure 2 Ammonia CI mass spectrum from scan 306, figure 1

ding LC/MS (64). Samples are examined by capillary GC/MS using EI and CI, then by LC/MS again using both forms of ionization and we also undertake molecular weight profiling of samples using field desorption, desorption chemical ionization and spotting of the sample onto a moving belt interface under CI conditions. These latter studies serve as a check for loss or decomposition of sample during chromatographic/mass spectrometric study. In our studies we have found this combined approach very valuable resulting in the identification of a large number of compounds which would not otherwise have been identified. Figure 1 shows the computer reconstructed total ion current trace obtained by ammonia CI LC/MS of a basic fraction from an extract of a test well sample from a landfill site. Figure 2 shows the ammonia CI spectrum obtained from scan 306. This component has not yet been identified and was not found by EI LC/MS, EI or CI GC/MS study.

4. CONCLUSIONS

Although sensitivity is not yet comparable with that obtained by capillary GC/MS recent advances in combined LC/MS have improved the range of compounds amenable to study and the detection limits. At the present time there is no universal LC/MS system; the choice of interface depending on the type of study to be undertaken and the possible types of compound to be investigated. Current systems have considerable potential for extending the range of identification of organic micropollutants in water.

5. ACKNOWLEDGEMENTS

We thank the SERC and Royal Society for assistance in the purchase of mass spectral equipment. M.A.M. thanks the SERC for financial support.

REFERENCES

1. P.J. Arpino and G. Guiochon, Anal. Chem., 1979, 51, 682A.
2. P.J. Arpino, Int. J. Mass Spectrom. Ion Phys., 1982, 45, 161.
3. W.H. McFadden, J. Chromatogr. Sci., 1979, 17, 2; 1980, 18, 9; Anal. Proc., 1982, 19, 258.
4. D.E. Games, Anal. Proc., 1980, 17, 110; 1980 17, 322; Biomed. Mass Spectrom., 1981, 8, 454; in H.R. Morris (Ed.) Soft Ionization Biological Mass Spectrometry, Heyden, London, 1981, p. 54; in J.C. Giddings, E. Grushka, J. Gazes and P.R. Brown (Eds.) Advances in Chromatography Vol. 21, Marcel Dekker, New York, 1983, p. 1.
5. J.H. Knox, Anal. Proc., 1982, 19, 166.
6. K. Levsen, Comm. Eur. Communities (Rep), 1982, 7623, 149.
7. Z.F. Curry, J. Liq. Chromatog. 1982, 5(Sup 2), 257.
8. R.C. Willoughby and R.F. Browner, Trace Anal, 1982, 2, 69.
9. D.I. Carroll, I. Dzidic, R.N. Stillwell, K.D. Haegele and E.C. Horning, Anal. Chem., 1975, 47, 2369.
10. H. Kambara, Anal. Chem., 1982, 54, 143.
11. B.A. Thomson and L. Danylewych-May, 23rd Annual Conference on Mass Spectrometry and Allied Topics, Boston, 1983, paper FOE 2.
12. B.A. Thomson, J.V. Iribarne and P.J. Dziedic, Anal. Chem., 1982, 54, 2219.
13. B. Shushan, J.E. Fulford, B.A. Thomson, W.R. Davidson, L.M. Danylewich, A. Ngo, S. Nacson and S.D. Tanner, Int. J. Mass Spectrom. Ion Phys., 1983, 46, 225.
14. M. Tsuchiya and T. Taira, Int. J. Mass Spectrom. Ion Phys., 1980, 34, 351.

15. M. Tsuchiya, T. Taira, H. Kuwabara and T. Nonaka, Int. J. Mass Spectrom Ion Phys., 1983, 46, 355.
16. J.D. Henion, B.A. Thomson and P.H. Dawson, Anal. Chem. 1982, 54, 451.
17. H. Yoshida, K. Matsumoto, K. Itoh, S. Tsuge, Y. Hirata, K. Mochizuki, N. Kokubun and Y. Yoshida, Fresenius Z. Anal. Chem., 1982, 311, 674.
18. J.R. Chapman, E.H. Harden, S. Evans and L.E. Moore, Int. J. Mass Spectrom. Ion. Phys., 1983, 46, 201.
19. C.R. Blakely and M.L. Vestal, Anal. Chem., 1983, 55, 750.
20. W.H. McFadden, Spectra, 1983, 9, 23.
21. A. Melera, Adv. Mass Spectrom., 1980, 8B, 1597.
22. P.J. Arpino, P. Krien, S. Vajta and G. Devant, J. Chromatogr., 1981, 203, 117.
23. M. Dedieu, C. Juin, P.J. Arpino, J.P. Bounine and G. Guiochon, J. Chromatogr., 1982, 251, 203.
24. M. Dedieu, C. Juin, P.J. Arpino and G. Guiochon, Anal. Chem., 1982, 54, 2372.
25. P.J. Arpino, J.P. Bounine, M. Dedieu and G. Guiochon, J. Chromatogr., 1983, 271, 43.
26. J.D. Henion, Anal. Chem., 1978, 50, 1687.
27. N. Evans and J.E. Williamson, Biomed. Mass Spectrom., 1981, 8, 316.
28. J.J. Brophy, D. Nelson and M.K. Withers, Int. J. Mass Spectrom. Ion. Phys., 1980, 36, 205.
29. E. Yamauchi, T. Mizuno and K. Azuma, Shibak Shitsuryo Bunski, 1980, 28, 227.
30. K.H. Schäfer and K. Levsen, J. Chromatogr., 1981, 206, 245.
31. A.P. Bruins and B.F.H. Drenth, J. Chromatogr., 1983, 271, 71.
32. J.D. Henion and G.A. Maylin, Biomed. Mass Spectrom., 1980, 7, 115; J.D. Henion and T. Wachs, Anal. Chem., 1981, 53, 1963.
33. R.D. Voyksner, J.R. Hass and M.M. Bursey, Anal. Lett., 1982, 15, 1.
34. R.G. Christensen, H.S. Hertz, S. Meiselman and E. White V, Anal. Chem. 1981, 53, 171.
35. R.G. Christensen, E. White, V.S. Meiselman and H.S. Hertz, J. Chromatogr., 1983, 271, 61.
36. N.J. Alcock, C. Eckers, D.E. Games, M.P.L. Games, M.S. Lant, M.A. McDowall, M. Rossiter, R.W. Smith, S.A. Westwood and H.-Y. Wong, J. Chromatogr., 1982, 251, 165.
37. W.H. McFadden, H.L. Schwartz and S. Evans, J. Chromatogr., 1976, 122, 389.
38. D.S. Millington, D.A. Yorke and P. Burns, Adv. Mass Spectrom., 1980, 8B, 1819.
39. P. Dobberstein, E. Korte, G. Meyerhoff and R. Pesch, Int. J. Mass Spectrom. Ion Phys., 1983, 46, 185.
40. D.E. Games, M.A. McDowall, K. Levsen, K.H. Schäffer, P. Dobberstein and J.L. Gower, Biomed. Mass Spectrom., in press.
41. R.D. Smith and A.L. Johnson, Anal. Chem., 1981, 53, 739.
42. E.P. Lankmayr, M.J. Hayes, B.L. Karger, P. Vouros and J.M. McQuire, Int. J. Mass Spectrom. Ion Phys., 1983, 46, 177.
43. D.E. Games, M.S. Lant, S.A. Westwood, M.J. Cocksedge, N. Evans, J. Williamson and B.J. Woodhall, Biomed. Mass Spectrom., 1982, 9, 215.
44. N.J. Alcock, L. Corbelli, D.E. Games, M.S. Lant and S.A. Westwood, Biomed. Mass Spectrom., 1982, 9, 499.
45. M.G. Foster, O. Meresz, D.E. Games, M.S. Lant and S.A. Westwood, Biomed. Mass Spectrom., 1983, 10, 338.
46. A. Benninghoven, A. Eicke, M. Junack, W. Sichtermann, J. Krizek and H. Peters, Org. Mass Spectrom., 1980, 15, 459; R.D. Smith, J.E. Burger

and A.L. Johnson, Anal. Chem., 1981, 53, 1603.
47. E.L. Hardin and M.L. Vestal, Anal. Chem., 1981, 53, 1492.
48. H. Jungclas, H. Danigel and L. Schmidt, J. Chromatogr., 1983, 271, 35.
49. J.F.K. Huber, T. Dzido and F. Heresch, J. Chromatogr., 1983, 271, 27.
50. D.F. Hunt, W.M. Bone, J. Shabanowitz, J. Rhodes and J.M. Ballard, Anal. Chem., 1981, 53, 1706.
51. I.A.S. Lewis and P.W. Brooks, 23rd Annual Conference on Mass Spectrometry and Allied Topics, Boston, 1983, paper FOE. 1.
52. C.G. Edmonds, J.A. McCloskey and V.A. Edmonds, Biomed. Mass Spectrom., 1983, 10, 237.
53. L.H. Wright, J. Chromatogr. Sci., 1982, 20, 1.
54. T. Cairns, E.G. Siegmund and G.M. Doose, Biomed. Mass Spectrom., 1983, 10, 24.
55. L.H. Wright, M.D. Jackson and R.G. Lewis, Bull. Environm. Contam. Toxicol., 1982, 28, 740.
56. T. Cairns, E.G. Siegmund and G.M. Doose, Anal. Chem., 1982, 54, 953.
57. D.E. Games and N.C.A. Weerasinghe, J. Chromatogr. Sci., 1980, 18, 106.
58. L.H. Wright, T.R. Edgerton, S.J. Arbes Jr. and E.M. Lores, Biomed.Mass Spectrom., 1981, 8, 475.
59. P. Dymerski, M. Kennedy and L. Kaminsky in H.S. Hertz and S.N. Chesler (Eds.) Trace Organic Analysis: A New Frontier in Analytical Chemistry, N.B.S. Washington, 1979, 685.
60. C.E. Parker, C.A. Haney, D.J. Harvan and J.R. Hass, J. Chromatogr., 1982, 242, 77.
61. C.E. Parker, C.A. Haney and J.R. Hass, J. Chromatogr, 1982, 237, 233.
62. K. Levsen, K.H. Schäfer and J. Freudenthal, J. Chromatogr., 1983, 271, 51.
63. A.D. Thruston Jr. and J.M. McGuire, Biomed. Mass Spectrom., 1981, 8, 47.
64. M. Glenys Foster, O. Meresz, D.E. Games, M.S. Lant and S.A. Westwood, Biomed. Mass Spectrom., 1983, 10, 338.

COMPUTERIZED DATA HANDLING IN GAS CHROMATOGRAPHY/MASS SPECTROMETRY.

P. GROLL
Institut für Heiße Chemie, Kernforschungszentrum Karlsruhe,
Postfach 3640, D-75 Karlsruhe, BRD.

Summary

A review on special problems of the data treatment of mass spectra of gas chromatography/mass spectrometry is given. Attempts for the characterization of the quality of mass spectra in a mass spectral data base are discussed. Due to the complex mixture of organic pollutants the automatic data processing of unresolved mixtures is reviewed. EPA published their experience of the analysis of aqueous samples by gc/ms. This problem, similar to that of Cost-action 64b is discussed in detail. Finally some aspects of the work of the working party "Computerized Data Processing" are given.

1. INTRODUCTION

The use of gaschromatography/mass spectrometry (gc/ms) has become the most used method for the identification of volatile pollutants in aqueous enviromental samples. The most commercially available gc/ms-systems are fitted with data systems for data acquisition and data processing. In data acquisition a computer system can handle easy large amounts of data for instance of repetitive scans. In data processing the computer system can extract much more informations out of the measured data as reconstructed mass chromatograms, background correction and library search. Many of the data systems are also controlling the mass spectrometer. In the following review I will focus only on data processing.

2. SPECTRA COLLECTION

Most of the mass spectra interpretation by computer is done by comparing the spectrum of the unknown with the spectra in a data base. To improve this technique we have collected during the past action of 64b and 64b bis the spectra of organic pollutants measured in the participating laboratories. For some substances duplicate spectra have been added to the collection. Similar to us the EPA had the same problem some years ago, which had been described by Heller and coworkers [1]. In 1978 McLafferty [2] published an algorithm which examines a mass spectrum for the occurrence of standard errors. He defined nine quality factors.

By computing this quality index (QI) of all spectra and omitting the spectrum with the smaller QI if dubletts have been present in the data base the data base had been reduced to about 78%. The average QI was about 500. 1353 spectra out of 33898 spectra in the NIH/EPA/MSCC Mass Spectra Data Base received a QI of zero. Before adding a new spectrum to the data base, the data base is scanned for the existence of this compound. The QI is computed and, if the compound is new, the spectrum and the QI are added to the file. If the substance is already in the file the spectrum with the

Quality factor	Feature tested
QF1	Electron voltage
QF2	Peaks above molecular weight
QF3	Illogical neutral losses
QF4	Isotopic abundances
QF5	Numbers of peaks
QF6	Lower mass limit
QF7	Sample purity
QF8	Calibration date
QF9	Similarity index of calibration mass spectrum

Table I: Quality factors [1].

higher QI becomes member of the data base. In 1978 3862 or 8,9% of the substances in the inventory of the TSCA (Toxic Substance Control Act) were represented in the Mass Spectral Data Base, which contained 32187 spectra. To change this situation a priority ordered list of chemicals in the TSCA Inventory not present in the MS Data Base was prepared. More than 1400 mass spectra of such substances have been prepared by using pure compounds.

Besides this qualifying procedure McLafferty is preparing a set of standards for defining the "quality class" of a spectrum.

3. MASS-SPECTRA INTERPRETATION

In the last years besides the most prominent matching procedures the Biemann search [3], PBM [4] and SISCOM [5] a number of new library search methods have been published. Terwinger has reported on a combination of forward and reverse search named STAR, which is designed for the identification of trace components in mixtures, using the eight most prominent peaks of each spectrum [6]. The method published by Kwiatkowski and Riepe is also a combination of forward and reverse search based on binary coded spectra [7]. Van Marlen and coworkers have recently studied the content of information of binary coded spectra, concluding that such an abbreviation does not contain enough information for retrieval purposes [8].

Another method for spectra interpretation is pattern recognition, which results in assigning the class of an unknown compound after training the system with a set of known training compounds. It is based on the tendency of compounds with similar structure to form clusters in a multidimensional space. On this basis the KNN(K nearest neighbours)-method has been developed [9-17], which has the disadvantage to be a very time consuming method. Therefore a number of other clustering methods have been developed. Another approach is the learning machine with the most extensive method STRIRS (Self Training Interpretative and Retrieval System) of McLafferty [18-21].

4. AUTOMATIC DATA PROCESSING OF MASS-SPECTRA OF UNRESOLVED MIXTURES

In the most cases the analyzed samples of this Cost-action are mixtures of organic compounds, partly volatile. When the resolution of the gaschromatograph is sufficient to separate the compounds of the mixture completely the interpretation can be done manually in many cases, if the operators are trained. Much more difficulties are arising if the compounds are not completely separated. A manual subtraction of mass spectra is very tedious and therefore not possible for routine work.

The 'Biller-Biemann method' [22] searches mass chromatograms for maxima and then takes the occurrence of these maxima together to indicate the presence of a discrete component not visible in the total ion current. For each component at least one specific mass is necessary and the components must be separated by at least three scan periods. The clean-up program by Dromey [23] is much more sophisticated but has similar requirements for specific masses. This method allows the calculation of purified spectra, which can then be much more successfully compared with spectra in a spectra collection.

Another method using a shape model of a gc-peak has been recently used by Harris [24]. In an iterative process the retention time and the number of components under the unresolved gc-peak are calculated. A separation of at least two scan periods between each component is necessary.

A further method is the so called factor analysis [25-30]. From successive scans a correlation matrix is constructed, relating all intensities at each m/z-values to that at all m/z-values of the scan. Successive factors are calculated accounting for the largest possible variance minimizing the residual variance. The number of factors represent the number of compounds under the unresolved gc-peak. As Kowalski demonstrated [29] the accuracy of the extracted spectra is high. A peak separation of at least two scan periods is necessary.

More simple than the above mentioned methods is the reverse library search. It provides good results for the identification of the main component if its intensity is high. For this method no separation of the components is necessary. The only two restrictions are that the components are well distinct and the procedure is restricted by the amount and the quality of the library data base. By this method it is possible to identify one component under a gc-peak, to calculate the difference between experimental spectrum and the data base spectrum and try to identify the remaining spectrum too [31].

Rosenthal [32] showed recently theoretically that using 2-sec scans over a gc-run of 60 minutes 60 components could be fully resolved to a level of 50% confidence. At a 2-sec/scan rate the number of components is reduced to 45. For a full resolution of a mixture of 200 components (with 50% success rate) over 20.000 mass spectra are necessary to be run during an 1-hour run of the ms. This means that a ms-scan rate of 0,2 sec/scan is necessary. This capability is entirely beyond the realm of current technology. He showed further for 200 components 82% of the components are measured as single mass spectra, assuming a 60-minute gc-run with a ms-rate of 2-sec/scan and an effective probable resolution of 2 scans. In a complex mixture, where the number of components exceed 100 it is a virtual certainty that it is not possible to identify more than a fraction of the components by first order technique.

5. EXPERIENCES WITH POLLUTION SAMPLES

Comparatively few components in any given sample of surface water are member in a preselected list, such as priority pollutants. Shakelford and coworkers of EPA reported about the evaluation of automated matching for survey identification of wastewater components by gc/ms [33]. Over a 2.5-year duration 22500 spectra have been submitted for spectrum matching. The method used was PBM [34,4]. Only 30% of the spectra submitted were matched with spectra in the reference library to a degree sufficient that the unknown was considered to be tentatively identified and entered into a historical library.

The spectra were extracted using the clean up-program of Dromey [23]. For complicated data files many of the extracted spectra have been contaminated by one or more other compounds. Matching of the unknown spectrum was done by PBM and followed by a retention data match with a historical library. Retention data were used instead of reliability ranking. Modifications as spectrum subtraction, spectrum tilting and reliability ranking were not employed. As internal standard for measuring retention data in gc for acid and base fractions anthracene-d_{10} and for purgeable fractions bromochloromethane and 1,4-dichlorobutane have been used.

To confirm the tentative identifications made by spectrum matching selected extracts were subjected to glass capillary column gc/ms. A standard of each of the compounds found in the re-examination have then been added to the extract for a second gc/ms experiment. Coelution confirmed then the compound identified before.

Three reference libraries have been used for this study: the Wiley library (30476 spectra of 30476 compounds), the EPA-NIH library (34363 compounds) and the EPA master data base (over 40000 spectra of about 32000 compounds).

The success of the matching program can be determined by the number of correct and incorrect answers as it has been reported by Shakelford of EPA [33]. Influences for a correct interpretation are arising from the difficulty of the software system to distinguish between chromatographic artefacts as noise and column bleeding and the compounds of interest.

A second difficulty was that the components were not separated well enough to obtain mass spectra of pure substances. In this case the operator had to decide if there were one or more compounds present.

The third difficulty for confirmation by glass capillary column gc was the possibility of degradation of components over time. Table 2 summarizes the overall effectiveness of the automated system.

Column	Fraction type	Spectra matched (%)	
		Samples	Standards
0,2% Carbowax 1500 on Carbopack C (packed)	Purgeable Acid	57	70
SP-1240 DA (packed)	Base/neutral	27	77
3% SP-2250 (packed)	Acid-base/	26	71
SE-54 fused silica (capillary)	neutral (combined) Fraction type	65	85

Table II: Comparison of g.c.-columns and fraction type with matching efficiency [33].

The results of the paper of Shakelford are:
1) The amount of matched spectra is greater for capillary column gc than for packed column.
2) The high yield of 57% in the purgeable fraction is due to the preponderance of compounds with distinctive spectra such as low-molecular-weight halogenated compounds.

Another conclusion of this paper is that the match quality parameters are differing more than by a factor of 2, whereas the standard deviation of the relative retention is very narrow (better than ±0,06 and for capillary columns better than ±0,007). The use of relative retention times improves matching efficiency. 620 compounds in the historical library gave an efficiency of 18, whereas 1553 compounds resulted in an efficiency of 31. For high ΔK-matches the reliability is increased remarkable by the use of retention data. For carboxylic acids and aldehydes Shakelford found a poor confirmation rate of about 30-40% compared with 70% and more for alkanes, alkyl benzenes, esters, alkyl substituted PAH, phenols, PAH, alcohols and phtalates. The overall reliability was 71% whereas specific isomer resolution was not required.

Another serious restriction to the efficiency of gc/ms is the time required for matching the spectra of a complex mixture against a data base. By this operation the instrument time for running the gc/ms can be increased by a factor of two or more. Dromey [35] and McLafferty and coworkers [36] proposed the use of an ordered file search procedure to reduce the necessary matches.

Dromey ordered the data base according to the most intense peaks in 6 ranges of 10 mass units starting at 40. Ambiguous peaks which are to weak in intensity or which are similar in intensity to another peak in this amu-region are marked. For 10.000 mass spectra about 61.000 representations are necessary. For retrieval less than 1% of the file has to be matched. For a spectrum with 5 uncertain peaks about 8% of the file had to be searched. For spectra with only very small peaks in the amu-region of 40 to 100 the ordering code would be 000000. For this substances another range (e.g. 100 to 159) has to be used.

McLafferty defined another method proposing a mass uniqueness and an abundance value. 15-27 peaks are selected of each spectrum. Also this method reduces the number of matches in a data base of 41.000 spectra to about 7.5%. The computertime was reduced by a factor of 10. Compared with the algorithm of Dromey this method has advantages because it enables a reverse search.

6. SUMMARY

Summarizing all these developments the application of the computer to mass spectrometry enlarges remarkable the advantages of this analytical technique. In future the knowledge of the fragmantation process and its implementation in the software system will refine the mentioned methods. An unresolved problem is till now a truly independent comparative study of the various software systems evaluating their strength and weakness.

The work of the working party 6 "Computerized Data Processing" was during the past period the collection of mass spectra of pollutants and not the development of a new interpretation technique. Most of the participating laboratories are equipped with ms-systems interfaced to a commercial hard- and software system. They can mostly not decide between the different software systems because the spectrometer companies mostly offer only one software system. On the other hand it is generally not so easy to install a commercial available software package on a given computer.

In addition to our recent work in which also the collection of gc-data should be included, we propose to install the above mentioned most prominent software systems as PBM, STIRS and SISCOM on our computer system and to try to run unidentified mass spectra of the participating laboratories. This would be helpful for the participating laboratories, if the substance of the spectrum can be identified, on the other hand it would be a step into the direction of a truly independent comparative study of the strength and the weakness of the used techniques.

REFERENCES

[1] Milne, G.W.; Budde, W.L.; Heller S.R.; Martinsen, D.P.; Oldham, R.G.
    Org.Mass Spectrom. 17, 547-552 (1982).
[2] Speck, D.D.; Venkataraghavan, R.; McLafferty, F.W.
    Org. Mass Spectrom. 13, 209-213 (1978).
[3] Hertz, H.S.; Hites, R.A.; Biemann, K.
    Anal.Chem. 43, 681-691 (1971).
[4] McLafferty, F.W.; Hertel, R.H.; Villwock, R.D.
    Org. Mass.Spec. 9, 690-702 (1974).
[5] Damen, H.; Henneberg, D.; Weimann, B.
    Anal.Chim.Acta 103, 289-302 (1978).
    Henneberg, D.
    Adv.Mass Spec. 8, 1511 (1980).
[6] Terwilliger, T.
    28th Annual Conference on Mass Spectrometry and Allied Topics,
    New York, May 30-June 2, 1980.
[7] Kwiatkowski, J.; Riepe, W.
    Anal.Chim.Acta 112, 219-231 (1979).
[8] van Marlen, G.; Dijkstra, A.; van't Klooster, H.A.
    Anal.Chim.Acta 112, 233-243 (1979).
    Anal.Chem. 51, 420-423 (1979).
[9] Kowalski, B.R.; Bender, C.F.
    J.Am.Chem.Soc. 94, 5632-5639 (1972).
[10] Crawford, L.R.; Morrison, J.D.
    Anal.Chem. 40, 1464-1469 (1968).
[11] Varmuza, K.
    Monats.Chem. 105, 1-10 (1974).
[12] Varmuza, K.
    Z.Anal.Chem. 268, 352-356 (1974).
[13] Kowalski, B.R.; Bender, C.F.
    Anal.Chem. 44, 1405-1411 (1972).
[14] Justice, J.B.; Isenhour, T.L.
    Anal.Chem. 46, 223-226 (1974).
[15] McGill, J.R.; Kowalski, B.R.
    J.Chem.Inf.Comput.Sci. 18, 52 (1978).
[16] Burgard, D.R.; Perone, S.P.; Wiebers J.L.
    Biochemistry 16, 1051 (1977).
[17] Ziemer, J.N.; Perone, S.P.; Caprioli, R.M.; Seifert, W.E.
    Anal.Chem. 51, 1732-1738 (1979).
[18] Kwok K.-S.; Venkataraghavan, R.; McLafferty, F.W.
    J.Am.Chem.Soc. 95, 4185-4194 (1973).
[19] Dayringer, H.E.; Pesyna, G.M.; Venkataraghavan, R.; McLafferty, F.W.
    Org.Mass Spectrom. 11, 529-542 (1976).
[20] Dayringer, H.E.; McLafferty, F.W.
    Org.Mass Spectrom. 11, 543-551 (1976).
[21] Dayringer, H.E.; McLafferty,.F.W.; Venkataraghavan, R.;
    Org.Mass Spectrom. 11, 895-900 (1976).
[22] Biller, T.G.; Biemann, K.
    Anal.Lett. 7, 515 (1974).
[23] Dromey, R.G.; Stefik, M.J.; Rindfleisch, T.C.; Duffield, A.M.
    Anal.Chem. 48, 1368-1375 (1974).
[24] Knorr, F.J.; Thorsheim, H.R.; Harris, J.M.
    Anal.Chem. 53, 821-825 (1981).
[25] Davies, T.E.; Shepard, A.; Stanford, N.; Rogers, L.B.
    Anal.Lett. 46, 821 (1974).
[26] Ritter, G.L.; Lowry, S.R.; Isenhour, T.L.; Wilkins, C.L.

Anal.Chem. $\underline{48}$, 591-595 (1976).
[27] Malinowski, E.R.
Anal.Chim.Acta $\underline{134}$, 129-137 (1982).
[28] Halket, J.M.
J.Chromatogr. $\underline{186}$, 443-455 (1979).
[29] Sharaf, M.A.; Kowalski, B.R.
Anal.Chem. $\underline{53}$, 518-522 (1981).
[30] Rasmussen, G.T.; Hohne, B.A.; Hieboldt, R.C.; Isenhour, T.L.
Anal.Chim.Acta $\underline{112}$, 151-164 (1979).
[31] Chapman, J.R.
Int. J. Mass Spec. and Ion Phys. $\underline{45}$, 207-218 (1982).
[32] Rosenthal, D.
Anal.Chem. $\underline{54}$, 63-66 (1982).
[33] Shakelford, W.M.; Cline, D.M.; Faas, L.; Kurth, G.
Anal.Chim.Acta $\underline{146}$, 15-27 (1983).
[34] Pesyna, G.M.; Venkataraghavan, R.; Dayringer, H.E.; McLafferty, F.W.
Anal.Chem. $\underline{48}$, 1362-1368 (1976).
[35] Dromey, R.G.
Anal.Chem. $\underline{51}$, 229-232 (1979).
[36] Mun, I.K.; Bartholomew, D.R.; Stauffer, D.B.; McLafferty, F.W.
Anal.Chem. $\underline{53}$, 1938-1939 (1981).

NEW DEVELOPMENTS IN SELECTIVE DETECTION IN CAPILLARY GAS CHROMATOGRAPHY

P. SANDRA
Laboratory of Organic Chemistry, State University of Ghent,
Krijgslaan, 281 (S.4), B-9000 GENT (Belgium)

Summary

The importance of "selectivity" in capillary gas chromatography is emphasized. Some recent developments in selective detection are discussed.

INTRODUCTION

Due to the complexity of "real world samples", i.e. environmental, biological etc., high resolution gas chromatography without selectivity would have only limited utility. Production of capillary chromatograms displaying a wealth of peaks is not difficult. A successful interpretation of the results strongly depends on the selectivity that can be introduced in several steps of the analytical procedure (sample preparation and chromatographic system). Therefore, new developments aimed at enhancing selectivity are of great importance. Progress in this aspect will further broaden the applicability of capillary gas chromatography.
In strictu sensu, selective detection in capillary gas chromatography implies the use of detection devices (selective detectors), which allow selective and preferably sensitive monitoring of chemical functionalities and/or physico-chemical properties of the solutes eluting from the capillary column.
In our opinion, selective detection must by discussed in a broader sense. All steps of the analytical procedure, which improve the selectivity for the compounds of interest, are benificial for both the qualitative elucidation and the quantitative determination. As a consequence, selectivity obtained by sample preparation and/or sample introduction through column and stationary phase selection, by data handling etc. should not be overlooked when discussing selective detection.
In this contribution a survey is given of the progress made in recent years to improve selectivity or specificity, with respect to several parts of the chromatographic system. Emphasis is given to selective detectors. It is however impossible to present a complete picture of the current situation. Some points were selected in view of the projected applicability in daily practice.

SELECTIVE DETECTORS

A large number of papers on selective detection has been published in the last two years. For developments before 1981, we refer to the excellent review article written by Prof. E.

Mantica on the occasion of the Second European Symposium on the Analysis of Organic Micropollutants in Water (1). Most of the papers deal with developments in theories and operation principles of well-known selective detectors. Their applicability in capillary gas chromatography is also well documented. Some new "potentially interesting" detection devices have been described but, at present, most are still laboratory curiosities, which need further development in order to increase their practical use for capillary gas chromatography.
A number of improvements on commonly used detectors are worthwhile mentioning. Moreover, the introduction of flexible fused silica capillary columns has led to the solution of most problems related to interfacing. The characteristics of well-known detectors are summarized in Table 1. Following comments can be added :

Flame ionization detection (FID)

The FID is the most universal detector for capillary gas chromatography. The linearity and the sensitivity can be strongly improved by adding nitrogen make-up gas (2). Flame optimization is often omitted but is necessary for real quantitative work. By slight modification, a FID detector can be modified in a silicium selective detector - HAFID : hydrogen atmospheric flame ionization detector (3). The response can further be enhanced by metal doping (4). The detector exhibits a sensitivity for silicium compounds in the nanogram range and a selectivity for these compounds over normal hydrocarbons of 2500. Silylation of oxygen-containing compounds results in indirect selective detection of oxygenated organic compounds.

Photoionization detector (PID)

By utilizing UV lamps with different photon energies, some degree of selectivity can be obtained with PID detection. The non-destructive character of the PID is however more important than its selectivity. Low-volume gas-tight PID detectors have been used in capillary gas chromatography in series with destructive detectors (5,6).

Electron Capture detector (ECD)

Electron capture detection is the most useful tool to obtain information on compounds with high electron affinity. The sensitivity of the ECD is very high. There is a strong relationship between the biological activity of organic compounds (mutagenic or carginogenic properties) and the ability to form negative ions (7). Of the 114 organic compounds of the US EPA-list of priority pollutants, only seven are not significantly el-capturing. The ECD detector is therefore highly appreciated for environmental research. Due to the high sensitivity of the ECD, the selectivity is often suppressed by the sample matrix. Selective sample preparation is therefore most important in utilizing ECD. A new detector, the tunable ion mobility detector, closely related to the ECD detector, has been described (8,9,10). The possibility of monitoring product ions offers a selectivity dimension, which is not achievable with ECD. It will be interesting to follow its development. The non-radioactive electron capture detector (11), based on a thermionic emitter (platinum wire coated with barium zirconate) as source of electrons, has not yet been evaluated for capillary gas chromatography, although a sensitivity to detect femtograms is claimed. Selective electron capture sensitization (12) by adding oxygen traces to the carrier gas or to the make-up gas is an analytical useful way to alter or enhance the sensitivity and selectivity of the ECD. Response enhancements of more than 100 are reported for monochloroalkanes, dichloroalkanes and polyaromatic hydrocarbons (7). It is advised to dope the make-up gas rather than the carrier gas for capillary columns (13). The ECD is non destructive and other detectors can be coupled in series (6).

Alkali Flame Ionization Detector (AFID)

The fields of application of the AFID or the thermionic nitrogen/phosphorus detector are continuously growing. A tunable NP detector has been developped, which shows excellent performance with capillary columns (14). By actuating specific detector configurations, N and P can unambiguously be distinguished without any modification to the chromatographic operating conditions.
The response of interfering carbon compounds can be suppressed or adjusted in positive or negative direction on the chromatographic trace. This development has drastically improved the selectivity of the AFID.

Hall Electrolytic Conductivity Detector (HECD)

The HECD detector, selective for nitrogen, sulfur and halogens, has been successfully combined with capillary columns (15). In the sulfur-specific mode, the detector shows some advantages over the FPD detector. The response quenching is considerably less noticeable, while the linearity of the HECD is 10^4 compared to 5.10^2 for the linearized FPD (16). Some efficiency loss has been noted when coupling the HECD to capillary columns and the operation of the detector seems to be not simple.

Flame Photometric Detector (FPD)

The possibilities and limitations of the FPD are well-known. The current N/P thermionic detectors are much more sensitive to phosphorus than the FPD detector. On the other hand each FPD needs careful characterization before placing any confidence in the data obtained when working in the sulfur mode. The selectivity for sulfur compounds is often quenched by other compounds and the sensitivity is limited by excessive flame noise. Higher sensitivity can be obtained by doping the flame gases with sulfur-containing gases (17).

Microwave-Induced Plasma Detector (MIPD)

Specific element gas chromatographic detection by plasma emission spectroscopy has received much attention in the last years (18,19,20,21,22,23). The atmospheric pressure microwave

induced and sustained helium plasma utilizing a Beenakker TM_{010} cavity is mostly used for gas chromatographic detection. The main advantages of the emission plasma detectors are :
- interfacing with capillary columns is easy.
- works in the universal mode or selective mode.
- several elements can be monitored simultaneously.
- the determination of the empirical formula of eluting compounds is possible.
- an approximately linear relationship has been found between the peak area of each compound determined by MIPD and the elemental content of the respective compound; therefore quantitation can be carried out without calibration.

The major limitations of the MIPD detectors are their relative high cost and their complexity. The multi-element determinations (C, H, D, S, F, Cl, Br, I, Hg), published in the literature, are very impressive. Problems are encountered with the determination of O, P and N.

Atomic Absorption Spectrometric Detection (AASD)

An atomic absorption spectrometer as mono-functional detector can easily be interfaced with a fused silica capillary gas chromatographic system (24); 100 pg of Hg can be detected very selectively without any problem.

Far-UV-absorbance Detector (FUV)

A new spectroscopic detector has been presented at the last Pittsburgh Conference. The FUV detector from HNU consists of a set of stable UV source lamps (120-150 nm), an absorption cavity and a photodiode. An emitted photon is absorbed by a molecule passing through the cell. The absorption causes a decrease in the photon flux at the photodiode. The detector can be used in combination with capillary columns; it is non-destructive and more sensitive and selective than the UV-VIS detector.

Mass Spectrometric Detection (MS)

The possibilities of high resolution gas chromatography combined with mass spectrometry are straightforward. The different modes of operation (electron impact, positive ion chemical ionization and negative ion chemical ionization) guarantee a very high degree of selectivity. Sensitivity in the pg range is obtained by applying mass fragmentography. The best way to interface CGC and MS is to use no coupling device at all; the column outlet needs only to be inserted into the ion source (25). The NICI-mode is very useful to search for mutagenic compounds (see ECD detector) and, by using CH_4/N_2O reaction gas mixtures, isomers can be differentiated (26). Two new MS detectors have recently been introduced : the HP 5970A Mass-Selective detector and the Finnigan Ion Trap detector. Both instruments cannot compete with sophisticated mass spectrometers, but they are simple in operation, perform quite well and are relatively cheap.

Operating modes are total ion current mode and selected ion monitoring. The scan range of the MSD is 10-800 amu and for the Finnigan 10-650 amu.

Infrared Detector (FTIRD)

Capillary GC-FTIR has developed into a powerful technique to provide functional group information of the eluting compounds (27,28). The sensitivity strongly depends on the nature of the compounds investigated. The inertness of the GC-IR lightpipe and transfer line has to be further improved to allow the analysis of (highly) polar compounds. The chromatographic efficiency of the combination CGC-FTIR is good. The combination CGC-FTIR-MS can find application in more difficult analytical problems (28).

MULTIPLE SIMULTANEOUS DETECTION

The main benefit of multiple detection is the simultaneous registration of specific and selective items by a single injection, reducing drastically the analysis time of multiple detection. Multidetector configurations can be arranged in series and/or in parallel. Destructive detectors can directly be mounted on non-destructive detectors (ECD, PID, FTIR, FUV). The combination of destructive detectors requires the installation of an effluent splitter in the CGC apparatus (6,29,30).

MULTIDIMENSIONAL GAS CHROMATOGRAPHY (31,32)

Multidimensional gas chromatography is a powerful technique to enhance selectivity. Interesting peak groups from a complex mixture, eluting from a first capillary column can be transferred to a second capillary column of higher selectivity. With the new developed live-switching system, heart-cutting in one second is possible. Trace compounds can be cut selectively from a precolumn, cold trapped and after enrichment by multiple injection transferred to the second capillary column. Multidimensional capillary gas chromatography will restrict the need for selective stationary phases.

SELECTIVE SAMPLE INTRODUCTION (33)

Complex mixtures can be separated according to the molecular characteristics of the compounds by HPLC. These selective preseparations facilitate structural elucidations and quantitative determinations. A similar result can be obtained by filling syringes used for on-column injection with HPLC material.

REFERENCES

1. MANTICA, E. (1981). Proc. Second Eur. Symp. on Analysis of Organic Micropollutants in Water, Killarney, Ireland. Ed. A. Bjorseth and G. Angeletti, D. Reidel Pub. Cie, Dordrecht, p. 61.
2. SANDRA, P., unpublished results.
3. OSMAN, M.A., HILL, Jr., H.H., HOLDREN, M.W. and WESTBERG, H.H. (1979). Anal. Chem., 51, 1286.
4. OSMAN, M.A. and HILL, Jr., H.H. (1981). J. Chromatogr. 213, 397.
5. SHUBHENDER KAPILA and CORAZON R. VOGT (1981). HRC & CC, 4, 233.

6. GAGLIARDI, P. and VERGA, G.R. (1983). Proc. Fifth Int. Symp. Cap. Chrom., Riva del Garda, Italy. Ed. J. Rijks, Elsevier Amsterdam, p. 418.
7. POOLE, C.F. (1982). HRC & CC, 5, 454 and ref. cited.
8. BAIM, M.A. and HILL, Jr., H.H. (1982). Anal. Chem. 54, 38.
9. BAIM, M.A. and HILL, Jr., H.H. (1983). HRC & CC, 6, 4.
10. BAIM, M.A. and HILL, Jr., H.H. (1983). Proc. Fifth Int. Symp. Cap. Chrom., Riva del Garda, Italy, Ed. J. Rijks, Elsevier Amsterdam, p. 809.
11. NEUKERMANS, A., KRUGER, W. and McMANIGILL, D. (1982). J. Chromatogr. 235, 1.
12. VAN DER WEIL, H.J. and TOMMASSEN, P. (1972). J. Chromatogr. 71, 1.
13. GROB, K. and HABICH, A. (1983). HRC & CC, 6, 11.
14. VERGA, G.R. (1983). Proc. Fifth Int. Symp. Cap. Chrom., Riva del Garda, Italy. Ed. J. Rijks, Elsevier Amsterdam, p. 835.
15. McCARTHY, L.V., OVERTON, E.B., MABERRY, M.A., ANTOINE, S.A. and LASETER, J.L. (1981). HRC & CC, 4, 164.
16. EHRLICH, B.J., HALL, R.C., ANDERSON, R.J. and COX, H.G. (1981). J. Chrom. Sci., 19, 245.
17. ZEHNER, J.M. and SIMONAITIS, R.A. (1976). J. Chrom. Sci., 14, 348.
18. YU WEI-LU, QU QING-YU, ZENG KE-HUI and WANG GUO-CHUEN (1981). Proc. Fourth Int. Symp. Cap. Chrom., Hindelang, GFR. Ed. R. Kaiser, Hüthig Verlag, Heidelberg, p. 445.
19. STIEGLITZ, L. and ZWICK, G. (1981). Proc. Second Eur. Symp. on Analysis of Organic Micropollutants in Water, Killarney, Ireland. Ed. A. Bjorseth and G. Angeletti, D. Reidel Pub. Cie, Dordrecht, 105.
20. ESTES, S.A., UDEN, P.C., RAUSCH, M.D. and BARNES, R.M. (1980). HRC & CC, 3, 471.
21. WASIK, S.P. and SCHWARZ, F.P. (1980). J. Chrom. Sci., 18, 660.
22. QU QING-YU, WANG GUI-CHUEN, ZENG KE-HUI and YU WEI-LU (1983). Spectrochemica Acta, 38B, 1,2., 419.
23. RISKA, G.D., ESTES, S.A., BEYER, J.O. and UDEN, P.C. (1983). Spectrochemica Acta, 38B, 1,2., 407.
24. DUMAREY, R., DAMS, R. and SANDRA, P. (1982). HRC & CC, 5, 687.
25. CRAMERS, C.A., SCHERPENZEEL, G.J. and LECLERCQ, P.A. (1981). J. Chromatogr., 203, 207.
26. OEHME, M., MANO, S. and STRAY, H. (1983). Proc. Fifth Int. Symp. Cap. Chrom., Riva del Garda, Italy. Ed. J. Rijks, Elsevier Amsterdam, p. 827.
27. SHAFER, K.H., BJORSETH, A., TABOR, J. and JAKOBSEN, R.J. (1980). HRC & CC, 3, 87.
28. SMITH, S.L. and ADAMS, G.E. (1983). Proc. Fifth Int. Symp. Cap. Chrom., Riva del Garda, Italy. Ed. J. Rijks, Elsevier Amsterdam, p. 800.
29. SHAFER, K.H., COOKE, M., DEROOS, F., JAKOBSEN, R.J., ROSARIO, O. and MULIK, J.D. (1981). App. Spectrosc., 35, 469.
30. MROWETZ, D. and STAN, H.J. (1983). Proc. Fifth Int. Symp. Cap. Chrom., Riva del Garda, Italy. Ed. J. Rijks, Elsevier Amsterdam, p. 203.

31. SANDRA, P., VERZELE, M. and VANLUCHENE, E. (1983). HRC & CC, Sept. 1983, in press.
32. SCHOMBURG, G., WEEKE, F., MULLER, F. and OREANS, M. (1982). Chromatographia, 16, 87.
33. MULLER, F. (1983). Int. Lab., July/August, 56.
34. SANDRA, P. (1983). J. Chromatogr., in press.

DETECTOR	UNIV	SELECTIVITY	SENSITIVITY	MOLECULAR INFORMATION	ELEMENTAL INFORMATION	CGC COMPATIBILITY
FID	+		++			+++
PID	+	+	+	+		+
ECD		+	+++	+		++
AFID		++	++		++ N,P	++
HECD	+	++	+		+ N,S,Cl	?
FPD		+	+		+ P,S	++
MIPD	+	++	++		++ C,H,D,F,Cl,Br,J,S ++ Hg,V,Cr,Mn,Fe,Co,Ni... + O,P,N	++
AAS		++	+		++ (Hg)	++
FUV	+	?	?	+		?
MS	+	+++	+++	++	+	+++
FTIR	+	+	+	++		+

Table 1

DETERMINATION OF ORGANIC WATER POLLUTANTS BY THE COMBINED USE OF HIGH-PERFORMANCE LIQUID CHROMATOGRAPHY AND HIGH-RESOLUTION GAS CHROMATOGRAPHY

W. GIGER, M. AHEL[*] and C. SCHAFFNER

Swiss Federal Institute for Water Resources and Water Pollution Control (EAWAG) and Swiss Federal Institute of Technology, CH-8600 Dübendorf, Switzerland

Summary

The combined use of high-performance liquid chromatography (HPLC) and high-resolution gas chromatography (HRGC) offers many advantages for the analysis of complex mixtures of organic chemicals commonly encountered in environmental samples. Of particular benefit is the complementary application of normal- and reversed-phase separation systems in HPLC and the use of HPLC as a preparative separation before detailed analysis by HRGC and HRGC/MS. Several examples of such applications are presented including determinations of polycyclic aromatic hydrocarbons, oil hydrocarbons, alkylphenols and alkylphenolpolyethoxylates.

1. INTRODUCTION

High-resolution gas chromatography (HRGC) and high-performance liquid chromatography (HPLC) are powerful techniques for the identification and quantitative determination of organic chemicals in environmental samples [1, 2]. Table I gives an overview of the different application purposes of HPLC and HRGC for the analyses of complex mixtures of organic compounds commonly encountered in environmental samples. Ratings are given for the feasibility and validity of the various application modes. The major advantage of HRGC lies in its high potential for complete analyses, specific determinations and, particularly when directly combined with mass spectrometry, for structure identifications. HPLC, on the other hand, is an excellent method for preparative group separations and trace enrichment.

---

[*] On leave of absence from the Center for Marine Research, "Rudjer Bošković" Institute, Zagreb, Yugoslavia.

	HPLC:	HRGC:
Analyses of complex mixtures		
complete analyses	−	+ +
specific determinations	+	+ +
Structure identifications		
mass spectrometry	+	+ +
UV/VIS-absorption/fluorescence	+ +	−
IR-absorption	+*	+
NMR	+*	−
(*: off line)		
Preparative operation		
individual components	+	−
group separation	+ +	−
trace enrichment	+ +	−

Table I

Feasibility and potential of HPLC and HRGC for the analysis of complex mixtures

++: highly useful

+ : feasible and useful

− : not feasible and of little use

Table I indicates that the potentials of two methods are often complementary to each other. Thus, a combined use of HPLC and HRGC is very promising, particularly if the two techniques are employed in their most powerful operation mode. A highly successful combination is the use of HPLC as a preparative group separation method followed by HRGC and HRGC/MS for complete analyses.

Figure 1 depicts schematically the different application modes of HRGC and HPLC for the determination of organic pollutants in water and in other environmental samples (sediments, sludges, biological materials). HPLC and HRGC can be employed to determine organic water pollutants by direct aqueous injections (--- in Fig. 1, refs. 3, 4). In most cases, however, the organic compounds have to be enriched by extraction and fractionation procedures (clean-up) before they can be determined by HPLC and HRGC (—— in Fig. 1).

HPLC is subdivided in Fig. 1 into normal-phase and reversed-phase operation because these two separation modes often provide different separations and can thus complement each other. The two HPLC modes can also be combined sequentially (—··— in Fig. 1).

In this paper, we emphasize the application of HRGC subsequent to a preparative separation either by normal-phase or by reversed-phase HPLC (—— in Fig. 1). A series of examples will be given including the determinations of polycyclic aromatic hydrocarbons, aliphatic hydrocarbons, alkylphenols and alkylphenolethoxylates.

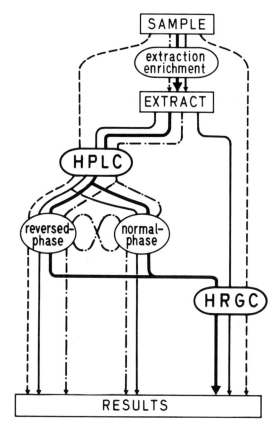

Fig. 1

Application modes of HPLC and HRGC for the determination of organic chemicals in water and in other environmental samples

––– direct aqueous injection HRGC or HPLC
——— HRGC or HPLC analysis after enrichment
–·– combined use of normal- and reversed-phase HPLC
▬▬ combined use of HPLC and HRGC

2. EXPERIMENTAL

HPLC solvent delivery was performed by Waters pumps and programmer. A Perkin Elmer LC-55 UV-VIS absorption detector and an Optilab refractive index detector were applied. All HPLC columns were purchased as packed columns from various suppliers. HRGC was carried out with Carlo-Erba gas chromatographs equipped with Grob-type split-splitless injectors. The glass capillary columns were kindly supplied by K. and G. Grob. For HRGC/MS analyses a computerised Finnigan system was employed.

The organic compounds were isolated from the aqueous and solid matrices by various extraction and fractionation procedures which are not discussed in this report.

3. RESULTS

3.1 Polycyclic aromatic hydrocarbons

Polycyclic aromatic hydrocarbons (PAH) are environmental trace pollutants which are predominantly associated with the particulates in air and water. HRGC has been extensively used for investigations of this compound class [5]. PAH almost always occur as complex mixtures of isomers and homologues with aromatic systems of up to seven rings. In recent lake sediments PAH have been found as common trace contaminants of pyrolytic origin [5,6]. In addition, natural PAH were detected which were formed in recent sediments as products of early diagenesis [7].

In a previous study, we have applied the combination of HPLC and HRGC/MS for the elucidation of diagenetic PAH in recent lake sediments [8]. Some sediment samples from the Greifensee, a highly eutrophic Swiss lake, contained a predominant pair of PAH which eluted very close to each other in HRGC analyses. The mass spectra indicated that these two components were isomers with molecular masses of 274 daltons. Possible structures of such PAH could contain a phenanthrene nucleus and seven additional carbon atoms including one saturated ring. The mass spectra, however, provided no evidence for the triaromatic ring structure. Thus, HPLC was applied to determine the structure of the aromatic ring system. At first, a reference mixture of phenanthrene, 2-methylphenanthrene and 3,6-dimethyl-phenantrene was analyzed by normal- and reversed-phase HPLC. In the normal-phase mode (SiO_2 column, hexane as eluting solvent), the phenanthrene and the two alkylphenanthrenes coeluted in one peak. However, with reversed-phase HPLC (SiO_2-C_{18}, methanol/water elution gradient) the homologues of the reference mixture were separated.

The HPLC analyses of the sedimentary PAH showed that in the normal-phase separation system the two major PAH coeluted with an elution volume corresponding to phenanthrene. In the reversed-phase operation the two 274-compounds eluted separately and much later than phenanthrene. Thus, the HPLC analyses helped to determine the phenanthrene-type structures of the two major PAH. The precise structures 1 and 2 were elucidated by the synthesis of reference compounds which was accomplished by Spyckerelle and coworkers [9] of the organic geochemistry group at the University of Strasbourg.

(1) 274 A

(2) 274 B

In our earlier investigation [8] we used preparative normal-phase HPLC to obtain PAH-fractions containing compounds with different aromatic ring structures (naphthalenes, phenanthrenes, chrysenes). Subsequent HRGC/MS allowed the identifications of a series of di-, tri- and tetraaromatic hydrocarbons of diagenetic origins. In this paper, we present combined

HPLC/HRGC characterizations of PAH isolated from sediments of the Bodensee (Lake Constance). Surface sediments (0 - 2 cm) and a deeper section of a sediment core (70 - 75 cm) were selected to demonstrate the results obtained by this approach.

Figure 2A shows the capillary gas chromatogram of the total PAH from the surface sediments. The complex pattern is typical for PAH which are of pyrolytic origin. Detailed structure assignments were presented elsewhere [5, 6]. One interesting question was whether this mixture also contained the two 274-PAH. One way to answer this question was by HRGC/MS, but the combined use of preparative normal-phase HPLC and HRGC/MS served as an excellent confirmatory technique.

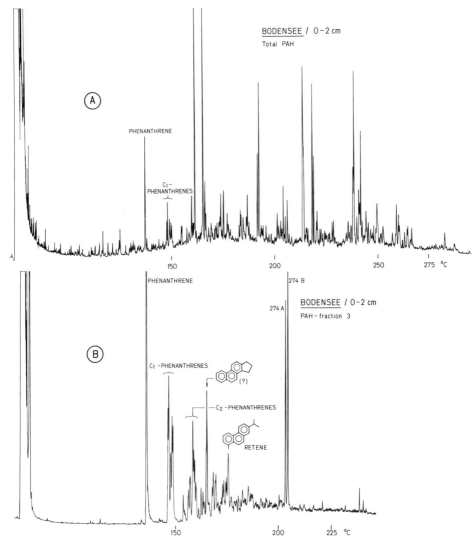

Fig. 2 HRGC determinations of PAH isolated from surface sediments (0 - 2 cm) of the Bodensee. A: total PAH, B: PAH-fraction 3.

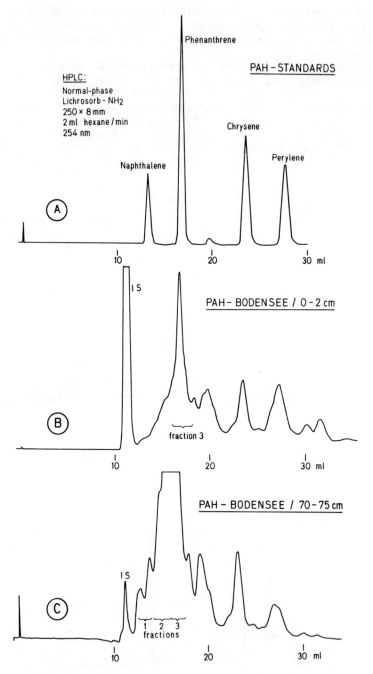

Fig. 3 Normal-phase HPLC analyses of PAH. A: reference mixture, B: PAH from surface sediments (0 - 2 cm) of the Bodensee, C: PAH from subsurface sediments (70 - 75 cm) of the Bodensee. IS: internal standard (n-decylbenzene).

Figure 3A shows the normal-phase HPLC analysis of a reference PAH mixture containing naphthalene, phenanthrene, chrysene and perylene. A bonded phase (SiO_2-NH_2), semipreparative (8 mm diameter) column and hexane as eluent were used. In Fig. 3B the chromatogram obtained from the total PAH fraction of the surface sediment is depicted. Group separation into PAH with tri, tetra- and pentacyclic aromatic ring systems was accomplished.

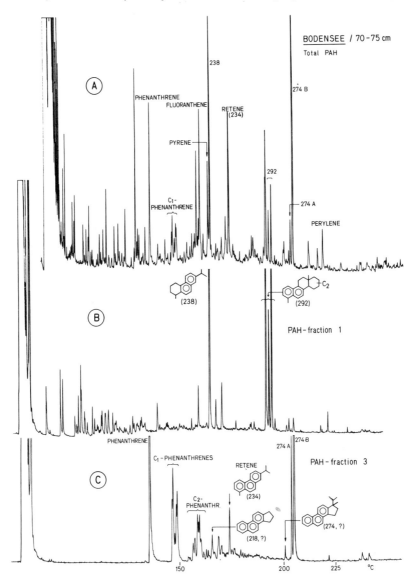

Fig. 4 HRGC determinations of PAH isolated from subsurface sediments (70 - 75 cm) of the Bodensee.

A: total PAH, B: PAH-fraction 1, C: PAH-fraction 3.

The result of the HRGC analysis of fraction 3 is shown in Fig. 2B. The peaks for 274A and 274B are well separated from the other phenanthrene homologues. In addition, retene and, more tentatively, cyclopentanophenanthrene could be identified.

Figure 4A depicts the capillary gas chromatogram of the total PAH isolated from the 70-75 cm section of a sediment core taken at the deepest point of the Bodensee. In this sample the pyrolytic PAH are accompanied by a series of diagenetic compounds. HRGC/MS could again be used to determine the molecular masses of the various constitutents, but more information on the aromatic ring structures was required. In Fig. 3C the preparative HPLC trace of the PAH from the deep sediment sample is shown and it is indicated which fractions were collected for further HRGC/MS analyses. Fig. 4B and 4C show examples of HRGC analyses of the fractions 1 and 3. Fraction 1 contained tetrahydrogenated retene (molecular mass: 238 daltons) and three isomeric tetracyclic hydrocarbons with two aromatic rings (molecular mass: 292 daltons, see ref. 7). The three 292-compounds have the same skeleton structure as the 274-PAH but with only a naphthalene-type aromatic system. Retene was the major component of fraction 2 which is not included in Fig. 4. Fraction 3 (Fig. 4C) had a similar composition as the corresponding fraction of the surface sediment. A third 274 PAH could be detected and was assigned the structure of a C_{21}-triaromatic steroid hydrocarbon, a compound which was identified in petroleum and bituminous sedimentary rocks [10].

It should be emphasized that the HRGC/MS analyses of the PAH subfractions provided much better spectra than could be obtained from the total PAH mixtures, in particular for minor trace components.

3.2 Aliphatic hydrocarbons

Aliphatic hydrocarbon mixtures isolated from environmental or geological samples usually contain components with boiling points varying over a very broad temperature range. Temperature programmed HRGC can accomplish the analyses of such mixtures but rather long analyses times are required particularly if long columns (>50 m) are employed. Thus, an easy possibility to separate such samples into fractions of smaller volatility ranges would be very valuable. Furthermore, HRGC/MS is often only required for parts of the hydrocarbon mixtures. If only a subfraction could be analyzed, analysis time and data storage space could be saved. In fact, reversed-phase HPLC offers a very practical means to fractionate aliphatic hydrocarbon mixtures of broad boiling ranges.

We report here the analysis of an oil shale extract by such a combined reversed-phase HPLC/HRGC approach. Figure 5A shows the HRGC analysis of the total aliphatic hydrocarbons extracted from the Triassic Serpiano oil shale which outcrops in Southern Switzerland. This hydrocarbon mixture is characterized by a typical bimodal distribution starting around n-dodecane and comprises, at the high boiling end, components in the volatility range of n-C_{33}-alkane. The lower boiling point range is dominated by n-alkanes and isoprenoid hydrocarbons such as pristane and phytane. In the higher boiling range, a series of homologous pentacyclic triterpanes (hopanes, see ref. 11) are the most abundant constitutents.

Figure 6 includes the reversed-phase HPLC analyses of a standard mixture of aliphatic hydrocarbons (pristane, phytane and cholestane) and of the oil shale extract. A refractive index detector was necessary to monitor the elution of these aliphatic, non-UV/VIS-absorbing hydrocarbons. The

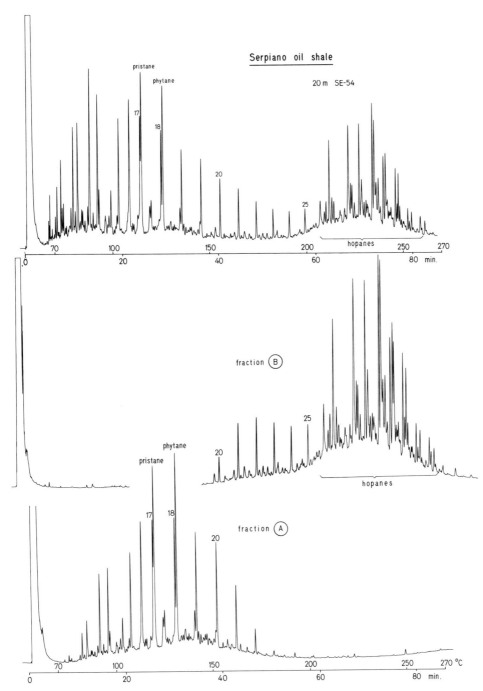

Fig. 5 HRGC determinations of aliphatic hydrocarbons isolated from Serpiano oil shale Fractions A and B refer to Fig. 6. The numbers assign n-alkane peaks.

bimodal composition of the oil shale extract was reflected also in the HPLC trace and a cut at 3.5 ml elution volume was set to separate the two fraction A and B. From the two lower gas chromatograms of Fig. 5 it can be seen that HPLC provided fractions with only little overlap. Figure 7 shows the HRGC analysis of fraction A with a 62 m glass capillary column. This analysis allowed the separation of the diastereomeric forms of pristane, phytane and of the C_{18}-isoprenoid hydrocarbon. Because of the long column and the slow temperature program, pristane eluted only after more than 40 min. Very tedious, time consuming operations would have been necessary, if the complete aliphatic hydrocarbon mixture would have had to be analyzed. Similarly, HRGC/MS determinations of the hopanes were much more practical if only the fraction B was analyzed.

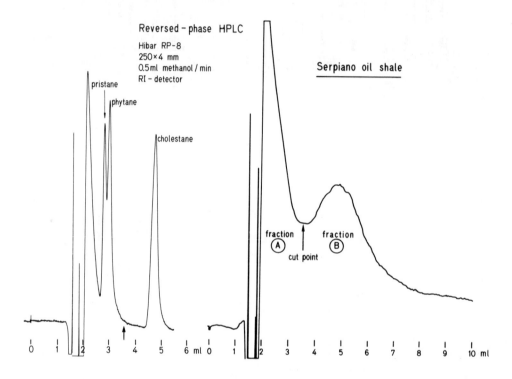

Fig. 6   Reversed-phase HPLC analysis of aliphatic hydrocarbons.

3.3 Alkylphenols, alkylphenolethoxylates and alkylphenolethoxy carboxylic acids

4-Alkylphenolpolyethoxylates (1 in Fig. 8) are widely used nonionic surfactants which are manufactured in several hundred thousand tons per year. Biological degradation during wastewater treatment occurs primarily via shortening of the hydrophilic polyethoxylate chain. More resistant 4-nonylphenolmono- and diethoxylates (2a) have been detected as major constituents in biologically treated wastewater effluents [12, 13, 14], and

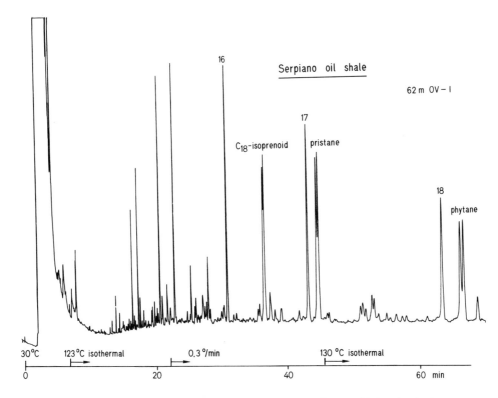

Fig. 7 Long column HRGC determination of low-boiling aliphatic hydrocarbons (fraction A) isolated from Serpiano oil shale.

in river waters [15]. In addition, alkylphenolpolyethoxy carboxylic acids (3) were identified in treated municipal wastewater [16]. Anaerobically stabilized sewage sludge was found to contain extraordinarily high concentrations of 4-nonylphenol (4a), ranging from 0.45 to 2.5 g/kg dry matter [17]. In this section, we present the determinations of alkylphenolic compounds (Fig. 8) in sewage effluents and in sewage sludges by the combined use of HPLC and HRGC.

Figure 9 shows the gas chromatograms of commercially available, technical 4-nonylphenol and of the extract of anaerobically stabilized sewage sludge [17]. The major peaks in both chromatograms were assigned by HRGC/MS to isomeric 4-nonylphenols which had variously branched nonyl substituents [12]. These isomers were present because the nonyl chains were synthesized from propylene. The technical 4-nonylphenol also contained approximately 10% of 2-nonylphenol which was absent in the sewage sludge extract. Furthermore, both samples contain 4-decylphenol, and in the sludge extract 4-octylphenol (4-(1,1,3,3-tetramethylbutyl)-phenol) was identified.

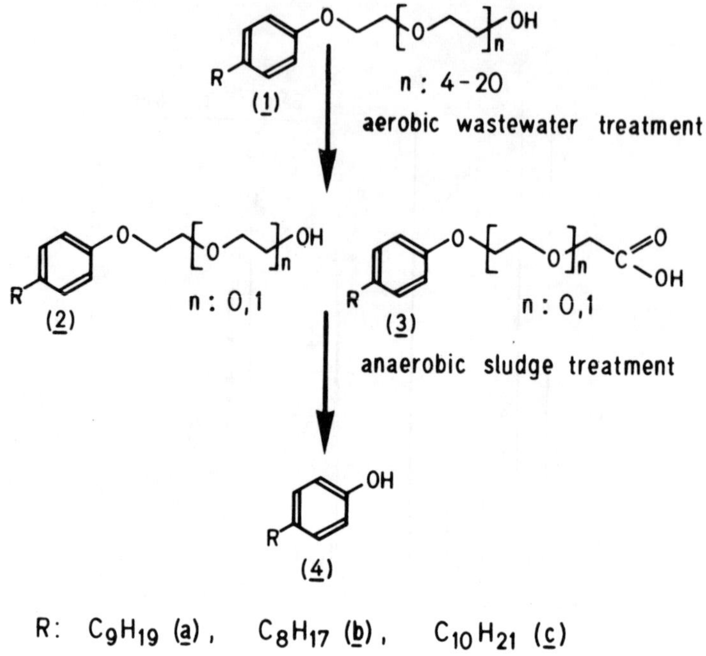

Fig. 8   Degradation scheme of alkylphenolpolyethoxylates.

Figure 10 contains the results of the determination of technical 4-nonylphenol by the combined use of normal-phase HPLC and HRGC. The HPLC trace showed two major and two minor peaks which were well separated from each other. These fractions could be easily collected for subsequent HRGC analysis. A similar characterization of technical 4-nonylphenol has been carried out by Gerhardt [18]. The early eluting fractions 1 and 2 contained ethers and dinonylphenols, respectively. In fraction 3 we detected 2-nonylphenol while 4-nonyl- and 4-decylphenol were found in fraction 4. Normal-phase HPLC as shown in Fig. 10 allowed the preparative separation of 2- and 4-nonylphenols in sufficient amounts for further characterization by proton magnetic resonance. These results confirmed the 2- and 4-position of the alkyl substituents [19].

Figure 11 depicts the combined normal-phase HPLC/HRGC analysis of a digested sewage sludge extract. The 4-substitution of the major nonylphenolic constituents and the absence of 2-nonylphenol were confirmed. A digested sewage sludge extract has been analyzed by combined reversed-phase HPLC and HRGC. As shown in Fig. 12, this HPLC technique separated the homologous octyl-, nonyl- and decylphenols (fractions 1, 2 and 3) which could be assured by HRGC/MS of the three fractions.

HPLC was also applied to directly determine alkylphenolic compounds in wastewater and sludge extracts. Examples of two sludge extract analyses are depicted in Fig. 13A and B. In comparison to HRGC [14, 17], the HPLC method is more rapid but also more susceptible to interferences. In the extract of an anaerobically treated sewage sludge an interfering compound

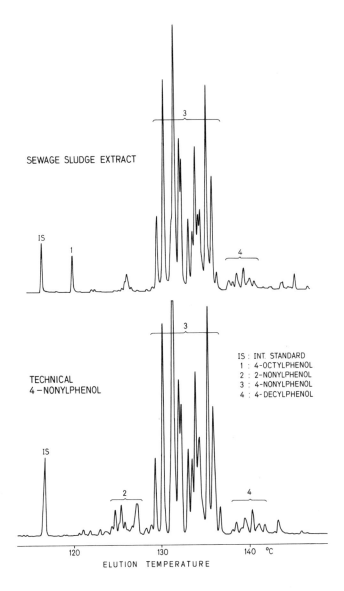

Fig. 9   HRGC analyses of the extract of an anaerobically stabilized sewage sludge and of technical 4-nonylphenol.

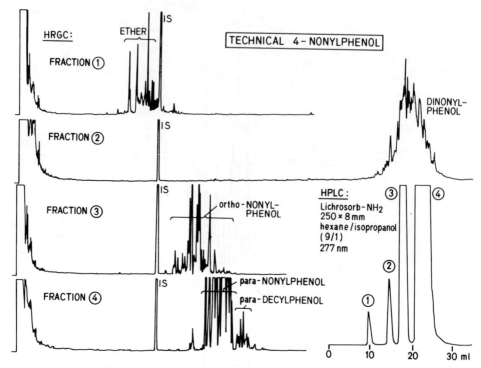

Fig. 10 Combined normal-phase HPLC/HRGC analysis of technical 4-nonyl-phenol.
IS: internal standard (n-nonylbenzene).

eluted very closely to the 4-nonylphenolmonoethoxylate (Fig. 13A and B). In fact, in the chromatogram of Fig. 13A this interference was not separated from 4-nonylphenolmonoethoxylate because the chromatographic performance was not as good as for Fig. 13B. The interfering peak was collected as fraction 1 and analyzed by HRGC/MS (Fig. 13C). 3-Methylindole (skatole) could be identified by its mass spectrum. This compound, a metabolite of the amino acid tryptophane, is encountered quite frequently in wastewaters and sludges. This example should demonstrate the use of HRGC/MS to elucidate the structures of chemicals interfering in HPLC analyses.

Nonylphenolmonoethoxy carboxylic acid (Fig. 8, 3a, n=0) and nonylphenoldiethoxy carboxylic acid (3a, n=1) in secondary sewage effluents could be well determined by HRGC (Fig. 14C). In primary, sewage effluents, however, large amounts of many other acids were present in the same fraction. Therefore, a reliable determination of the nonylphenolethoxy acids was not feasible (Fig. 13A). For this purpose HRGC/MS and single ion monitoring techniques would have been necessary. In this case HPLC enabled the specific detection of the nonylphenolethoxy carboxylic acids as can be seen in the HPLC traces Fig. 14B and D.

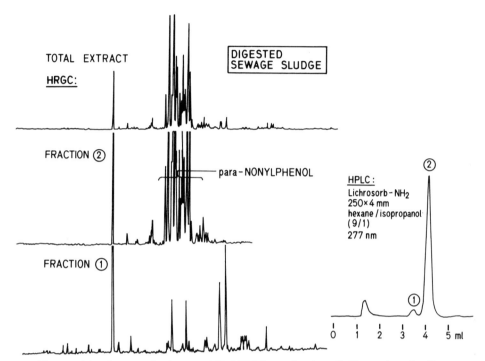

Fig.11: Combined normal-phase HPLC/HRGC analysis of the extract of a digested sewage sludge.

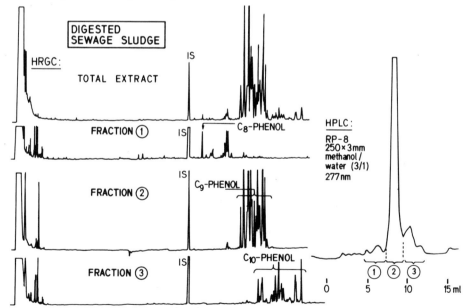

Fig.12: Combined reversed-phase HPLC/HRGC analysis of the extract of a digested sewage sludge.
IS: internal standard (n-nonylbenzene).

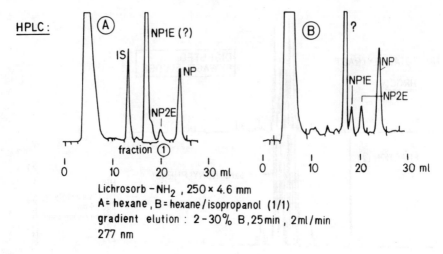
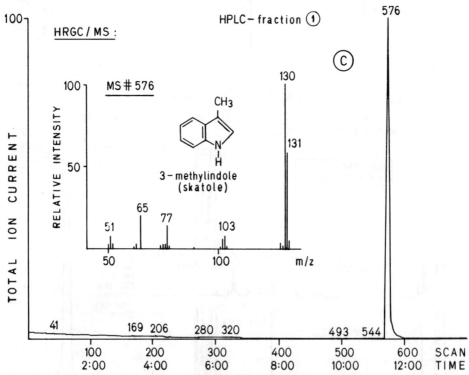

Fig. 13 Combined HPLC/HRGC/MS analyses of the extract of an anaerobically treated sewage sludge.
NP: 4-nonylphenol (4a in Fig. 8); NP1E: 4-nonylphenolmonoethoxylate (2a, n=0 in Fig. 8); NP2E: 4-nonylphenoldiethoxylate (2a, n=1 in Fig. 8).

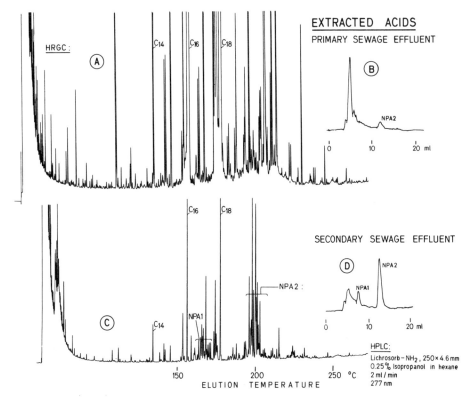

Fig. 14　HRGC and HPLC determinations of acids isolated from primary and secondary sewage effluents.
NPA1: 4-Nonylphenolmonoethoxy carboxylic acid
(3a, n=o in Fig. 8)
NPA2: 4-Nonylphenoldiethoxy carboxylic acid
(3a, n=1 in Fig. 8)
C_{14}, C_{16}, C_{18}: saturated fatty acids.

Figure 15 shows the gas chromatogram of a fraction collected from the HPLC analysis of a primary sewage effluent (see inserted HPLC trace). In contrast to the HRGC analysis in Fig. 14A, many of the major acids have been eliminated and the nonylphenolmono- and diethoxy carboxylic acids (NPA1 & NPA2) were easily recognized.

ACKNOWLEDGMENT

This work was in part supported by the Swiss Department of Commerce (Project COST 64b).
We thank K. and G. Grob for supplying glass capillaries for gas chromatography.

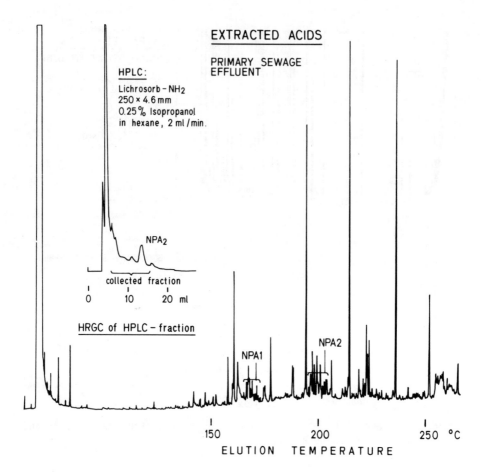

Fig. 15 Combined HPLC/HRGC analysis of acids isolated from a primary sewage effluent.
NPA1, NPA2: see legend of Fig. 14.

REFERENCES

1. Bjoerseth A., Angeletti G., eds., "Analysis of Organic Micropollutants in Water", Reidel Publishing Company, Dordrecht, 1982.
2. Keith L.H., ed., "Advances in the Identification and Analysis of Organic Pollutants in Water", vol. 1 and 2, Ann Arbor Science, Ann Arbor, 1981.
3. Kummert R., Molnar-Kubica E., Giger W., Anal. Chem. 50, 1637-1639 (1978).
4. Grob K., Habich A., HRC & CC, 6, 11-15 (1983).
5. Giger W., Schaffner C., Anal. Chem., 50, 243-249 (1978).
6. Wakeham S.G., Schaffner C., Giger W., Geochim. Cosmochim. Acta, 44, 403-413 (1980).

7. Wakeham S.G., Schaffner C., Giger W., Geochim. Cosmochim. Acta, 44, 415-429 (1980).
8. Wakeham S.G., Schaffner C., Giger W., in Adv. Org. Geochem., 1979, 353-363, Douglas A.G., Maxwell J.R., eds., Pergamon Press, Oxford, 1980.
9. Spyckerelle C., Greiner A.C., Albrecht P., Ourisson G., J. Chem. Res. (M) 1977, 3746-3777; (S) 1977, 330-331.
10. Mackenzie A.S., Hoffmann C.F., Maxwell J.R., Geochim. Cosmochim. Acta, 45, 1345-1355 (1981).
11. Ourisson G., Albrecht P., Rohmer M., Pure Appl. Chem., 51, 709-729 (1979).
12. Giger W., Stephanou E., Schaffner C., Chemosphere, 10, 1253-1263 (1981).
13. Schaffner C., Stephanou E., Giger W., in reference [1], pp. 330-334.
14. Stephanou E., Giger W., Environ. Sci. Technol., 16, 800-805 (1982).
15. Ahel M., Giger W., Molnar-Kubica E., Schaffner C., this volume.
16. Reinhard M., Goodman W., Mortelmans K.E., Environ. Sci. Technol., 16, 351-362 (1982).
17. Schaffner C., Brunner P.H., Giger W., in "Proceedings of the Third International Symposium on Processing and Use of Sewage Sludge", Brighton, England, September 1983, in press.
18. Gerhardt W., Much H., Tenside Detergents, 18, 120-123 (1981).
19. Giger W., Brunner P.H., Schaffner C., in preparation.

ANALYSIS OF ORGANIC MICROPOLLUTANTS BY HPLC

J.C. KRAAK
Laboratory for Analytical Chemistry, University of Amsterdam

Summary

The potential of HPLC as analytical method for the analysis of non-volatile organic compounds in complex mixtures is discussed. The state of art to optimize the analysis via the number of theoretical plates of the separation column, via multicolumn operation and by applying selective detection is reviewed.

1. INTRODUCTION

When analyzing complex mixtures at low concentration levels by HPLC one is faced with two main problems:
i) the limited separation power of the chromatography
ii) the detectability of the solutes of interest.

The separation power of a phase system for two solutes is best reflected in the resolution R_{ji}, which can be expressed in terms of the chromatographic parameters:

$$R_{ji} = (r_{ji} - 1) \left(\frac{k'_i}{1+k'} \right) \left(\sqrt{\frac{L}{H_i}} \right) \qquad (1)$$

in which
r_{ji} = selectivity factor (= k'_j / k'_i)
k'_i = capacity factor
L = column length
H_i = theoretical plate height

$\frac{L}{H_i}$ = N = number of theoretical plates.

The detectability of solutes depends on their sensitivity towards the applied detection system and on the dilution of the solutes in the chromatographic set up. The maximal concentration of a solute at the end of the column (e.g. the peak height) $(c_{im}^{max})_L$ is given by the following expression:

$$(c_{im}^{max})_L = \frac{V_{inj} \cdot c_i^{inj}}{\varepsilon_m \cdot A_s \cdot (1+k'_i) \sqrt{2\pi} \cdot \sqrt{L \cdot H_i}} \qquad (2)$$

in which
V_{inj} = injection volume

c_i^{inj} = concentration of i in the sample
ε_m = interstitial porosity of the column
A_s = cross sectional area of the column.

When including the contribution of external peak broadening effects into the discussions via:

$$\sigma_{total}^2 = \sigma_{column}^2 + \sigma_{det.}^2 + \sigma_{inj.}^2 + \ldots \quad (3)$$

where σ = standard deviation in volume or time units and further $\sigma_{V_{inj}} \sim V_{inj}$, the state of art of HPLC systems for trace analysis in complex mixtures under common practical experimental conditions such as an upper pressure limit of 500 bar, flow rates in the range of 50-2000 μl/min etc. can be reviewed straightforwardly.

When comparing the influence of the parameters L, k_i', V_{inj} of equations 1 and 2 it will be obvious that the optimization of R_{ji} and $(c_{im}^{max})_L$ contradicts each other and will ask for a compromise in practice.

2. APPLICATION OF COLUMNS WITH LARGE PLATE NUMBERS

According to equation 1 the resolution increases with the square root of the number of theoretical plates N. Although in practice a large number of separations can be carried out on the now commonly available 6000-10.000 plate columns, in some instances more plates are necessary in order to obtain a desired resolution. These situations occur when despite optimization of the phase system for some components of the sample the selectivity factor r_{ji} is still very small or when the k' of the solutes of interest are very small and cannot be increased and further when the mixtures to be separated are so complex that usually the optimization of the phase system (e.g. r_{ji}) for all solute pairs is virtually impossible.

There are three ways to generate a large number of theoretical plates:
i) by decreasing the theoretical plate height
ii) by recycling
iii) by increasing the length of the column.

It will be obvious that due to experimental limitations (for instance pressure), there will be trades off between analysis time, dilution and the number of theoretical plates.

Apart from working in the minimum of the H versus the linear velocity (<v>) curve, the plate height can be decreased according to theory by decreasing the particle size below the now common used 5-10 μm. Recently column packings of 3 μm have become available and its potential to create very small plate heights (6-9 μm) is shown in a number of papers (1-3). However, these 3 μm particles require a significant larger pressure drop which limits the length of the column. Moreover, the heat dissipation due to frictional forces in such 3 μm columns is large and gives rise to rather steep H/<v> curves if no precautions are made (4). Another problem with these highly efficient 3 μm columns arises from the very small peak volumes (σ_V) which become too small compared to the external peak broadening effects of most common HPLC systems and thus might lead to a significant loss in plates. Fortunately, there is a trend to adapt the detection and injection systems for ultra efficient columns including microbore columns (5).

The other way of increasing the number of theoretical plates is by recycling the sample a number of times over the same column (6). Although very useful for simple mixtures, this technique is not suitable for complex mixtures as the less and more retained peaks are merged with increasing number of cycles.

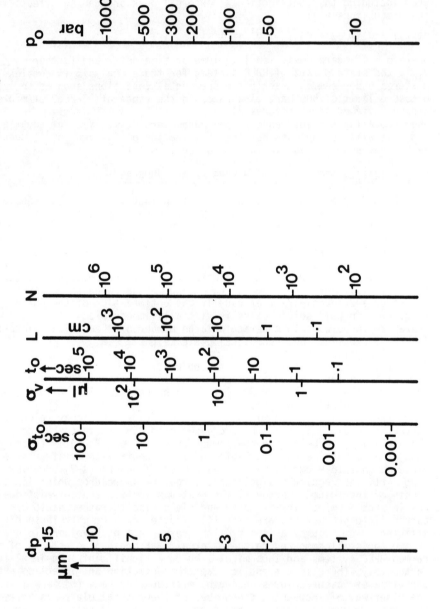

Fig. 1. Nomogram representing the relation between d_p, N, L, P, t_{R_o}, σ_{t_o} and σ_{v_o}.

The simplest way to increase the plate number is by lengthening of the column. Because of the pressure limitation very large plate numbers, for instance 10^6, can only be obtained if somewhat larger particles (10-20 µm) are used, as is shown in Fig. 1. As experience has learned that with the applied present packing techniques longer columns (50-100 cm) can be packed less efficiently than shorter ones (10-25 cm), it will be obvious that coupling of efficient short columns with equal plate heights is the way to go to increase the number of theoretical plates. This was demonstrated previously by Kraak et al. (7) who coupled five 25 cm, 4.6 I.D. columns filled with 7 µm silicagel and thus were able to create a column with 50.000 plates. The loss in plate number due to the coupling was found to be less than 10%. Recently, Verzele et al. (8) demonstrated a similar experiment with reversed phase columns. The separation potential of this 50.000 plate column is demonstrated in Fig. 2, showing a chromatogram of a TLC fraction (a spot) of an extract of a refuse incinerator. It can be seen from this figure that although the column discriminates a large number of peaks, still

Fig. 2. *HPLC separation of an extract of a* refuse incinerator *on a 125 cm silica column of 3 mm I.D., having 50.000 theoretical plates.*

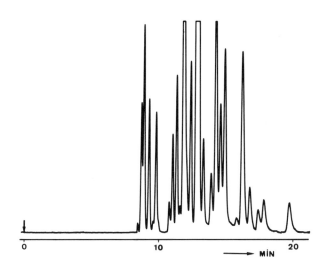

significant overlap occurs. In order to resolve all peaks in this particular sample a much larger plate number than 50.000 plates is required. However, with the practical requirements put on the pressure limitation it will be obvious that this can only be realized by accepting excessively long retention times and significant dilution. This latter might give problems to detect the solutes.
Although large plate numbers can be obtained by coupling of commonly used 4-5 mm I.D. columns at the cost of the separation time, another aspect, namely the worse detectability of the solutes due to the increasing dilution with increasing column length, often determines the maximal column length (e.g. plate number). In order to diminish the dilution (e.g. keeping reasonable $(c_{1m}^{max})_L$) and for economic reasons (eluent consumption, amount of packing material) columns with a significant smaller diameter (0.5-1 mm) are in focus nowadays (9,10).

According to equation 2, the peak height $((c_{im}^{max})_l)$ is reverse proportional with the column cross sectional area. When decreasing the diameter from 4.6 to 1 mm the dilution will be a factor of 20 less, providing both columns can be packed equally efficient. This latter is the case, as has been demonstrated by various workers (9-11). The coupling of these microbore columns seems to be more efficient than for the larger diameter columns. Coupled 1 mm I.D. microbore columns up to 14 m having 750.000 theoretical plates have been demonstrated by Scott and Kucera (10,11). Recently, packed micro glass capillaries have been applied (12) showing excellent efficiencies. Because of the pressure limitations the separation times on coupled microbore columns are of the order of 12-50 hours (see Fig.3). In order to

Fig. 3. *Separation of bergamot oil on a 14 m microbore column of 1 mm ID, filled with 5 μm silicagel, having 750.000 theoretical plates* (10).

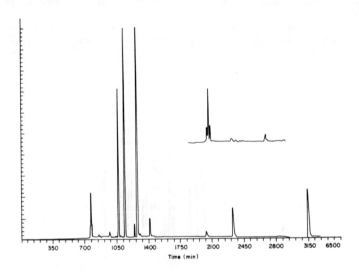

generate more plates with a given pressure drop, attempts are now made to apply open tubular columns, as used in GC, in LC (13). Theory predicts (14) that these open capillaries can only compete with packed columns if the column diameter is of the same order as the particle sizes as used in HPLC, namely 5-10 μm. Apart from manufacturing difficulties, such capillaries lead to serious problems arising from the contribution of the external peak broadening caused by the injection and detection system. The volume of such capillaries is very small and prescribes detection systems with cell volumes of a few nanolitre. For the same reason very small injection volumes have to be applied which prescribes detection systems capable to measure very small amounts of solutes. It will be obvious that the present available detection systems, even when miniaturization is applied, are not suitable for such capillary columns. However, there are some detection systems which have some future potential in this area like laser fluorescence, mass spectrometry. The exploration of these relatively expensive detection systems is now subject of research in many laboratories and these efforts hopely lead to the full exploration of open capillaries in LC in the coming decade.

Although columns with large plate numbers can be useful to create resolution, in practice it is usually found that the full profit of the peak capacity on such columns is distressingly small. Usually, peaks cluster in a small elution volume leaving other parts of the chromatogram empty. In order to improve the resolution of these clustered peaks with the selected phase system a manifold of, for a greater part wasted, theoretical plates is required. Recently, Giddings et al. (15) showed that when chromatographing a randomly distributed sample on an arbitrary phase system maximal 18% of the solutes in the sample can be eluted as a single peak if 50% of the peak capacity is exploited (50 compounds when the peak capacity is 100). Or even more disappointing in order to elute a solute as a singlet with a probability of 90%, the chromatogram must be empty for 95%! These statistical calculations by Giddings show, that the separation of a complex mixture can hardly be realized on one column and needs multicolumn systems (16). If one is interested in just the analysis of a few solutes in a complex mixture, the situation can be less serious if for these solutes a very specific detection system (for instance post-column reaction detection) can be applied. However, if one wants to collect the peaks for identification purposes or further experiments (for instance biological tests), multi-column operation will be the route to follow.

3. MULTI-COLUMN OPERATION

In the previous section it was discussed that the complete separation of a complex mixture on one column can seldomly be obtained. Usually the selected column discriminates quite a number of peaks which, however, will consist, with a large chance, of more than one solute. When optimizing the phase system (r_{ij}), for instance via the composition of the mobile phase, the separation of some solute pairs can be improved at the cost of other pairs. Such optimization procedures make clear that there is no universal phase system which can separate all types of solutes. For each solute pair or a limited number of solutes an optimal phase system can be composed. In multi-column operation now the separation system consists of a number of different columns, each of these being suitable for the separation of a particular group of solutes as present in the mixture (17,18). By means of switching valves unresolved solutes (peaks) on one column can be transferred (heart-cutting) to another column with more favourable separation conditions for these particular solutes. Usually, one main column serves to fractionate the mixture into various groups of solutes, which then are transferred to one of the second columns. It must be noticed that similar fractionation, although much less effective, can sometimes be realized by pretreatment of the sample by other separation techniques such as liquid-liquid and liquid-solid extraction, gel permeation, etc. How many of such "tailor made" columns are needed depends on the complexity of the mixture. In some cases it might be even necessary to transfer a peak from the second column to a third one in order to realize elution of the solutes of interest as singlets.

The whole multi-column operation can be executed on- or off-line. The on-line set up has definite advantages with respect to automation, but unfortunately a number of serious problems still delay its full exploration. The most serious problem arises from the fact that in HPLC the composition of the sample solution is not allowed to deviate much from that of the mobile phase. As usually the composition of the mobile phase on the selected columns differs significantly, only in a very few cases a direct transfer of a peak to another column can be applied. In most cases, however, the mobile phases are not tuned to each other (compare reversed and normal phase) and thus a kind of interface is needed to transfer a "heart-cut"

from one column to another. However, such on-line interfaces are not available at this moment. In this respect the interface principle as used by technicon (EDM system) (19) to transfer an organic extract on-line into an aqueous solvent might be a way to design the desired interfaces for on-line multi-column operation.
Another problem arises from the peak broadening. The peaks coming from the first column start by transference to the second column with a certain width which will influence the column efficiency. In order to have full profit of the second column, the peak has to be compressed before entering the column. In some particular cases it is possible to realize this by diluting the peak with a diluent of low elution strength before entering the second column (20). By this the peak enters the column in a solution of low elution strength and thus concentration on the top of the second column occurs.
 It will be clear that due to these aforementioned problems on-line multi-column operation is not widely applicable at this moment. The offline method, although more laborious than the on-line system, is easier to implement and must be the method of choice to explore fully at this moment the separation power of multi-column systems. After collection of a "heartcut" the analytes can be isolated by evaporation, dry freezing, etc. and then redissolved in a small volume of eluent for the selected second column.
 Although at this moment a large number of phase systems can be composed with the few available types of stationary phases, the potential of multi-column operation can be still significantly enlarged when more "tailor made" stationary phases become available. In this respect the application of liquid-liquid systems as phase systems (21) might be worth to be investigated again.

4. SELECTIVE DETECTION SYSTEMS

If one is only interested in the analysis of a few specified compounds in a sample and not in the isolation of the solutes for identification purposes or other tests, selective detection is another solution to diminish the interference problem. By applying selective detection one can reduce significantly the number of registered peaks and thus increases the chance to detect the solutes as singlets.
 Of course selective detection can be applied also with columns with very large plate numbers or in multi-column operation. Of the available detection systems the following can be considered as more or less selective (22):
- fluorescence
- electrochemical detection
- homogeneous post-column reaction
- extraction detection
- atomic spectrometry
- electron capture
- photochemical
- phosphorescence.
 Apart from the selectivity towards a limited class of compounds, the detection system must show adequate sensitivity towards the compounds of interest to allow analysis on the ppb level as is common for organic micropollutants in water.
 The *fluorescence detector* fulfils both demands of high selectivity and high sensitivity. The detection system is very useful in environmental analysis as quite a number of micropollutants such as PAHs show fluorescence. Recently, fluorescence detectors have become commercially available with

cell volumes of 3 µl without significant loss in sensitivity (23). These detectors are suited to be connected with packed microbore columns and form a powerful combination for ultra trace analysis.

The *electrochemical detection* systems based on oxidation and reduction processes at electrodes also show high selectivity and sensitivity towards limited classes of compounds, some of them are found as micropollutants in water. With the solid electrode mode (24,25) detection limits in the low picogram range and for the mercury drop systems (26) in the low nanogram level can be obtained. The anodic version allows miniaturization of the cell without loss in sensitivity (27) and thus can be made suitable as detection system for packed microbore columns.

Post-column reaction detection is probably the most versatile selective detection system as specific chemical reactions are carried out on the eluting peaks by which the compounds are converted into derivatives which are highly sensitive for the common detection systems (28). By proper choice of the reaction and reaction conditions the selectivity of detection can be very specific, as is known from classical organic analysis. A very powerful detection system is obtained when the post-column reaction can be combined with a selective detector like fluorescence. A disadvantage of post-column reaction systems is the relatively large extra peak broadening caused by the mixing Tees and reactor (29). This certainly will lead to a loss of resolution and peak height. However, with normal sized columns the extra peak broadening can be kept at an acceptable level. Recently, it was shown that post-column reaction systems can be miniaturized to such an extent that they become suitable for connection with packed microbore columns (30).

In *post-column extraction detection* the effluent is segmented with an immiscible solvent and transported to a wound extraction coil. Before entering the detector the segmented stream is separated again by means of a phase separator into an aqueous and organic phase. One of these phases, containing the solutes of interest, is fed to the detector (31). Such extraction systems have definite advantages compared to the homogeneous reaction systems. First, the effluent is segmented, which can significantly reduce the extra peak broadening depending on the reaction time. Secondly, the formed derivatives and reagent usually distribute differently between the two phases. This allows the use of labelling reagents which cannot be used in the homogeneous system because of their response towards the detection system. For instance ion-pair extraction using pairing ions with fluorophores or chromophores as "label" for non-detectable ionizable substances can then be used as is demonstrated for alkylsulfonates (32) in Fig. 4.

Atomic spectroscopic detection has been applied in HPLC (33). The simplest but also the most expensive system to be interfaced with HPLC is inductively coupled argon plasma(ICAP). Such HPLC-ICAP combination can be a powerful tool for investigations of metal speciation in the environment. The interfacing of microwave induced plasma (MIP) with HPLC promises a tremendous possibility for detection of traces of halogen, N, S, P containing compounds. However, until now this is still not yet realized.

Electron capture detection in which the mobile phase is completely vaporized has been applied as HPLC detector (34). The detection system can only be used with aqueous free and very pure solvents of very low polarity such as hexane. This causes a tremendous restriction on the choice of the phase system, for instance it excludes reversed phase chromatography.

By *photochemical reaction detection* the effluent is exposed to a high UV irradiation (20-40 s) before entering a fluorescence or UV detector. Due to the irradiation the solutes are converted into fluorescent or UV active products (35). This detection principle can be highly selective and sensitive for certain classes of compounds.

Fig. 4. *HPLC separation and post-column ion-pair extraction detection of a commercial sulfactant formulation of secondary alkylsulphonates (left) and a water sample (right) spiked with c_{10}-c_{14} alkylsulphonates (20 ppb of each) after on-line preconcentration of 50 ml on a small reversed phase column (32).*

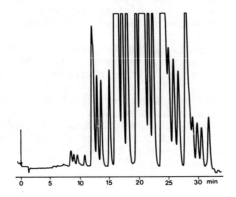
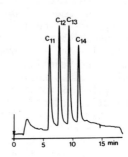

When applying *sensitized room temperature phosporescence* as detection system the analytes are being exited and then transfer their triplet energy to an acceptor added to the mobile phase (for instance biacetyl) which in turn phosphoresces (36). For halogenated naphtalenes and biphenyls detection limits in the subnanogram range are obtained.

All these aforementioned more or less selective detection systems show excellent sensitivity and detection limits. Unfortunately, most of them can only be used in combination with a limited number of phase systems (e.g. mobile phases). Also in many cases one can use them only for known compounds in the sample.

5. CONCLUSION

The significant improvement of the column technology, the stationary phases and detection systems during the last decade turned HPLC into a suitable method for trace analysis of organic micropollutants. However, the separation power (the peak capacity) of common HPLC columns is insufficient for very complex mixtures. Columns with very large plate numbers, multi-column operation and/or selective detection are the ways to attack such separation problems. The applicability of packed microbore columns of about 0.5-1 mm I.D. as a means to lower detection limits and for economical reasons is now subject of large interest. The full exploration of such microbore columns will suit HPLC even better for trace analysis such as micropollutants in water. Moreover, it will bring the reliable on-line LC-MS combination closer to reality.

REFERENCES

1. COOK, N.H.C., ARCHER, B.G., OLSEN, K., and BERICK, A., Anal.Chem. 54 (1982) 2277.
2. MELLOR, N., Chromatographia 16 (1982) 359.
3. VERZELE, M., DYCK, J. van, MUSSCHE, P. and DEWAELE, C., J.Liq.Chrom. 5 (8) (1982) 1431.
4. POPPE, H., KRAAK, J.C., HUBER, J.F.K. and BERG, J.H.M. van den, Chromatographia 14 (1981) 515.
5. SCOTT, R.P.W. and KUCERA, P., J.Chromatogr. 169 (1979) 51.
6. HENRY, R.A., BYRNE, S.H. and HUDSON, D.R., J.Chrom.Sci. 12 (1974) 197.
7. KRAAK, J.C., POPPE, H. and SMEDES, F., J.Chromatogr. 122 (1976) 147.
8. VERZELE, M. and DEWAELE, C., HRC&CC 5 (1982) 245.
9. SCOTT, R.P.W. and KUCERA, P., J.Chromatogr. 169 (1979) 51.
10. SCOTT, R.P.W., J.Chrom.Sci. 18 (1980) 49.
11. KUCERA, P. and MANIUS, G., J.Chromatogr. 216 (1981) 9.
12. HIRATA, Y. and GINNO, K., HRC&CC 6 (1983) 196.
13. TIJSSEN, R., BLEUMER, J.P.A., SMIT, A.L.C. and KREVELD, M.E. van, J.Chromatogr. 218 (1981) 137.
14. KNOX, J.H. and GILBERT, M.T., J.Chromatogr. 186 (1979) 405.
15. DAVIS, J.M. and GIDDINGS, J.C., Anal.Chem. 55 (1983) 418.
16. HUBER, J.F.K. and EISENBEISS, F., J.Chromatogr. 149 (1978) 127.
17. ERNI, F., KELLER, H.P., MORIN, C. and SCHMITT, M., J.Chromatogr. 204 (1981) 65.
18. APFEL, J.A., ALFREDSON, T.V. and MAJORS, R.E., J.Chromatogr. 206 (1981) 43.
19. DOLAN, J.W., WAL, Sj. van der, BANNISTER, S.J. and Snyder, L.R., Clin.Chem. 26 (1980) 871.
20. LINDNER, W., RUCKENDORFER, H. and LECHNER, W., J.Chromatogr., in press.
21. CROMBEEN, J.P., HEEMSTRA, S. and KRAAK, J.C., J.Chromatogr., in press.
22. POPPE, H., Proceedings of the second European Symp. on Analysis of micropollutants in water, Killarney, 1981, p. 141.
23. PERKIN ELMER (Fluorescence detector LS-4).
24. KISSINGER, P.T., Anal.Chem. 49 (1977) 447A.
25. LANKELMA, J. and POPPE, H., J.Chromatogr. 125 (1976) 375.
26. HANEKAMP, H.B., VOOGT, W.H., BOS, P. and FREI, R.W., Anal.Chim.Acta 118 (1980) 81.
27. SLAIS, K. and KOURILOVA, D., J.Chromatogr. 258 (1983) 57.
28. LAWRENCE, J.F. and FREI, R.W., Chemical derivatization in LC, Elsevier, Amsterdam (1976).
29. HUBER, J.F.K., JONKER, K.M. and POPPE, H., Anal.Chem. 52 (1980) 2.
30. APFEL, J.A., BRINKMAN, U.A.Th., and FREI, R.W., Chromatographia 17 (1983) 125.
31. LAWRENCE, J.F., BRINKMAN, U.A.Th. and FREI, R.W., J.Chromatogr. 185 (1979) 473.
32. SMEDES, F., KRAAK, J.C., WERKHOVEN-GOEWIE, C.F., BRINKMAN, U.A.Th. and FREI, R.W., J.Chromatogr. 247 (1982) 123.
33. GAST, C.H., KRAAK, J.C., POPPE, H. and MAESSEN, F.J.M.J., J.Chromatogr. 185 (1979) 549.
34. KOK, A. de, GEERDINK, R.B. and BRINKMAN, U.A.Th., J.Chromatogr. 252 (1982) 101.
35. SCHOLTEN, A.H.M.T., and FREI, R.W., J.Chromatogr. 176 (1979) 349.
36. DONKERBROEK, J.J., EIKEMA HOMMES, N.J.R. van, GOVYER, C., VELTHORST, N.H., and FREI, R.W., Chromatographia 15 (4)(1982) 218.

IDENTIFICATION OF NON-VOLATILE ORGANICS IN WATER USING

FIELD DESORPTION MASS SPECTROMETRY AND HIGH

PERFORMANCE LIQUID CHROMATOGRAPHY

C.D. WATTS, B. CRATHORNE, M. FIELDING and C.P. STEEL
Water Research Centre,
Henley Road, Medmenham,
Marlow, Buckinghamshire
ENGLAND.

SUMMARY

Results obtained from the use of high-performance liquid chromatography and field desorption mass spectrometry for the separation and identification of non-volatile organics isolated from river and drinking waters are presented.

Complex mixtures of non-volatile organics with a wide diversity of structural types have been shown to be present in these water samples.

Among the non-volatile organics identified are poly chlorinated terphenyls, non-ionic and cationic surfactants, pharmaceuticals, pesticides and epoxy resin components. The majority of the identified compounds are anthropogenic in origin and many of them are biologically active.

Only a small proportion of the non-volatiles present have been identified and the others await further application of the techniques used here and the development of new techniques.

1. INTRODUCTION

The need to develop methods for identifying non-volatile organics in water has become apparent over the last few years. It is a problem which has been highlighted by the fact that the large number of volatile organics which have been identified in water represent, at most, only 20% of the total organic material. The remaining 80% is generally referred to as the non-volatile organic fraction. The identification of volatile compounds was facilitated by the application of high-resolution gas chromatography-mass spectrometry and the availability of suitable mass spectral data banks. This technique provides an ideal method for separation and identification of complex mixtures of volatile organics. No similar combined separation-identification method or mass spectral data bank is available for non-volatile organics and this has proved a major obstacle to more widespread research into non-volatile organics in water. As a direct consequence, the composition of the non-volatile organic fraction is largely unknown. Undoubtedly, a large proportion consists of humic material (humic, fulvic and hymatomelanic acids), which is intractable to most methods of molecular structure elucidation. The remainder probably consists of discrete organic compounds which are non-volatile by virtue of polarity, thermal instability or high molecular weight. In fact, the available evidence (1-3) indicates that the non-volatile organics present in water consist of mixtures at least as complex as those found for volatiles. This data has shown that modern liquid chromatographic and mass spectrometic techniques can provide an insight into the nature of the non-volatile organics present in water. Results from the analysis of river and drinking water samples using these techniques are reported in this paper.

Although identification of unknown compounds using the survey type approach is successful using modern techniques, it is undoubtedly slow. One way of overcoming this limitation is to use a "target" compound approach. This requires the selection of particular compounds or classes of compounds which could pose problems with respect to water quality. The criteria we adopted for "target" compounds are that they must be:-

(a) non-volatile organics,
(b) in widespread use,
(c) produced in large amounts,
(d) of potential toxicological concern,
(e) likely to enter the water cycle.

Examples of compounds which meet all of these criteria can be found among the most commonly used pesticides and pharmaceuticals. We have developed methods for certain pesticides and pharmaceuticals selected with these criteria in mind and preliminary results are presented from the analysis of "target" pharmaceuticals in a river water.

The importance of identifying the non-volatile organics in water in addition to the volatiles has been stressed previously (3). It is worth repeating here that concern over possible health effects resulting from long-term exposure to organics in drinking water has arisen mainly from the volatile organics identified. Furthermore, most water quality standards referring to individual organic compounds are concerned with volatile organics (4). In order to make a properly balanced judgement of the possible health effects of organics in water, more information on the identity of non-volatile organics is essential.

2. EXPERIMENTAL

2.1 Isolation Techniques

Extracts of non-volatile organics from river and drinking water were obtained by either XAD-2 resin adsorption and elution with a suitable solvent or freeze drying and extraction of the solid residue with a suitable solvent. Details of these procedures have been previously reported (3).

2.2 Separation Techniques

Separation and fractionation of the extracts was carried out using either normal- or reversed- phase high-performance liquid chromatography (HPLC) as appropriate. Details of the separation conditions have been published elsewhere (3).

2.3 Identification Techniques

Total extracts and HPLC fractions were examined using field desorption mass spectrometry (FDMS), with peak matching and collision activated dissociation (CAD) as ancillary techniques. A limited number of samples were examined using electron impact (EI) MS. Details have been published previously (3). In general, total extracts were used for the "target" compound approach and HPLC fractions for the broad survey work.

The only change from previously established procedures concerns preparation of FD emitters. These were prepared using indene as the activation vapour (5).

3. RESULTS AND DISCUSSION

HPLC analysis of extracts of both river and drinking waters invariably results in complex chromatograms. Typical examples of HPLC chromatograms obtained for extracts (XAD-2/diethylether; normal phase) from river water and drinking water are presented (Figure 1). In fact, the mixtures present in these extracts are more complex than is indicated on HPLC chromatograms with uv detection, since peaks on the HPLC chromatogram apparently consisting of a single component are frequently found to consist of mixtures on subsequent analysis by FDMS. This emphasises the necessity of using an HPLC system with the highest possible resolution and an identification technique capable of handling mixtures.

As mentioned earlier, FDMS is carried out both on HPLC fractions and total extracts and typical examples of FD mass spectra obtained from extracts of a river water (freeze-drying dichloromethane) and a drinking water (XAD-2/diethyl ether) are shown (Figures 2 and 3). Generally, FD yields only molecular ions and hence these mass spectra indicate both the complexity and the wide molecular weight range of the mixtures isolated from water samples.

The nature of FDMS is such that different compounds desorb at different emitter heating currents and with different absolute sensitivities. For a complex mixture, the emitter heating current is increased linearly (with or without integrated ion current feedback) and the mass spectrometer scanned repetitively. Thus the FD mass spectra shown represent single spectra selected at the point of the highest intensity integrated ion current. As a result of this, only approximate quantitative information can be obtained from this data. Accurate quantitative information can be obtained from FDMS by the use of multichannel analysers, photoplate detectors or single ion monitoring. The examples here show compounds with molecular weights up to the maximum acquired into the data system, ~ 1300 Daltons. Using manual scanning of the mass spectrometer, ions can be observed up to the maximum mass available on the instrument used, i.e 1750 Daltons.

A list of some organics identified in river and drinking waters using both the survey and "target" approaches is given in Table 1. The certainty of identification of these compounds is variable since it was not possible to obtain authentic standards for confirmation in all cases. It is clear from this data that although the techniques used were developed specifically for identification of non-volatile organics, they are not exclusive of volatile organics. Thus, many of the compounds identified, e.g alkanes and phthalates, are amenable to analysis by GC and GCMS and have been previously identified in river and drinking waters using those techniques. Discussion will concentrate on some of the more interesting non-volatile organics that have been identified.

Polychlorinated terphenyls (PCTs) were found in XAD-2 extracts of drinking water derived from both groundwater and river water sources. Concentrations were probably < 1 µg.l^{-1} for the most abundant PCT perchlorinated terphenyl (C_{18}, Cl_{14}) was always the most abundant and the relative proportion of the other PCTs decreased with decreasing chlorine content. To our knowledge, PCTs have not been previously identified in drinking waters but have been reported in a water sample from the River Rhine (6). They are undoubtedly anthropogenic in origin and are used commercially as a substitute for PCBs in certain applications (7). However, there is little evidence to suggest that there is any significant difference in toxicity between PCBs and PCTs (7). The relative abundances of the individual PCTs in commercial PCT mixtures (e.g Arochlor 5460) are such that the maximum individual concentration occurs for the Cl_9 or Cl_{10} isomers and decreases for the higher and lower molecular weight compounds. The difference in

TABLE 1   ORGANICS IDENTIFIED IN WATER EXTRACTS

COMPOUND	WATER TYPE
5-Chlorouridine	Drinking
5-Chlorouracil	"
4-Chlororesorcinol	"
5-Chlorosalicylic acid	"
2,4,6-Trichlorophenol	"
Monochloro-diphenyl-propane isomer	"
Dichloro-nonane isomer	"
Bromo-chloro-hydroxy-dihydro-benztriazole isomer	River
Polychlorinated terphenyls (Cl_9-Cl_{14})	Drinking
Pentadecanoic acid	"
Palmitic acid	"
Linolenic acid	"
Linoleic acid	"
Oleic acid	"
Stearic acid	"
Myristic acid	"
Eicosanoic acid	"
$\underline{n}$-Alkanes (C_{24}-C_{31})	Drinking, River
$\underline{n}$-Alkanes (C_{32}-C_{50})	Drinking
Squalene	River
Di-butyl phthalate	Drinking, River
Di-hexyl phthalate	" "
Di-octyl phthalate	" "
Di-nonyl phthalate	" "
Di-decyl phthalate	" "
Poly di-methyl siloxanes $R_1 \cdot [(CH_3)_2\ Si - O -]_n\ R_2; n = 8-16$	" "
Poly(oxyethylene) nonyl phenylethers $C_9 H_{19} \cdot C_6 H_4 \cdot O\ (C_2H_4O)_n; n = 6-15$	" "
Tetra-alkyl ammonium compounds $(CH_3)_2\ {}^+N\ R_1\ R_2$ $R_1 = R_2 = C_9 H_{19} - C_{24} H_{49}$	Drinking

TABLE 1 (continued)   ORGANICS IDENTIFIED IN WATER EXTRACTS

COMPOUND	WATER TYPE
Salicylic acid	Drinking
Trimethylhexa-methylene diamine	"
Isophorone diamine	"
p-Toluene sulphonamide	"
Tetradecyl trimethyl-ammonium ion	"
Bisphenol F diglycidyl ether	"
Bisphenol A diglycidyl ether	"
Bisphenol A diglycidyl ether/Bisphenol F diglycidyl ether dimer	"
Bisphenol A diglycidyl ether dimer	"
Bisphenol A diglycidyl ether trimer	"
Triamino cyclohexane isomer	"
C_{30} Alkyl diamine	"
C_{18} Alkyl diamine	"
Diemethly-naphthyl-carbamate	"
Thiomethyl-benzothiazole-N-oxide	"
Methoxythiomethyl-benzothiazole-N-oxide	"
2-Hydroxybenzthiazole	"
2-Phenylbenzimidazole	"
p-Isopropyldiphenylamine	"
Di-tolyl ethane isomer	"
Linuron	River
Metbromuron	"
Monolinuron	"
Chlorotoluron	"
Buturon	"
Carbaryl	"
Phenmediphan	"
Diuron	"
Metoxuron	"
Chloroxuron	"
Methabenzthiazuron	"

TABLE 1 (continued)   ORGANICS IDENTIFIED IN WATER EXTRACTS

COMPOUND	WATER TYPE
Fenuron	River
Cycluron	"
Erythromycin	"
Tetracycline	"
Theophylline	"

chlorine atom distribution between commercial mixtures and mixtures found in water extracts is difficult to rationalise, but may relate to physical differences e.g water solubility of the PCTs.

The polydimethyl siloxanes identified in some drinking and river water samples covered a molecular weight range of $\sim$ 600-1200 Daltons. While the compound type has been identified, the particular end groups associated with them are unknown. These compounds are anthropogenic in origin and are widely used as both plasticisers and anti-foaming agents (8). They possess negligible toxicity and are used in both foodstuffs (9) and pharmaceutical preparations (10).

Poly (oxyethylene) nonyl phenyl ethers covering a molecular weight range of $\sim$ 500-900 were found in both river and drinking waters. They have been previously identified in river water using FDMS (11,12) but not in drinking water. These man-made polymers are used as non-ionic surfactants in a wide range of applications. Identification of polymeric materials is facilitated by the regular repeating mass differences which are observed in their FD mass spectra. Thus, the dimethyl siloxanes produce an envelope of peaks with a spacing of 74 Daltons and the poly oxyethylene surfactants have a spacing of 44 Daltons. The only other polymeric species which have been identified in drinking water are poly vinyl acetates (13). These were identified by a Japanese research group using the same procedures used for the work reported here.

Another class of compounds we have frequently identified in drinking water extracts are the tetra-alkyl (quaternary) ammonium compounds. As an example of the type of data that can be obtained by FDMS of HPLC fractions of water extracts, the FD mass spectrum of an HPLC fraction containing these compounds is presented (Figure 4). This shows a range of dialkyl dimethyl ammonium compounds from dinonyl- to di-nonadecyl with the major contribution from the dioctadecyl- ion at $\frac{m}{z}$ 550. A distribution of this type is frequently observed in drinking water extracts, although tetra-alkyl ammonium ions up to dimethylditetracosanyl [$(CH_3)_2$ $^+N(C_{24}H_{49})_2$] have been observed during the course of this work. The tetra-alkyl ammonium compounds are man-made organics and are used both as surfactants and mild bactericides in a wide range of applications.

The "target" compound approach originally started as a way of quickly screening total extracts for the presence or absence of compounds we had previously identified in water extracts. Thus mass chromatography using the nominal mass of the "target" compound was used to screen FDMS data derived from total extracts. An extension of this approach to enable initial tentative identification of unknown compounds was developed during the analysis of a River Main extract as part of COST 64b BIS collaboration. Thus all of these 'uron' type herbicides found in the River Main sample

(Table 1; Linuron to Cycluron) had been previously reported in extracts of water from the River Rhine by the use of HPLC and FDMS (14). In the work reported here, the molecular weights of these compounds were used for mass chromatography of FDMS data from the total extract (Freeze-dried/dichloromethane) of a River Main sample to provide tentative evidence of their presence. These identifications were then confirmed by the use of additional techniques, e.g the presence of 37 Cl isotope peaks for chlorine-containing compounds. The presence of isoproturon, linuron, monolinuron and chlorotoluron in the same River Main sample, has also been confirmed using LCMS (15). Some of the phenylurea herbicides reported here have been previously reported in other river water samples by the use of FDMS (16). Although these compounds are not truly non-volatile, their analysis by GC and GCMS is difficult due to thermal instability.

The "target" compound approach was also used to identify various organics in extracts (freeze-dried/dichloromethane) of drinking water samples taken from water mains relined with epoxy resin (17). These compounds (dimethyl dioctadecyl and dimethyl diheptadecyl ammonium and salicylic acid to bisphenol A diglycidyl ether trimer in Table 1) were originally identified in the epoxy coating resins and hardeners. The identification of the bisphenol A resin oligomers was facilitated by a previous FDMS study of epoxy resin pre-polymer (18). Most of the compounds are inherently non-volatile and were not detected in a parallel study using GCMS (17).

More recently the "target" compound approach has been further refined by the use of internal standards and peak matching to improve the quality of the data obtained. Methods have been developed for the analysis of non-volatile pesticides (carbendazim, tridemorph and paraquat) and pharmaceuticals (sulphamethoxazole, erythromycin, tetracycline and theophylline) down to microgram litre^{-1} levels in raw and drinking water. The preliminary use of these methods led to the identification of erythromycin, tetracycline and theophylline in a river water sample. These pharmacologically active compounds were present at the $\mu g.l^{-1}$ level and have not been previously reported to occur in river.

4. CONCLUSIONS

The use of HPLC and FDMS affords a solution to the long-standing problem of analysing non-volatile organics in river and drinking waters. It has been demonstrated that river and drinking waters contain extremely complex mixtures of non-volatile organic compounds of diverse structural types. Many of these compounds have been identified, some for the first time, in such samples.

Many of the non-volatile organics identified are anthropogenic in origin and possess biological activity. The possible toxicological significance of the presence of low amounts of these compounds in river and drinking waters requires further consideration.

The majority of the non-volatile organics still await indentification. This can be accomplished by the application of the techniques used here, in particular by expansion of the "target" compound approach and by the use of newer techniques. Thus, future work at the Water Research Centre will investigate improved isolation/concentration methods, higher resolution HPLC using coupled columns and an alternative method of MS ionisation using fast atom bombardment (FAB).

ACKNOWLEDGEMENTS

The work carried out at WRC is funded by the Department of the Environment. Permission from the Department of the Environment and the Water Research Centre to publish this work is gratefully acknowledged.

REFERENCES

1. CRATHORNE, B., WATTS, C.D. and FIELDING, M. Analysis of Organic Micropollutants in Water. Proceedings of the 1st European Symposium. Berlin, 1979, p. 76-90.

2. WATTS, C.D., CRATHORNE, B., FIELDING M. and KILLOPS, S.D. Environmental Health Perspectives, 46, (1982), p. 87-99.

3. CRATHORNE, B. and WATTS, C.D. Analysis of Organic Micropollutants in Water. Proceedings of the 2nd European Symposium, Killarney, 1981, p. 159-173.

4. BEDDING, N.D., McINTYRE, A.E. and LESTER, J.N. Science of the Total Environment, 27, (1983), p. 163-200.

5. RABRENOVICH, M., AST, T. and KRAMER, V. International Journal of Mass Spectrometry and Ion Physics, 37, (1981), p. 277-307.

6. FREUDENTHAL, J. and GREVE, P.A. Bulletin of Environmental Contamination and Toxicology, 10, (1973), p. 108-111.

7. JENSEN, A.A. and JØRGENSEN, K.F. Science of the Total Environment, 27, (1983), p. 231-250.

8. KLEINERT, J.C. and WESCHLER, C.J. Analytical Chemistry, 52, (1980) p. 1245-1248.

9. MINISTRY OF AGRICULTURE, FISHERIES AND FOOD, "Food Standards Committee, Report on Emulsifying and Stabilising Agents". Appendix A, 1956, p. 11.

10. MARTINDALE, THE EXTRA PHARMACOPOEIA, Editor A. Wade, The Pharmaceutical Press, London, 27th Edition, 1977.

11. OTSUKI, A. and SHIRAISHI, H. Analytical Chemistry, 51, (1979), p. 2329-2332.

12. YASUHARA, A., SHIRAISHI, H., TSUJI, M. and OKUNO, T. Environmental Science and Technology, 15, (1981), p. 570-573.

13. SHINOHARA, R., KIDD, A., ETO, S., HORI, T., KOGA, K. and AKIYAMA, T. Environment International, 4, (1980), p. 31-37.

14. STÖBER, I. and SCHULTEN, H.R. Science of the Total Environment, 16, (1980), p. 249-262.

15. LEVSEN, K., SCHÄFER, K.H. and FREUDENTHAL, J. Journal of Chromatography, (1983), 271, 51-60.

16. YAMAMOTO, Y., SUZUKI, M. and WATANABE, T. Biomedical Mass Spectrometry, 6, (1979), p. 205-207.

17. WATTS, C.D., JAMES, H.A., GIBSON, T.M. and STEEL, C.P. Environmental Technology Letters, 4, (1983), 59-64.

18. SAITO, J., TODA, S. and TANAKA, S. Bunseki Kagaku, 29, (1980), p. 462-467.

Column = Spherisorb-CN (25cm x 4.6mm I.D.)
Eluent = 1% - 50% Isopropanol in n-hexane over 30 min.
Detection = u.v Absorption at 254 nm
Flow = 1 ml/min

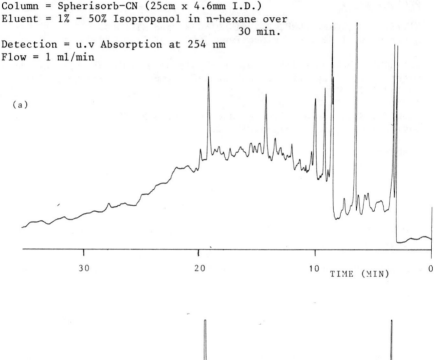

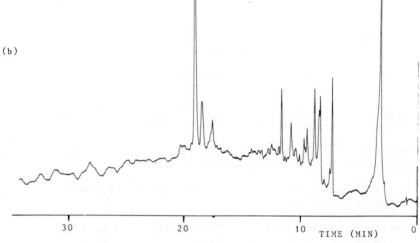

Figure 1 HPLC separation of XAD-2/diethyl ether extracts from (a) a drinking water sample and (b) a river water sample

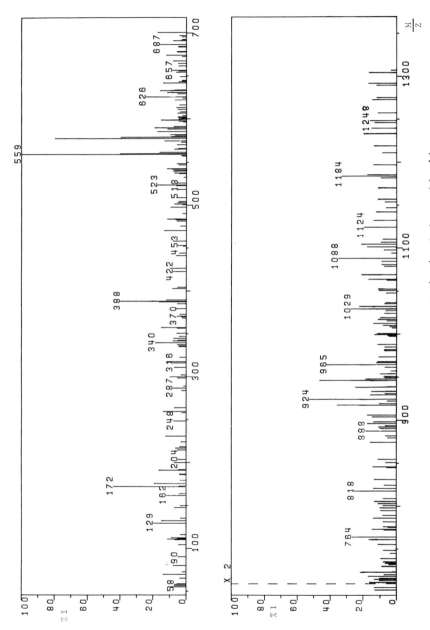

Figure 2 FD Mass spectrum from a freeze drying/methylene chloride extract of a River Maine sample

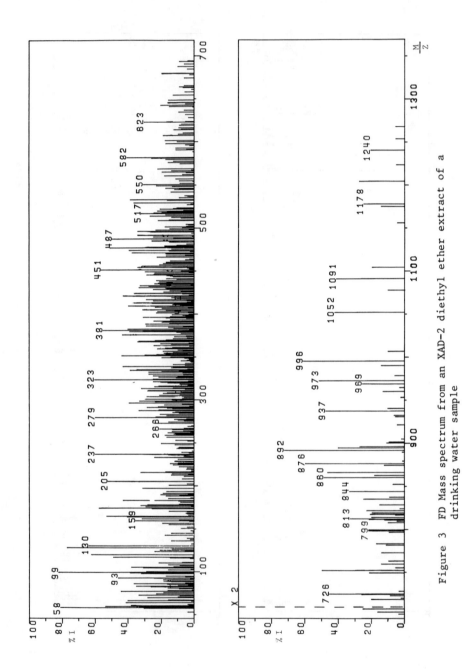

Figure 3 FD Mass spectrum from an XAD-2 diethyl ether extract of a drinking water sample

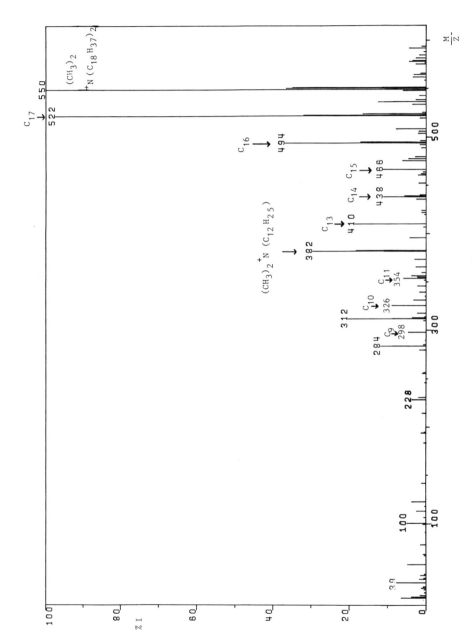

Figure 4 FD Mass spectrum of an HPLC fraction (from a drinking water extract) containing tetra-alkyl ammonium surfactants

MASS SPECTROMETRIC IDENTIFICATION OF SURFACTANTS

K. LEVSEN, E. SCHNEIDER, F.W. RÖLLGEN AND P. DÄHLING
Department of Physical Chemistry, The University of
Bonn, Germany

AND

A.J.H. BOERBOOM, P.G. KISTEMAKER; AND S.A. MCLUCKEY
FOM Institute for Atomic and Molecular Physics,
Amsterdam, The Netherlands

Summary

Various types of pure cationic and anionic surfactants have been analyzed by mass spectrometry using field desorption (FD) and "fast atom bombardment"(FAB) as ioization techniques. The FD spectra of all surfactants are dominated by quasimolecular ions while structure specific fragments are usually missing or of low abundance. In the contrary, the FAB method yields both quasimolecular ions and structure specific fragments. If these ionization techniques are combined with the method of collisionally activated decomposition (CAD) in a tandem mass spectrometer a direct mixture analysis is possible. It will be demonstrated that this approach can be used to identify surfactants in surface water.

1. INTRODUCTION

Surfactants are the major organic constituents of detergents. They are mostly classified by the nature of their ionic charges. Thus one distinguishes between three types of surfactants: (a) cationic (such as quaternary ammonium salts), (b) nonionic (formed e.g. by condensation reactions between ethylene oxide and fatty alcohols or alkylphenol), and (c) anionic (e.g. alkylbenzene sulfonates or alkyl sulfonates). Surfactants are typically manufactured as a mixture of homologous compounds, the constituents of which differ in the length of the alkyl or ethoxylate chain.
The release of surfactants to the environment (in particular to surface water) has required the development of sensitive analytical methods for their determination at trace levels. These methods have been reviewed recently (1). While organic compounds at trace levels in various sample matrices are most readily identified by combined gas chromatography/mass spectrometry, this method cannot be applied directly to the analysis of ionic surfactants as these compounds are too polar to be amenable either to gas chromatography or to conventional electron impact mass spectrometry.
In recent years a variety of new mass spectrometric ionization techniques have been developed which also allow the analysis of strongly polar compounds. Thus mass spectrometric equipment for FD and FAB measurement is now commercially available. In addition it has been demonstrated that tan-

dem mass spectrometry, now commonly referred to as mass spectrometry/mass spectrometry (MS/MS), can be used for direct mass spectrometric mixture analysis. Thus we have recently reported FD and FAB spectra of various types of surfactants (2,3) and emphasized that the combination of the FD and CAD technique in a tandem mass spectrometer is particularly well suited for the analysis of surfactants (4). FD and FAB spectra of long-chain quaternary amines have also been reported by Cotter et al. (5).

In the present study we compare the potential of the FD and FAB methods as well as the MS/MS technique for the analysis of commercial surfactants and wish to demonstrate that this approach can be used to identify surfactants in surface water. Experimental details have been presented earlier (2-4).

2. Field desorption and fast atom bombardment spectra of commercial surfactants

Fig. 1a and b compare the FD and FAB spectra of a textile softener (Lenor) containing cationic surfactants. In the upper mass range of the positive ion FD spectrum, Fig. 1a, (m/z 450-600) signals due to the intact dimethyldialkyl ammonium cations with two hexadecyl and/or octadecyl substituents are observed at m/z 494, 522 and 550 while three less abundant signals can be detected in the medium mass range at m/z 284, 312 and 326. The ions at m/z 284 and 312 are assigned as trimethylalkyl ammonium cations, i.e. $(CH_3)_3NC_{16}H_{33}^+$ and $(CH_3)_3NC_{18}H_{37}^+$. The FD spectra are avoid of any fragment ions.

On the contrary both quasimolecular ions and fragment ions are observed in the FAB spectrum of this surfactant (Fig. 1b). It is apparent that signals due to trimethyl ammonium cations are much more abundant than those due to dimethyl ammonium cations which is in contrast to the FD spectrum. It is known that it is often difficult to analyze mixtures using the FAB method as some constituents may be largely or completed suppressed. Thus we assume that the abundance ratios observed in the FD spectrum reflect the true composition of the mixture more realistically than those observed in the FAB spectrum. The fragments observed at m/z 184, 268 and 296 are formed by loss of one substituent as alkane. Moreover, loss of one substituent as alkyl radical followed by α-cleavage of the second substituent leads to a prominent fragment at m/z 58, not shown in the figure. The FD and FAB spectra of an anionic surfactant, an alkylsulfonate, are contrasted in Fig. 2. The positive ion FD spectrum (Fig.2a) shows abundant quasimolecular ions formed by sodium attachment (m/z 323, 337, 351) and by protonation (m/z 301, 315) while fragment ions are again missing. If the FD spectrum is recorded in the negative ion mode (Fig. 2b) signals due to the intact anions, [M-Na]⁻, are observed. It has been shown previously that the negative ion FD mode leads to a selective ionization of anionic surfactants (6).

The FAB spectrum of this surfactant is shown in Fig. 2c. The spectrum displays both abundant quasimolecular ions (formed by sodium attachment) and fragments. Structure specific fragments are, however, of low intensity.

The results demonstrate that both the FD and FAB method

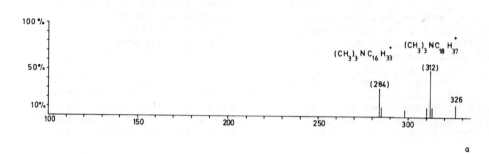
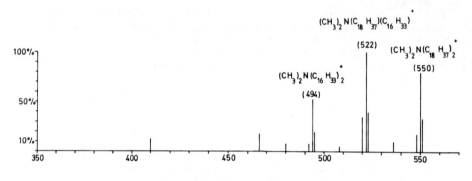
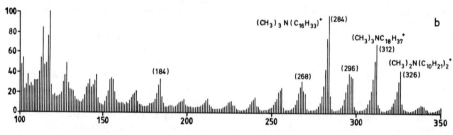
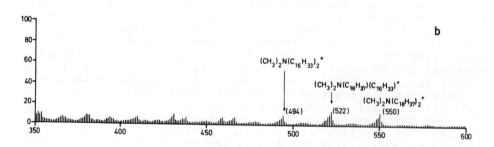

Fig. 1. FD (a) and FAB spectrum (b) of a textile softener

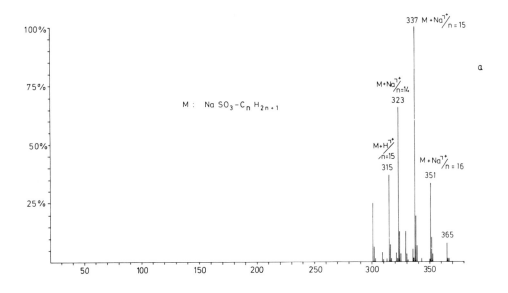
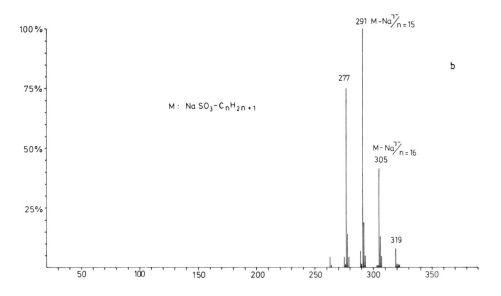

Fig. 2. Positive ion FD (a) and negative ion FD spectrum (b) of a linear alkylsulfonate

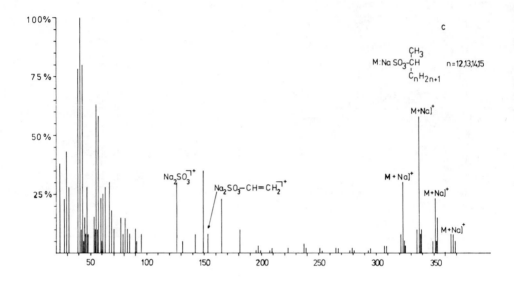

Fig. 2c. FAB spectrum of a linear alkylsulfonate

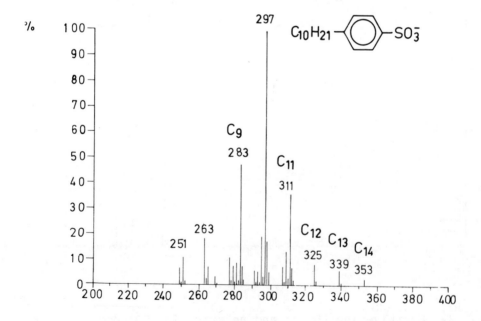

Fig. 3. Negative ion FD spectrum of a sewage water sample

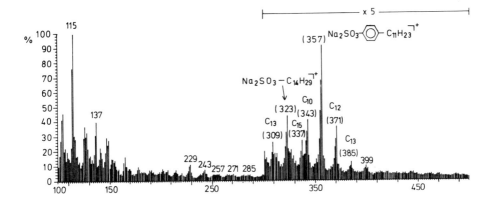

Fig. 4. FAB spectrum of a sewage water sample

allow to analyze intact cationic and anionic surfactants.

3. Identification of surfactants in surface water

Samples from a sewage plant and river water (river Rhine) have been analyzed. The surfactants were stripped from the water samples using the Wickbold procedure (7). No further separation of the various types of surfactants is necessary if the positive ion FD method is used as they desorb at different emitter temperatures. Alternatively, the negative ion FD mode can be used for a selective ionization of anionic surfactants. Such negative ion FD spectrum from a sewage water sample is shown in Fig. 3. It is obvious that the sample contains predominantly alkylbenzenesulfonates.

Fig. 4 shows the FAB spectrum of a waste water sample which again contains alkylbenzenesulfonates. It is noteworthy that in addition to the quasimolecular ions a structure specific fragment is observed at m/z 229 (vide infra). It is apparent that both methods allow identification of surfactants in surface water.

We have previously shown that the combination of the FD and CAD technique in a tandem mass spectrometer is particularly well suited for the identification of surfactants, as this method allows a direct mixture analysis and yields very detailed structural information (4). We have now studied several sewage water samples by this approach using the tandem mass spectrometer at the FOM institute in Amsterdam as described earlier (4).

Fig. 5 shows part of the positive ion FD spectrum of such a sewage water sample at medium emitter temperatures. The observed signals are due to cationic surfactants (see Fig. 1). M/z 550 has been selected by the first stage of the tandem mass spectrometer, subjected to collision with a neutral target gas and its CAD spectrum recorded with the second stage as shown in Fig. 6. The CAD spectrum shows two abundant fragments at m/z 296 (loss of one substituent as alkane) and m/z 58 (loss of one substituent as alkyl radical,

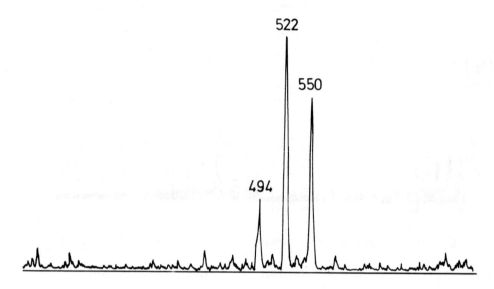

Fig. 5. Positive ion FD spectrum of a sewage water sample at medium emitter temperatures

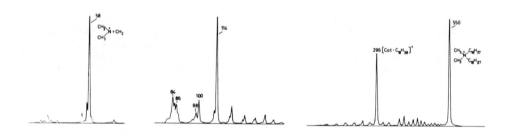

Fig. 6. CAD spectrum of m/z 550 in Fig. 5

followed by α-cleavage of the second substituent). These two fragments allow an unambiguous and straightforward identification of the cation.

At higher FD emitter temperatures signals due to cationic surfactants disappear, while those due to anionic surfactants are observed in the normal FD spectrum as shown in Fig. 7. (The signals in the higher mass range from m/z 677 to 719 are due to cluster ions). The CAD spectrum of the precursor at m/z 357 is shown in Fig. 8. Again this CAD spectrum contains very detailed structural information. The abundant fragment at m/z 126 characterizes the functional group while the intense fragment at m/z 229 gives information on the main branching of the alkyl chain.

The above results demonstrate that ionic sufactants can

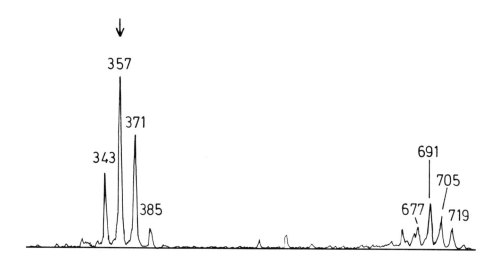

Fig. 7. Positive ion FD spectrum of a sewage water sample at higher emitter temperatures

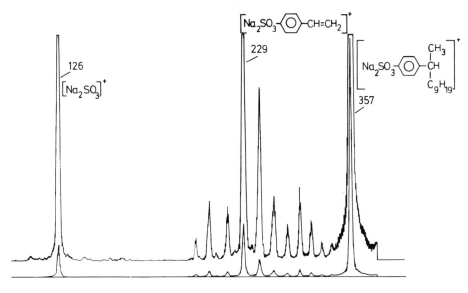

Fig. 8. CAD spectrum of m/z 357 in Fig. 7

be analyzed by newer mass spectrometric techniques. Particularly powerful is the combination of FD and CAD in a tandem mass spectrometer. This method gives very detailed structural information including information on the branching of the alkyl chain. The method can be used to identify ionic surfactants in surface water.

REFERENCES

1. LIENADO, R.A. and NEUBECKER, T.A. (1983). Anal. Chem. Vol. 55, 93 R.
2. SCHNEIDER, E., LEVSEN, K., DÄHLING, P. and RÖLLGEN, F.W. (1983), Fresenius 7. Anal. Chem., in press.
3. SCHNEIDER, E., LEVSEN, K., DÄHLING, P. and RÖLLGEN F.W. (1983), Fresenius 7. Anal. Chem., in press.
4. WEBER, R., LEVSEN, K., LOUTER, G.J., BOERBOOM, A.J.H., and HAVERKAMP, J. (1982), Anal. Chem., Vol. 54, 1458.
5. COTTER, R.J., HANSEN, G. and JONES, T.R. (1982), Anal. Chimica Acta, Vol. 136, 135.
6. DÄHLING, P., RÖLLGEN, F.W., ZWINSELMAN, J.J., FOKKERS, R.H. and NIBBERING, N.M.M. (1982), Fresenius 7. Anal. Chem., Vol. 312, 335.
7. WICKBOLD, R. (1972), Tenside Vol. 9, 173.

CHARACTERISATION OF NON-VOLATILE ORGANICS
FROM WATER BY PYROLYSIS-GASCHROMATOGRAPHY-MASS SPECTROMETRY

L. Stieglitz, W. Roth
Nuclear Research Centre Karlsruhe

Summary

The main fraction of the organic pollutants in water are non-volatile, high molecular compounds, from which humic acids and ligninsulfonic acids are the most important classes in the river Rhine. The combination of pyrolysis gas chromatography mass spectrometry is applied to the characterisation of these non-volatile compounds. The material is isolated from water and pyrolysed in a special inlet system for 10 sec. at 600/800°C. The pyrolysis products are identified and analysed by glass capillary GC-MS. From the pyrolysis fragments the presence of substructures in the macromolecules may be evaluated. As example the pyrolysis of ligninsulfonic acid yields characteristic fragments like o-methoxy-phenol, o-dimethoxybenzene, m-methoxyacetophenone and isoeugenol. These sructures could be found in extracts of surface water and paper mill effluents and prove the presence of lignine compound.
From humic acids of various molecular weight fractions characteristic patterns of substituted phenols, cresols, styrenes are obtained. The detection limit is about 10 micrograms, the ratios of the most important peaks are reproducible with standard deviations of 10-15%.
The technique is generally applied to isolated organic material from surface waters. Characteristic peak patterns can be assigned to samples of different origin.

1. INTRODUCTION

According to our knowledge up to now only 10 to 12 percent of the dissolved organic carbon in water may be identified by gas chromatographic techniques. The information about the remaining ninety percent is scarce; compound classes with no specific structural information are assumed such as proteins, carbohydrates, lipids, humic substances but also coal tar products, paper mill wastes and refuse from the organic polymer industry.
General characterisation methods are used such as measurement of UV-absorption, of molecular size and weight distribution. The determination of substructures is attempted by degrading the polymeric compounds chemically or thermally into smaller structural units, which can be identified by various chromatographic methods. The pyrolysis has been

repeatedly applied to the characterisation of organics in water. So far two different approaches can be distinguished: the pyrolysis mass spectrometry (Py-MS) and the pyrolysis gas chromatography mass spectrometry (Py-GC-MS). With Py-MS the data are obtained by directly inserting the probe into the ionisation chamber of the mass spectrometer. The data are processed by multi-variate statistical analysis and lead to a classification of complex organic substances (1). With Py-GC-MS the sample is pyrolized in a reactor and the products are then analysed by GC-MS. In our investigation the latter technique is used.

2. EXPERIMENTAL

Materials: The following samples were investigated: ligninsulfonic acid (Fa. Roth), humic acid (Fa. Roth), various humic acids prepared from ground water samples by extraction into alamin and separated into molecular weight fractions. For the study special concentrations were also prepared from surface waters (Rhine, Danube) by adsorption on XAD2/XAD4. The loaded columns (2 g resins, pretreated) are eluted with diethylether and then extracted with methanol. These extracts contain the high molecular material. Ca. 10 µg of material is used for the pyrolysis.
Instrumentation: The measurements were carried out with a gaschromatograph (Mod. 1400/1700 Varian Aerograph) with the conventional injection port replaced by a special Curie point pyrolysator as described by H.L.C. Meuzelaar (2). The sample is transferred to a disposable pyrolysis wire and positioned in a glass reaction tube coupled to a GC-column. Pyrolysis temperature 800°C, time 10 sec. For separation of the pyrolysis products OV101 and Se 54 glass capillary columns (50 m length, 0.3 mm i.d.) were used. Quantitative measurements were made with a flame ionisation detector. The identification of the degradation products was carried out under identical conditions with a GC-MS combination (MAT-112, magnetic field instrument) operated with a cyclic exponential scan (1 sec/decade).

3. RESULTS AND DISCUSSIONS

In order to facilitate the interpretation a number of reference compounds of ligninsulfonic and humic acids was investigated.
Ligninsulfonic acids: In the pyrogram of the commercial ligninsulfonic acid (fig.1) the main peaks were identified as phenol, o-methoxyphenol, o-dimethoxybenzene, m-methoxyacetophenone and isoeugenol. Less abundant peaks were inden, o and p-cresol, 2,4 dimethylphenol and eugenol. The higher volatile products were identified as CO/CO_2 cyclopentadiene, carbondisulfide, benzene, thiophen and toluene. The products are typical for the structure of the lignin with its cross linked methoxylated phenylpropane units. The technique was also successfully applied to papermill effluent.
Humic acids: The pyrograms of the humic acids (fig.2a,b) show a great difference, depending on origin and isolation proce-

dure. In table I are typical results of humic acid isolated from bog water with the standard variation (five runs) of retention times and normalized peak area. The main peaks are toluene, octane, octene, xylenes, styrene, nonane, nonene, isopropylbenzene, methylstyrene, phenol, cresols and xylenols. Tentatively identified were also 2,5 dimethyl 2,4 hexadiene and a dihydroxyoctane. Qualitatively the same pattern is obtained from humic acids isolated from ground water. Only minor changes are observed with material divided into molecular weight fractions (>30000, 10000-30000, <10000). A completely different pattern is obtained from commercial humic acid (fig. 2b). The characteristic feature is the appearance of a series of hydrocarbons saturated and unsaturated from C-5 to C-28. Besides the aliphatic hydrocarbons only an abundant peak of toluene and minor peaks of xylene and phenol is observed. This result indicates that the sample is to be considered rather as coal product than as humic acid.

Table I: Reproducibility of pyrolysis capillary GC
Sample: Humic acid (Holohsee), 10 ug, results from five runs.

Retention time min x 10	area normalized^x	pyrolysis product
33 ± 1	2.0 ± 0.5	toluene
37 ± 2	1.1 ± 0.2	C_8H_{16}
39 ± 2	0.7 ± 0.1	C_8H_{18}
55 ± 1.5	0.3 ± 0.1	xylene
58 ± 2	0.6 ± 0.1	xylene
64 ± 2	0.8 ± 0.3	styrene
66 ± 2	0.3 ± 0.05	C_9H_{18}/C_9H_{20}
101 ± 2	1.0 ± 0.2	cumene, methylstyrene
106 ± 1	1.4 ± 0.1	dimethylhexadiene
115 ± 1	1.6 ± 0.4	phenol
139 ± 1	0.5 ± 0.1	cresol
150 ± 1	0.4 ± 0.1	cresol
176 ± 4	0.2 ± 0.05	xylenol
190 ± 0.6	1.00	octanediol

Concentrates from Water Samples

a) River Rhine, XAD-2, methanol eluate. The main pyrolysis products are shown in fig.3. The most prominent peaks are styrene, naphthalene, methylnaphthalene and a homologue series of alkylbenzenes; as minor but characteristic peaks are registered methyl- and dimethylfuran, benzonitril and phenol, indicating lignine type biomers associated with carbohydrate and peptides.

b) River Danube, XAD-2, methanoleluate. The pyrograms are different insofar as only aromatic hydrocarbons like toluene and substituted styrenes appear, with divinylbenzene as the major compound. In contrast to the humic acid samples neither phenols, cresols aliphatic hydrocarbons or hetero-

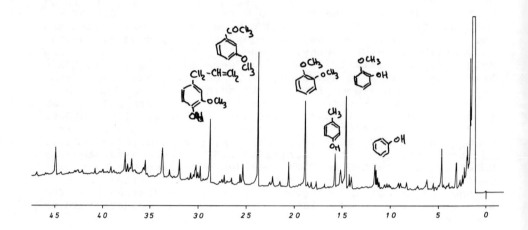

fig.1: Pyrogram of ligninsulfonic acid, 800°C

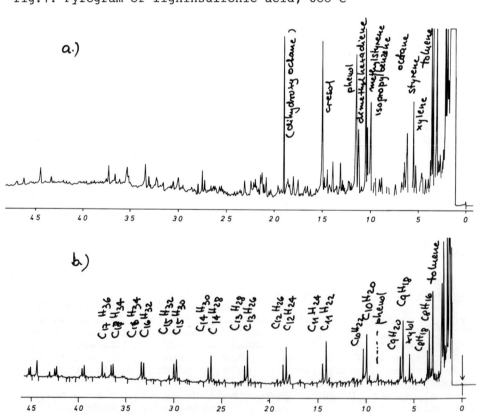

fig.2: Pyrogram of humic acids, 800°C
a) from bog water, b) commercial sample

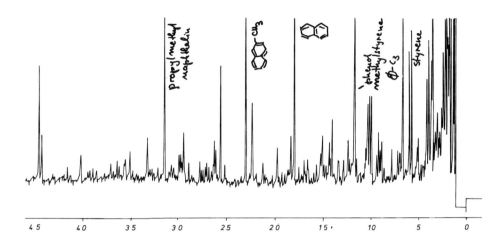

fig.3: Pyrogram from XAD-2 concentrate, methanol eluate riverwater Rhine; 800°C

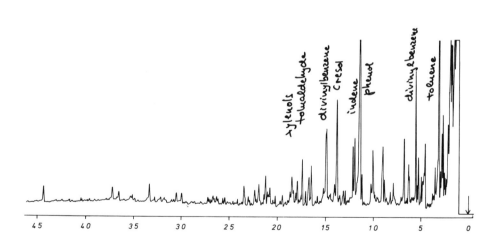

fig.4: Pyrogram from ionexchangeregenerate, riverwater, Mississippi, 800°C

cyclic compounds were detected as main products.

c) River Mississippi, anion exchange regenerate (fig.4). The sample was obtained from the Department of Civil and Mineral Engineering of the University of Minnesota as concentrate. The most prominent peak is phenol, accompanied by cresols and dimethylphenols. Besides furan derivatives (dimethylfuran, methylfurfural) a series of alkylbenzenes is present. From the pyrolysis pattern this sample has a close resemblance to the bog water humic acid. From our experiments and from literature data (3) the following compound classes can be characterised by their pyrolysis products:

humic acids: and lignines	benzene, toluene, benzonitrile, alkylbenzenes, naphthalene, methoxyphenol, methylmethoxyphenol vinyl-methoxyphenol
carbohydrate:	furan, 2-methylfuran, furfural, methylfurfural, pyrrol, dihydropyrone.
lipids:	n-alkanes, n-alkenes
peptides:	benzonitril, phenol, indol methylindol, alkylnitriles, pyrrol.

From this point of view all pyrolysis products observed in the samples isolated from water can be explained as originating from natural compounds such as humic acids, lignines, lipids and peptides.

4. CONCLUSIONS

It is demonstrated that Py-GC-MS may give useful information about substructures of non-volatile organics isolated from various water samples. In order to extend this range of application, however, a number of developments are necessary: improvement of isolation procedures for organics with recoveries up to ninety percent, avoiding any deterioration of the material; availability of standardized reference compounds, especially humic acids. Improvements are also necessary of the instrumentation for a versatile pyrolysor, using capillary columns and a multi-stage pyrolysing temperature.

REFERENCES

1. WINDIG, W. KISTEMAKER, P.G, HAVERKAMP, J, MEUZELAAR, H.L.C., J. Analyt.-Appl. Pyrolysis 1(1979), 39-52.
2. MEUZELAAR, H.L.C., FICKE, H.G., J. Chromatogr. Sci. 13, 12(1975)
3. VAN DE MEENT, DIK, DE LEEVW, J.W.; SCHENCK, P.A., J. Analyt.-Appl. Pyrolysis 2(1980) 249-263.

AUTOMATED GC/MS FOR ANALYSIS OF AROMATIC HYDROCARBONS
IN THE MARINE ENVIRONMENT

S. SPORSTØL and K. URDAL
Central Institute for Industrial Research
P.O. box 350 Blindern, Oslo 3, Norway

Summary

The method described for the determination of selected petrogenic aromatic hydrocarbons in environmental samples is based on computerized GC/MS. Quantitation is carried out by integrating the ion current profiles of the moleculare ions of the different compounds. The advantages of this analytical approach are discussed in terms of specificity, sensitivity, reproducibility and accuracy. Full automation of the instrumental analyses as well as of the data calculation and presentation is also described.

1. INTRODUCTION

During the last decade oil related activities increased significant in the North Sea. In Norway this has led to increased efforts of pollution research and monitoring, especially concerning hydrocarbons. An integral part of most studies is the analysis of water, sediments and organisms, and from the North Sea hundreds of samples have been analysed during the past years.

As part of monitoring programs our laboratory has analysed approximately 350 samples from the North Sea during the last two years. The samples are normally analysed for their total hydrocarbon content (THC) and for the amount of selected aromatic compounds. Among the aromatics it has become common practice in Norway to report the NPD as a separate parameter. NPD is the sum of naphthalene (N), phenanthrene/ anthracene (P), dibenzothiophene (D) and their C_1-, C_2, and C_3-alkyl homologs.

The THC is determined using a conventional GC technique, and the method i only slightly modified compared to the method described by the Intergovernmental Oceanographic Commission (1). The NPD's are analysed using computerized GC/MS, and the method is a modified version of the method outlined by Grahl Nielsen et al (2). Also this method is now recommended by the Intergovernmental Oceanographic Commission as reference method for determination of petroleum in sediments (1).

Although the GC/MS method for NPD determination is accepted as a reference method there is a great need of quality data. In this paper we report on how the NPD's are determined in our laboratory, and we present results from our quality assurance programs, which are run regularly. In addition we describe some of our work on automating the analysis, both by outlining the hardware involved and by describing the function of specially written computer programs for data aquisition and handling.

2. EXPERIMENTAL

2.1. Sample processing
Trace analysis of hydrocarbons in environmental samples requires control of background levels in chemicals and equipment. All chemicals are solvent washed and solvents are distilled over rectifying columns and kept in precleaned all-glass containers. Reagent blanks are checked by concentrating 100 ml of solvents to a volume of 50 µl followed by gas chromatographic analysis. The equipments are rinsed with dichloromethane, heated at 600 degr. C overnight, and kept in aluminium foil. To avoid contamination from the air, all samples are prepared in a hydrocarbon free atmosphere (laminar flow cabinet).

2.2. Water
Water samples (1 l) are extracted at sea with dichloromethane (normally 3 x 40 ml). The extracts are kept in precleaned bottles with teflon caps at approx. 4 degr. C or lower, and transferred to the laboratory. Prior to use internal standars (biphenyl-d10, anthracene-d10 and pyrene-d10) are added. The extracts are dried over Na_2SO_4 and evaporated under vacuum to approx. 5 ml. Further evaporation of solvent (normally to 150-500 µl) is done under a light stream of nitrogen keeping the sample at 30 degr. C (warming block).

2.3. Sediments
To approx. 50 g (wet weight) of each sediment sample internal standards (biphenyl-d10, anthracene-d10 and pyrene-d10) are added. The samples are saponified in 80 ml of 0.5 N methanolic KOH/NaOH under reflux for 2 hours.

The mixture is filtrated under suction and washed with 50 ml of 1N HCl in methanol (1:3), methanol, and finally with n-pentane. The combined filtrates are extracted twice with 50 ml n-pentane, washed with water and dried over Na_2SO_4. Polar components are removed by chromatographing on fluorisil and the eluate is concentrated.

3. INSTRUMENTAL ANALYSIS

Quantitation of the NPD's is performed by using computerized gas chromatography/mass spectrometry (GC/MS). The instrument is a Finnigan 9610 gas chromatograph interfaced to a Finnigan 4023 quadrupole mass spectrometer and an Incos 2300 data system with a 96-megabyte CDC disk drive. The gas chromatograph is supplied with a Varian 8000 autosampler. Mass spectra are recorded with 1 scan/sec in full scan mode (35-250 amu) or 1 scan/0.6 sec in MID mode (molecular ion of compounds). The analyses are performed using capillary columns (SE-54 glass capillary, SE-54 fused silica, DB-5 chemically bonded fused silica). Using fused silica the columns are introduced directly into the ion chamber. Typical injector temperatures are 280 degr. C, the interface oven is held at 240 degr. C. and the ion chamber at 250 degr. C. The oven is held at 30 degr. C for 1 min and then programmed by 4 or 5 degr. C/min up to 300 or 315 degr. C, which is held for 15 min.

The NPD's are determined in principally the same way as described previously (1, 2, 3). The quantitation is based on integration of chromatograms corresponding to the molecular ions of the compounds. Specific peaks in these profiles are not integrated if their spectra do not correspond to those of the compounds of interest or if they do not have characteristic retention times. The values are corrected by using response factors due to differences in extraction behaviour, volatility and GC/MS response of compound to be analysed and that of the internal standars. If reference compounds are not available, the response factors are determined indirectly using ratios between values of homologs obtained from reference compounds.

4. RESULTS AND DISCUSSION

The analytical methods are regularly tested with respect to specificity, sensitivity, reproducibility, accuracy, stability of GC/MS instrument, and linearity in GC/MS response.

4.1. Specificity

The specificity might be defined as to which degree the mean value of the measurements is due to the substance to be determined and not to other substances that may be present in the sample being analysed (1). The critical point in the described GC/MS method is whether the samples contain interfering compounds which disturb the chromatograms corresponding to the molecular ions of the compounds. Initially this has been examined carefully by inspection of the mass spectrum of each compound which gives response in the ion chromatograms within defined retention windows. Our general experience is that very few samples contain interfering compounds, and if such compounds are found they are most prominent in backgrounds samples with low concentration of anthropogenic hydrocarbons. Overestimation of peak areas is, however, prevented because ion chromatograms are visually inspected prior to area calculation. A skilled person will easily recognize and make area correction for new peaks in the ion chromatograms. An example of ion chromatograms of C_1-alkyl phenanthrenes/anthracenes in a petroleum polluted sediment sample is given in Figure 1.

4.2. Sensitivity

By using a mixture of available NPD standards typical detection limits at 50 pg/μl (2 μl injection volume) is obtained (MID mode, 19 ions). This corresponds to 10 ng/l for water samples and 150 ng/kg for sediment samples.

4.3. Reproducibility

The reproducibility of the analytical procedure is determined regularily. For water samples the determination is made using samples which are spiked with known amounts of standards and deuterated internal standards. Typical average relative standard deviation in the calculated concentrations are 3-6 %, except for the most volatile compounds (naphthalenes) which have values in the range of 5-12 %.

The reproducibility of sedimental analysis is determined both from background samples spiked with standards and from petroleum contaminated samples. The samples are divided into three parallels after saponification and extraction. Typical average standard deviations in these three parallels are in the range of 5-10 %.

4.4. Accuracy

The accuracy of the calculated NPD values are calculated by using a shale oil (Standard Reference Material 1580, National Bureau of Standards) that contains a certified value for pyrene (104 plus/minus 18 µg/g). Non-contaminated samples are spiked with known amounts of the shale oil and deuterated internal standards. From two different experiments, each with three parallels, values of 103 plus/minus 7 µg/g and 118 plus/minus 8 µg/g were obtained. These values are within the limit of variation in the certified value.

4.5. Stability of GC/MS system

Response factors are usually determined for each 10 samples analysed. The stability of the GC/MS system is regularly controlled by determination of short term response factors. From 5-7 measurements over a period of 24 hours typical average relative standard deviation of 1.5 - 2.0 % are obtained. Long term stability has been controlled from 9 measurements over a period of 11 days. Average relative standard deviation of 8.5 % was obtained.

4.6. Response factors

Response factors are determined from background samples spiked with known amounts of standards and deuterated internal standards. The response factors thereby represent the differences in extraction behaviour, volatility and GC/MS response of compounds to be determined to that of the deuterated internal standards.

In petroleum products there are great differences in concentration between the different NPD components. It is therefore not possible to operate at the same concentration level both for internal standard and all compounds to be analysed. Using one-level response factors it is therefore necessary to make sure that we are operating within the range of linear response. By spiking background samples with a fixed amount of deuterated internal standards and varying amounts of standards we have found approximately the same response factor in the range of 100 pg/µl - 50 ng/µl extract (2 µl injection volume, MID mode). Below 100 pg a drop in response is observed. The quantitation limit is therefore set to 100 pg/µl.

4.7. Automation

Due to the large amount of samples analysed great efforts have been made in automation of the analysis. All analyses are run on computer controlled autosampler, ensuring that maximum utilisation of instrument time is derived by round-the-clock operation. About 20 samples can be analysed within 24 hours, and with the 96 Mbyte disk drive all data from about 200 samples can be stored (MID mode).

The computer programmed handling of data is based on Finnigans quantitation software, but has been expanded and adapted to suit our specific requirements. A flow chart of the programs is given in Figure 2. The main advantages of the system can be summarized as follows : The quantitation is based on areas of ion chromatograms of molecular ions of compounds to be determined. For each compound/group of isomers visual display of the mass chromatograms of the molecular ions within given retention windows are given automatically. At this stage possible interfering compounds will be detected. Further, we select the retention window to be integrated. The area, calculated by the computer, is stored in a quantitative list. At this stage it is possible to make area corrections for possible interfering compounds.

After the area calculation steps are complete response factor reports or specially designed quantitation reports are generated. Some of the most important quantitative parameters from the quantitative report can be stored on the disk and the results from a whole series of analysis can be summarized in a final report. At present the program is designed for 25 compounds/group of isomers (NPD's plus some other aromatics) and the whole quantitation program is run in about 5 minutes per sample.

4.8. Application

The described method is found well suited both for water and sediment samples. Satisfying results are obtained for samples covering a 5 decade concentration range. With a few modifications the method can also be applied on biological samples.

5. LITERATURE

1. INTERGOVERNMENTAL OCEANOGRAPHIC COMMISSION.
   "The determination of Petroleum Hydrocarbons in Sediments", Manuals and Guides No. 11, Unesco 1982.
2. Grahl-Nielsen, O., Sundby, S., Westrheim, K. and Wilhelmsen, S. "Petroleum hydrocarbons in sediment resulting from drilling discharges from a production platform in the North Sea". Symp. on Research on Envir. Fate and Effects of Drilling Fluids and Cuttings, Lake Buena Vista, Florida (1980) p. 541.
3. Sporstøl, S., Gjøs, N., Lichtenthaler, R.G., Gustavsen, K.O., Urdal, K., Oreld, F. and Skei, J. " Source Identification of Aromatic Hydrocarbons in Sediments Using GC/MS". Envir. Sci. & Technology 17 (1983) 282.

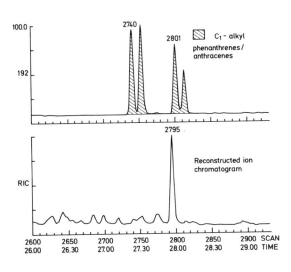

Figure 1 Example of ion chromatogram of C_1- alkyl phenanthrenes/anthracenes

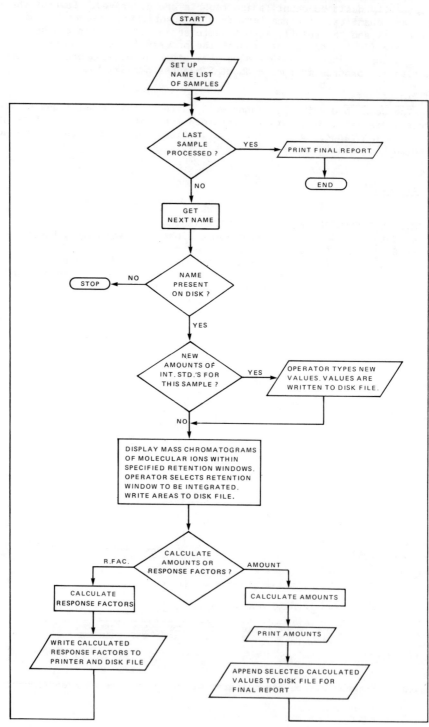

Figure 2. Flowchart of automatic quantitation programs for selected aromatics.

ANALYSIS OF SULPHONIC ACIDS AND OTHER IONIC ORGANIC COMPOUNDS
USING REVERSED-PHASE HPLC

K. J. CONNOR and A. WAGGOTT
Water Research Centre,
Stevenage Laboratory, UK

Summary

The determination of ionic organic compounds in aqueous samples by reversed-phase liquid chromatography poses problems not encountered with more non-polar organic compounds eg polynuclear aromatic hydrocarbons. The ionisation of sulphonated organic compounds must be suppressed either by the use of an ion-pairing reagent, by rendering the eluted solvent acidic, or by the use of more polar reversed-phase packings. The techniques employed for ionic suppression can be a problem, particularly when an in-situ preconcentration of the organic compound is required before HPLC separation.
These problems are discussed and compromise solutions are described which vary depending upon the types of organic compound to be separated. It is recommended that paired-ion reagents are used only when other techniques fail. Satisfactory separations have been achieved using a styrene-divinylbenzene copolymer reversed-phase medium in the absence of either paired-ion reagents or buffer solutions.

1. INTRODUCTION

The analysis of polar organic compounds in water samples by high performance liquid chromatography (HPLC) can give rise to severe problems at any stage of the analytical procedure. Extraction of polar, water soluble, organic compounds is in general difficult and conventional and routine methods can give low extraction efficiencies. The problem may be aggravated by the complex nature of the sample matrix. Further problems arise in chosing the most appropriate separation mode providing the desired degree of reproducibility and resolution and in the availability of a suitable detector giving the required selectivity and sensitivity.
These problem areas have been investigated in the course of work at WRC and those connected with sampling and chromatographic separation are discussed here.

2. EXTRACTION AND CONCENTRATION

The application of HPLC for the determination of organic compounds in water samples is frequently limited by inadequacies of available extraction techniques. This limitation can be overcome by taking advantage of the order of elution imposed by reversed-phase partition (RPP) chromatography. This allows the extraction and concentration of dissolved organic compounds from the water sample onto the media itself. The operation may be carried out by direct injection of the aqueous sample onto the analytical column or more usually by incorporating into the system a small precolumn containing the same stationary phase. When sampling is completed the precolumn is connected to the analytical column in the reverse direction and entrapped material separated by applying a reversed-phase solvent gradient. Particular advantages gained by employing this system are the potential integration of sampling and analysis, fewer possibilities for introduction of artefacts, and the achievement of high concentration factors during sampling (depending upon capacity of the pre-column). The technique has potentially wide application as a screening technique and for the quantitative determination of known pollutants, but a potential weakness is the limited polarity range of organic compounds normally separable on a RPP column.

Direct injection of the aqueous sample on to the analytical column has proved generally effective for relatively small volumes (typically around 5 ml) where relatively large concentrations of the target compound (upper µg to mg/l levels) were present in a simple sample matrix. Precolumn concentration techniques have proved effective at lower concentrations of the target compound (down to lower ng/l levels) and with more complex sample matrices.

Small columns for extraction of trace components from liquids, by displacement column liquid chromatography, have been frequently used. However the technique has not been systematically investigated theoretically or experimentally as an enrichment procedure until quite recently[1]. Many workers have established useful analytical techniques for determination of trace concentrations of water pollutants using the concentration of non-polar compounds which occurs on the head of a RPP column when relatively large amounts of aqueous sample are introduced. This procedure was first discussed by Kirkland in 1974[2] and shortly after, Little and Fallick[3], using octadecylsilyl (ODS) RPP columns, demonstrated the practicality of injecting very large volumes of up to 200 ml for preconcentration of relatively non-polar compounds without serious band broadening. Similar on-column enrichment techniques have been employed for determination of phthalate esters[4], and tetrachloroethylene[5], and for monitoring general pollution levels in various types of natural and process waters[6,7]. Schauwecker et al.[8] and Frei[9] have discussed quantitative aspects and the influence of very large injection volumes on peak broadening, retention period, and base-line shift. Frei[9] also considers the problems and potential of the technique in relation to environmental analysis.

If when applying the method to environmental water samples, the enrichment stage is carried out using the analytical column itself, a serious problem can occur involving the gradual irreversible contamination of the stationary phase by components of the sample, with a resulting decline in efficiency of the column. If this eventually necessitates replacement of the column, use of the technique can become prohibitively expensive. It is better, therefore, to employ short replaceable precolumns or 'guard' columns for the concentration stage and subsequently to connect these to the analytical column. Methods for the determination of phthalate esters[4,10,11] and polynuclear aromatic hydrocarbons[12] applicable to water pollution analysis have been developed using separate stages of sampling and analysis. A similar method has also been used for the determination of sea-water-soluble fractions of oil[13]. The precolumn is loaded with sample either locally or at a remote site before it is introduced into the chromatographic system. Methods for the determination of polynuclear aromatic hydrocarbons[14,15], phthalate esters[10], photographic chemicals[16], optical brighteners[16], and acid blue dyes[16] using an integrated enrichment column have also been developed. The system shown in Fig. 1 was developed at this laboratory[16]. It requires a second pump and suitable high-pressure low dead volume

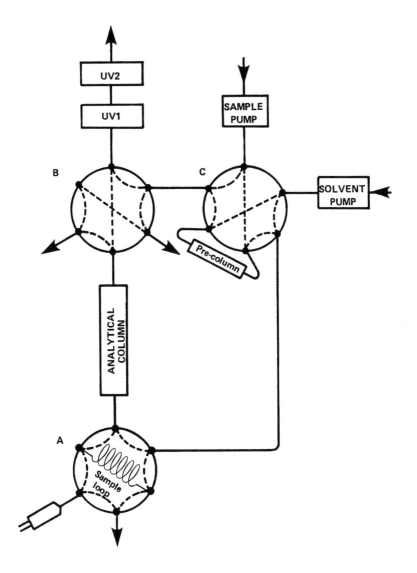

Fig. 1. Integrated sample concentration and separation HPLC system: precolumn backflush and analysis position shown

switching valves and has the advantage that it does not require the
routine and extensive replumbing of the systems required by other equivalent techniques. It also allows direct monitoring of the concentration
stage.

Earlier work employing paired ion chromatography (PIC) reagents
suggested that if they are added to the sample and used to condition the
precolumn, the scope of the technique may be extended to include polar
organic compounds not normally retained on a RPP medium[16]. These
findings have been confirmed in the course of further work to develop
analytical techniques for the analysis of optical brightners and acid
blue dyes. It has also been found that increased efficiency of recovery
may be obtained by using simple buffers or by using a more polar stationary phase. Ionic organic compounds of a basic nature eg. basic dyestuffs,
have been found to be strongly held on reversed-phase packings. They may
be eluted by making the mobile phase strongly acidic (pH 2).

3. COLUMN SEPARATION

Ionic organic compounds may be separated using HPLC by employing ion
exchange chromatography, by using RPP media with PIC reagents, and by
employing RPP media with pH-adjustment (ion suppression) using buffers.

3.1 Ion exchange chromatography

Classical ion exchange resins based on divinylbenzene - polystyrene
structures have a limited application in high pressure work because they
lack rigidity and can be seriously compressed at pressures greater than
1000 psi. More efficient and rigid media have been obtained by chemically
bonding the appropriate functional groups to the stationary phase. As
with adsorption media, coating can be carried out with both pellicular and
microparticulate media. It has been the general experience at WRC that it
is not always easy to predict the behaviour of ion exchange columns in
certain separations. This is probably due to spurious adsorption and
partition effects which may play some part in the separation process.
For this reason ion exchange chromatography has tended to be employed as
a last resort.

3.2 Reversed phase partition chromatography using PIC reagents

In recent years partition chromatography has become so popular that
it is adopted as the mode of separation in approximately 70 per cent of
all reported methods. Its popularity is due to its relatively wide
applicability, the speed at which columns may be re-equilibrated, and the
relatively simple choice of mobile phases. In fact the most difficult
aspect of this mode of chromatography is choice of the correct
stationary phase since the nature of the silica support and the technique
used to produce the bonded phase can cause significant differences in the
properties of similarly labelled commercial columns. The silica support
material may differ in its particle shape and surface area and its
particle size and distribution as well as pore size and distribution.
The silylating reagents used to coat the silica particles with a bonded
mono-layer may differ in their alkyl chain length, the end functional
group, eg. methyl, cyanide, primary amide, phenyl etc, as well as the
degree of crowning or capping of un-reacted silanol groups. Unless
reaction conditions are closely defined and controlled, considerable
differences in properties of bonded phase partition columns may be
unwittingly produced. This is generally so in the case of columns of
similarly coated phases from different manufacturers. All such columns
should be characterised chromatographically before use. It has been

found that the choice of a suitable RPP column with a higher degree of polarity than normal can considerably help in the separation of ionic compounds. However in general ionic organic compounds will elute straight through the RPP media and emerge with the solvent peak. One method of increasing the retention time is the application of PIC reagents. PIC chromatography relies on the supposed neutralisation of polar functional groups in the solute molecules by a suitable ionic species so that they may be separated by partition chromatography as neutral species. A suitable PIC reagent which does not interfere with the detection system employed must however be available. For purely acidic or basic species quaternary amines and alkyl sulphonates, respectively, are suitable counter ions for formation of neutral ion pairs. For compounds with both acidic and basic functional groups, a suitable procedure frequently adopted is to employ a cationic PIC reagent eg. tetrabutyl ammonium chloride (TBAC) to neutralise acidic groups. In order to suppress the ionisation of the basic function, PIC reagent is dissolved in a slightly basic mixed phosphate buffer solution. The nearly neutral species thereby produced will probably have a significant retention time on a RPP column, whereas the parent molecule would undoubtedly be washed through with the solvent peak.

The sample-ion/counter-ion adduct retains a relatively high degree of polarity and may be quickly eluted from the RPP column. Its retention may be enhanced by using a more polar packing than the usual ODS phase eg. an octyl or a tetramethylsilyl (TMS) phase.

Several disadvantages have been found when using PIC reagents. A major practical problem is the lack of solubility of PIC reagents in organic solvents. TBAH for example is only sparingly soluble in methanol and acetonitrile particularly in the presence of inorganic buffers eg. 0.01M TBAH in a 0.01M potassium dihydrogen phosphate/disodium hydrogen phosphate mixed buffer will begin to precipitate out if the percentage methanol exceeds 60% in a water-methanol mixture. This means that only very limited solvent gradients can be applied in the presence of TBAH.

If gradient elution is required in the presence of a PIC reagent then the concentration of the reagent and buffer must be kept constant while varying the solvent composition. The two reasons for this are that the PIC reagent and buffer mixture usually give a UV response which will cause unacceptable base line variation during the gradient elution and that the mechanism involved in PIC chromatography requires an equilibration time which is impossible to synchronise with the rate of change in TBAH concentration.

A full theoretical discussion of the mechanism of action of PIC reagents in HPLC is available[17]. It is a subject about which there remain some uncertainty and controversy[18]. It has been claimed that the basic mechanism is a straightforward liquid/liquid partition between the column packing phase and the sample/counter-ion adduct. It has also been claimed that the PIC reagent acts by forming an almost irreversible bond with the RPP packing so that the separation is achieved by an ion exchange mechanism. Our work would indicate that the latter mechanism is certainly occurring even though it is not the predominant one. All RPP columns through which PIC reagents have been passed retain these reagents quite strongly. Mass spectrometric analysis of column eluate have shown the presence of PIC reagents even after prolonged cleaning with polar solvents eg. dimethylsulphoxide.

It has been found that the retention of PIC reagent is not completely irreversible however. A slow bleed of PIC reagent during subsequent use of the column has been observed. It has been found necessary therefore to dedicate columns used with PIC reagent to a specific analysis.

Finally, the use of PIC reagents contaminates any fractions collected after the HPLC separation. Subsequent identification and characterisation procedures therefore become more difficult to apply.

3.3 Reversed-phase liquid partition chromatography using buffers (ionic suppression)

Because of the above mentioned disadvantages the use of PIC reagent is now avoided in our work wherever possible. It has generally been found that with the correct choice of RPP medium and an appropriate buffer solution much simpler systems can be evolved for the separation of most ionic organic compounds. Elution with buffer solution has the advantage that it is cheaper and improvements in peak separation and retention as well as peak shape are frequently observed. Buffers are also easily flushed out after use so that columns may be used for different separations. There is also no problem with long term stability provided algal growths can be prevented. The biggest advantages however are the swift equilibration of buffered solutions with the stationary phase and the useful characterisation work which may be applied to collected fractions.

A problem in using inorganic buffers for elution from RPP columns is the limited pH range which may be applied. The chemically bonded phases used in RPP chromatography are stable only in the range 2-8. Outside this range hydrolysis is likely to occur. This places a restriction on separations by ionic suppression using conventional RPP media. However, recently a new stationary phase which is claimed to effect separations by a partition mechanism and which can be used in the pH range 1-12, has been introduced. It is based upon a styrene-divinylbenzene co-polymer structure and, it is claimed, can also be used as a gel permeation medium.

3.4 Experimental work

In order to demonstrate the effectiveness of simple buffer solutions for separating ionic compounds on RPP media, three stationary phases covering a range of properties were chosen and used to separate three acid blue dyes in the presence and absence of a buffer solution. The three phases chosen were an ODS, a TMS, and the new styrene-divinylbenzene co-polymer medium. Results are shown in Table 1 and Fig. 2.

Table 1. Resolution of acid blue dyes on RPP phases in the presence and absence of buffer

Phase	Resolution without buffer		with buffer	
	AB9-AB1	AB1-AB3	AB9-AB1	AB1-AB3
ODS	0	0	4.5	2.9
TMS	2.3	1.0	11.5	8.0
Copolymer	2.6	1.3	2.2	2.3

In the absence of buffer the ODS phase was unable to resolve the three dyes. However the more polar TMS phase allowed the three dyes to be separated. The best separation was achieved by the copolymer phase even though this column was only half the length (15 cm) of the other two.

In all cases the application of a potassium dihydrogen orthophosphate buffer adjusted to pH 3 with phosphoric acid improved the resolution. The TMS phase gave the best and the ODS phase the poorest performance.

Fig. 2.

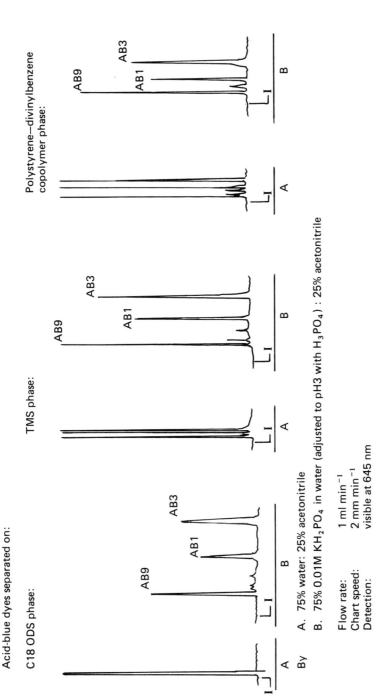

Acid-blue dyes separated on:

C18 ODS phase: TMS phase: Polystyrene–divinylbenzene copolymer phase:

A. 75% water: 25% acetonitrile
B. 75% 0.01M KH$_2$PO$_4$ in water (adjusted to pH3 with H$_3$PO$_4$) : 25% acetonitrile

Flow rate: 1 ml min^{-1}
Chart speed: 2 mm min^{-1}
Detection: visible at 645 nm

4. CONCLUSIONS

It is appreciated that for the separation of some ionic organic compounds eg. amphoterics, by RPP chromatography the use of PIC reagents is essential. However it is concluded that the use of these compounds is to be avoided if separation can be achieved by other means. The use of inorganic buffers and/or more polar RPP packings is recommended. The application of styrene-divinylbenzene copolymer phases used with and without buffers and with pH-gradients should be considered.

REFERENCES

1. HUBER, J.F.K. et al. (1971). Evaluation of dynamic chromatographic methods for the determination of adsorption and solution isotherms. J. Chromat., 58, 137.
2. KIRKLAND, J.J. (1974). Preferred experimental conditions for trace analysis by modern liquid chromatography. Analyst, 99, 859.
3. LITTLE, J.N. et al. (1975). New considerations in detector-application relationships. J. Chromat., 112, 389.
4. OTSUKI, A. (1977). Reversed-phase adsorption of phthalate esters from aqueous solutions and their gradient elution using a high-performance liquid chromatograph. J. Chromat., 133, 402.
5. KUMMERT, R. et al. (1978). Trace determination of tetrachloroethylene in natural waters by direct aqueous injection high-pressure liquid chromatography. Anal. Chem., 50, 837.
6. CREED, C.G. (1976). LC simplifies isolating organics from water. Research/development, 27, 40.
7. WALTON, H.F. (1978). Trace organic analysis of wastewater by liquid chromatography. National Bureau of Standards Special Publication 519. Trace Organic Analysis: A New Frontier in Analytical Chemistry. Proc. of the 9th materials research symposium, NBS, Gothenburg, 185.
8. SCHAUWECKER, P. et al. (1977). Trace enrichment techniques in reversed-phase high-performance liquid chromatography. J. Chromat., 136, 63.
9. FREI, R.W. (1978). Trace enrichment and chemical derivatisation in liquid chromatography; problems and potential in environmental analysis. Int. J. Environ. Anal. Chem., 5, 143.
10. ISHII, D. et al. (1978). Studies of micro high-performance liquid chromatography. III. Development of a "micro-pre-column method" for pre-treatment of samples. J. Chromat., 152, 341.
11. VAN VLIET, H.P.M. et al. (1979). On-line trace enrichment in high-performance liquid chromatography using a pre-column. J. Chromat., 185, 483.
12. OYLER, A.R. et al. (1978). Determination of aqueous chlorination reaction products of polynuclear aromatic hydrocarbons by reversed phase high performance liquid chromatography-gas chromatography. Anal. Chem., 50, 837.
13. SANER, W.A. et al. (1979). Trace enrichment with hand-packed CO:PELL ODS guard columns and Sep-Pak C18 cartridges. Analyt. Chem., 51, 2180.
14. EISSENBEISS, F. et al. (1978). The separation by LC and determination of polycyclic aromatic hydrocarbons in water using an integrated enrichment step. Chromat. Newsl. 6, 8.
15. OGAN, K. et al. (1978). Concentration and determination of trace amounts of several polycyclic aromatic hydrocarbons in aqueous samples. J. Chromat. Sci., 16, 517.

16. WAGGOTT, A. et al. (1979). An enrichment technique for integrated extraction and determination of organic compounds in water using high performance liquid chromatography. Concerted Action Analysis of organic micropollutants in water (COST project 64b bis). Proc. of the 1st European Sym., Berlin.
17. TOMLINSON, E. et al. (1978). Ion-pair high-performance liquid chromatography. J. Chromat., 159, 315.
18. KISSINGER, P.T. (1977). Comments on reserve-phase ion-pair partition chromatography. Anal. Chem. 49, 883.

CLOSED LOOP STRIPPING AND CRYOGENIC ON-COLUMN FOCUSING OF TRIHALOMETHANES AND RELATED COMPOUNDS FOR CAPILLARY GAS CHROMATOGRAPHY

M. TERMONIA, J. WALRAVENS, P. DOURTE and X. MONSEUR
Institut de Recherches Chimiques, Administration de la recherche agronomique, Secrétariat d'état à l'agriculture.
Museumlaan, 5 - B1980 Tervuren, Belgium.

Summary

The activated carbon filters, commonly used for sampling and enrichment of organic volatiles in the closed loop stripping system, are thermally desorbed by inserting them into the injection port of a gas chromatograph-mass spectrometer-data system. The desorbed substances are trapped at liquid nitrogen temperature onto a "U-shaped" glass capillary, packed with Chromosorb and then transfered to the analytical column by rapid heating of the trap. Using this system, highly volatile halocarbons in drinking water were analyzed, using a glass capillary column provided with a stationary phase film thickness of 2.5 um.

1. INTRODUCTION

Highly volatile organic micropollutants in water are commonly analyzed according different methods, including liquid-liquid extraction (1,2), purge-and-trap techniques using adsorption on a porous polymer (3), direct aqueous injection with ECD detection (4), headspace analysis (5) and recently, gas phase stripping followed by adsorption on activated charcoal (6,7).
We want to report over the use of a modified "closed loop stripping" procedure (8) involving thermal desorption of activated charcoal filters directly in front of a cryo-focusing trap, connected to a glass capillary column.

2. METHODS

A closed loop stripping (CLS) apparatus, build according the descripttion in ref.8, was equipped with 5 mg activated carbon filters (Brechbuhler AG). 1 liter of water sample (tap water) was submitted to stripping for periods ranging from 15 to 60 minutes. The charcoal filters were thermally desorbed at 200 C directly in the injection port of a Carlo Erba 2900 gas chromatograph (GC) with FID detection, or of a Carlo Erba 4160 GC coupled to a Finnigan-MAT 4500 mass spectrometer, interfaced to an Incos 2300 data system.
In the oven of the GC, a "U-shape", 15 cm long glass capillary (int. diam.:0.32mm), deactivated by persilylation, was connected between the injector and the analytical column. The connection of the trap with the column was accomplished by

means of shrinkable Teflon. A 2 cm long portion of this capillary was packed with persilylated Chromosorb W 120-140 mesh.

The analytical column (Pyrex, 50 m long, int.diam : 0.32mm) was coated with SE-54. The thickness of the stationary phase film was 2.5 um.

Immediately before use, the filters were submitted to conditioning, at 200 C in the injection port of the GC, by flushing with hydrogen at the analytical flow rate. This procedure was repeated until the blank chromatograms showed no or very little contamination.

After CLS, the activated carbon filters were introduced in the cold injection chamber, allowing the air to be flushed from the GC system.

Cryo-focusing of the desorbed sample was performed by immersing the trap in liquid nitrogen for 10 min. In this situation, a H2 pressure of 5 bar, measured in the injection port of the GC-FID, gave a flow rate of 3ml/min at the column outlet. Similarly, a flow rate of 1ml/min was obtained with a He pressure of 6 bar for the GC-MS system.

Injection occured when the liquid nitrogen was replaced by a hot water (approx. 80 C) bath. Temperature programming of the GC oven was then started (from 30 C to 200 C at 4 C/min).

3. RESULTS

In a first series of experiments, the sampling efficiency of the CLS system was estimated, using tap water as a sample system. A series of extractions of the same sample showed that for 15 min stripping periods, the recovery rate varied from 50 to 60%. For 45 min stripping time, a recovery rate of more than 90% was measured for all the components present in our sample (methylene chloride, chloroform, trichloroethylene, dichlorobromomethane, bromoform, chlorodibromomethane, tetrachloroethylene).

On the other hand, the adsorption efficiency of the activated carbon filters was controled by placing 2 filters in series. The corresponding analyses showed that the first filter efficiently adsorbs all the halocarbons of interest.

For quantitative measurements at high sensitivity, the same samples were submitted to GC-MS-DS analysis, with the MS working in the full scan mode (see Fig.1) or in the single-ion-monitoring mode (see Fig.2). Contamination of the CLS system by acetone, pentane and carbon disulfide are due to the washing procedure.

The ion chromatograms represented in fig.2 show that very little breakthrough occurred, when a packed capillary trap is used for the cryo-focusing. The same results have been reported previously for the determination of components of very high volatility (8).

4. REFERENCES

1. G.EKLUND, B.JOSEFSSON and C.ROOS, J.HRC & CC, 1 (1978) 34.
2. J.F.J. VAN RENSBURG and J.HASSET, J.HRC & CC, 5 (1982) 574.
3. W.BERTSCH, E.ANDERSSON and G.HOLZER, J.Chromatogr., 112 (1975) 701.

4. K.GROB and A.HABICH, J.HRC & CC, 6 (1983) 11.
5. G.J.PIET, P.SLINGERLAND, F.E.de GRUNT, M.P.M.VAN DEN HEUVEL and B.C.J.ZOETEMAN, Anal. Letters, 11 (5) A (1978) 437.
6. W.GIGER, COST 64b bis, Workshop of the working party 8 "Specific Analytical Problems", Dubendorf, 18-19 may 1982.
7. J.W.GRAYDON and K.GROB, J. Chromatogr., 254 (1983) 265.
8. K.GROB and F.ZURCHER, J. Chromatogr., 117 (1976) 285.

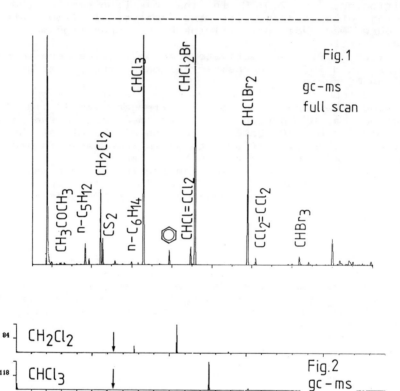

OBSERVATIONS WITH COLD ON-COLUMN INJECTION

P. SANDRA
Laboratory of Organic Chemistry, State University of Ghent,
Krijgslaan, 281 (S.4), B-9000 GENT (Belgium)

Summary
Problems prohibiting the application of on-column injection to some "real world" samples are discussed.

Cold on-column injection in capillary gas chromatography is well-known for its high analytical precision (1,2). However some practical problems prohibit the application of cold on-column injection to some "real world" samples. The most serious drawbacks can be summarized as follows :

- Peak distortion and splitting when injecting large samples of apolar solvents (> 2-3 μl) or small samples of polar solvents (< 1 μl) such as methanol on apolar polysiloxane columns, make qualitative and quantitative analysis impossible.

- The inlet section of the capillary column can easily be contaminated with non-volatiles, particles, metal ions, etc. This results in a loss of efficiency and inertness of the capillary column.

- The column has to be cooled down below the boiling point of the solvent or, when secondary cooling is available, to 10-30 degrees above the boiling point of the solvent. For high temperature capillary gas chromatography (e.g. the analysis of triglycerides, steroids, etc.) this is a waste of time and energy. Moreover due to baseline drift by fast temperature programming, less precise and accurate results are obtained in comparison with the moving needle injector. The analysis of pesticides with ECD detection can serve as an example.

To overcome these problems, several solutions has been advanced.
Peak distortion and peak splitting can be avoided through a "retention gap" i.e. the capillary column inlet is free of stationary phase (3,4).
The simplest way to obtain a retention gap is to wash out the column over the length needed (50 to 60 cm per μL injected). Washing has to be carried out after coating and before the conditioning or immobilization. This technique can only be applied by a limited number of chromatographers who are making their own columns. Commercial columns are mostly preconditioned and, at present, cross-linked or immobilized, so that the stationary phase film can no longer be removed by washing. The proposal of Grob that manufacturers should produce columns with a retention gap of 3 m is therefore sound. It is however doubtful that this will be realized in the near future. The connection of a pre-column retention gap to the

separating column is a more realistic approach. This offers
the advantage that when the column inlet, i.e. the retention
gap, is contaminated, replacement of the retention gap is a-
chievable. Due to the problems with straightening glass columns, a retention gap must be made with fused silica tubing,
which has to be deactivated by silylation. In our hands, the
best and easiest way to connect the retention gap to the capillary column is by means of a polyimide seal (5,6).
Peak distortion effects of cold on-column injection are however not as general as currently believed. A retention gap is
more particularly needed when polar solvents such as methanol
are injected on apolar columns. During experiments with normal and immobilized polyethyleneglycol (Superox 20M) films
we observed that neither apolar nor polar solvents produced
peak distortion (6).

Fig. 1 and 2 show the analysis of 1 and 3 μL samples
sizes of n-hexane and methanol solutions of nonanol-1 and the
methyl esters of decanoic, undecanoic and dodecanoic acid on
a 20 m x 0.3mm glass capillary column coated with 0.12 μm immobilized Superox 20M. The TZ values decrease with increasing
sample size (the increase of the flooding zone accounts for
this) but injection peak splitting phenomena are absent.
In general, somewhat lower TZ values are found for methanol
solutions. As visually observed the length of the flooding
zone for methanol injections is 1.5 to 2 times the length of
the flooding zone for n-hexane solutions and this can explain
the small decrease.

Therefore we must conclude that the nature of the stationary phase is an important parameter in on-column injection
phenomena. Due to the absence of peak splitting on PEG films,
a PEG precolumn can be connected to an OV-1 column for the direct analysis of water samples.

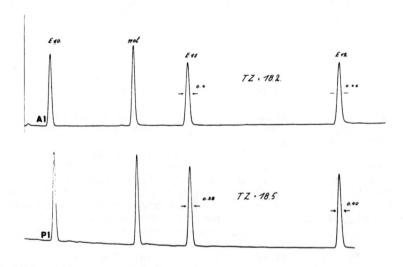

Fig. 1 : A_1 : 1 μl hexane solution
        P_1 : 1 μl methanol solution

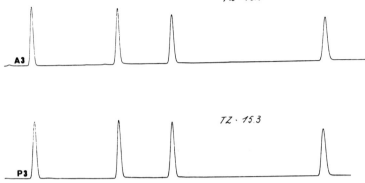

Fig. 2 : A_3 : 3 µl hexane solution
         P_3 : 3 µl methanol solution

- Contamination of the inlet section of the capillary column is also avoided through a retention gap. When a decrease in efficiency and inertness is observed, the retention gap is replaced. Capillary columns with immobilized phases can be washed out. In our experience however we state that capillary column inlet contaminants cannot always be properly removed.

- To overcome the problem of cooling down the gc oven below the boiling point of the solvent, a movable on-column injector has been developed for high temperature work (7,8). The idea of removing only the injection zone of the capillary column out of the oven (capillary column remains at high oven temperature) and thus applying natural cooling, has been patented by the Carlo Erba Company.
A home-made movable on-column injector has been described in ref. 7 and 8. The main advantages of the system are its simplicity, the precise and accurate results obtained and the gain of time in routine capillary gas chromatography. The use of the movable on-column injector is not restricted to high temperature work and can also be applied to normal temperature capillary GC.

References

1. GROB, K. and GROB, K. Jr. (1978), J. Chromatogr., 151, 311.
2. GALLI, M., TRESTIANU, S. and GROB, K. Jr. (1979), J.HRC & CC, 2, 366.
3. GROB, K. Jr. (1982), J. Chromatogr., 237, 15.
4. GROB, K. Jr. and MÜLLER, R. (1982), J. Chromatogr., 244, 185.
5. SANDRA, M., SCHELFAUT, M. and VERZELE, M. (1982), J.HRC & CC, 5, 50.
6. SANDRA, P., VAN ROELENBOSCH, M., VERZELE, M. and BICCHI, C. (1983). Proc. Fifth Int. Symp. Cap. Chrom., Riva del Garda, Italy. Ed. J. Rijks, Publ. Elsevier Amsterdam, pg. 315.

7. GEERAERT, E., SANDRA, P. and DE SCHEPPER, D. (1983). Proc. Fifth Int. Symp. Cap. Chrom., Riva del Garda, Italy. Ed. J. Rijks, Publ. Elsevier Amsterdam, pg. 346.
8. GEERAERT, E., SANDRA, P. and DE SCHEPPER, D. (1983). J.HRC & CC, 6, 387.

POLYMETHYLPHENYLSILICONES IN FUSED SILICA CAPILLARY COLUMNS FOR GAS CHROMATOGRAPHY

M. VERZELE, F. DAVID and P. SANDRA
Laboratory of Organic Chemistry, State University of Ghent,
Krijgslaan, 281 (S.4), B-9000 GENT (Belgium)

Summary

A polymethylphenylsilicone, synthesized in the laboratory of the authors was successfully immobilized in fused silica capillary columns. The possibilities of the polymethylphenylsilicone columns are illustrated with applications of environmental research.

Much effort is presently directed at the preparation of fused silica columns with immobilized films of polar polysiloxanes. Because of their high polarity and unique selectivity, the phases most interesting to investigate are the methylphenylsilicones with a high phenyl content (OV-17, OV-22, OV-25), the cyanopropylsilicones (Silar 10C, OV-275) and the methyltrifluorosilicones (QF 1, OV-210). The availability of efficient, inert and thermally stable columns with the phases mentioned, will further broaden the applicability of capillary gas chromatography. This however, only if it is possible to manufacture the columns in a reproducible way with a well-defined polarity and selectivity. This last point is very important because in order to make high quality columns, capillary compatible stationary phases have to be synthesized. In practice this means that the above mentioned phases have to be synthesized in gum form. Moreover the phases should preferably be immobilized. These two requirements could be fulfilled by synthesizing a viscous methylphenylsilicone polymer which is terminated with silanol functions. The columns are statically coated with methylphenylsilicone polymer. Gummification in the column is then carried out by heat-curing. An immobilized and solvent resistant methylphenylsilicone film is obtained without needing addition of peroxides.

The procedure offers the advantage that no groups other than methyl and phenyl are present in the polymer finally deposited on the column wall. Therefore the polarity and selectivity are well-defined and equal to that of OV-17. The yield of immobilization is 100 %. The chromatographic efficiency is above 80 % and the columns show a thermal stability to at least 280°C.

Details on the procedure have been described at the Fifth Int. Symposium Cap. Chrom., Riva del Garda, April 1983 (1). Columns prepared according to the procedure described are commercially available through Alltech, Europe.

Some applications are shown in Fig. 1, 2 and 3. Fig. 1 shows the analysis of a polyaromatic hydrocarbon (PAH) standard mixture. The benzopyrene isomers are better separated than on a polydimethylsilicone stationary phase. Note also the low bleeding at high temperatures.

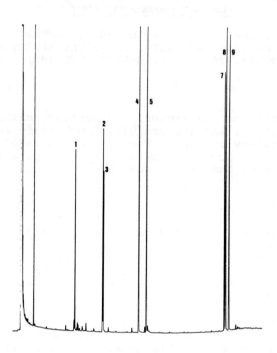

Fig. 1 : PAH standard mixture.
Column 25 m x 0.32 mm FSOT, 0.12 µm film of methylphenylsilicone gum. Injection on column 50°.
Temperature 50° ↑ 100° - 5°/min - 270°.
Peaks : 1. naphtalene; 2. fluorene; 3. phenantrene; 4. antracene; 5. fluoranthene; 6. pyrene; 7. benzo|e| pyrene; 8. benzo|a|pyrene; 9. perylene.

Since dibenzothiophene derivatives are very resistant to physical, chemical and biological processes and are not synthesized by biota, it is logical that these compounds can be used as pollution markers in environmental analyses (2,3). Each oil (petroleumbatch) contains different concentration ratios of dibenzothiophene isomers. By monitoring the concentration ratio of the C_1-dibenzothiophene isomers, it becomes possible to indicate the origin of oil spillage (4).

This attractive potential method, however, was weakened by the fact that 2-methyldibenzothiophene and 3-methyldibenzothiophene could not be separated on several stationary phases (4) (OV-1, SE-54, SE-52, OV-73, OV-1701, OV-225 and Carbowax 20M). We now show the separation of these compounds on a 15 m x 0.32 mm ID FSOT column coated with the immobilised polymethylphenylsilicone discussed in the present paper.

This example illustrates the interesting selectivity of polymethylphenylsilicone.

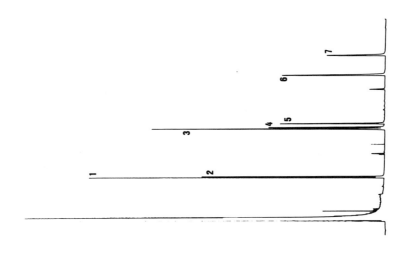

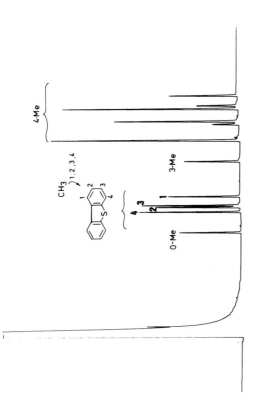

Fig. 2 Dibenzothiophene derivatives. Column 15 m x 0.32 mm FSOT; 0.15 μm film of methylphenylsilicone gum. Injection on column. Temperature 50° ↑ 120° - 3°/min-280°. Peaks: 1.2.3.4. methyldibenzothiophene isomers.

Fig. 3 Underivatized phenols. Column 20 m x 0.32 mm FSOT, 0.24 μm film of polymethylphenyl silicone gum. Injection on column. Temperature 25° ↑ 70° - 3°/min-200°. Peaks: 1. phenol; 2. 2-chlorophenol; 3. O-nitrophenol; 4. 2,5-dimethylphenol; 5. 2,4-dichlorophenol; 6. 4-choro-m-cresol; 7. 2,4,6-trichlorophenol

The analysis of dibenzothiophene derivatives is illustrated in fig. 2.

The inertness of the columns is demonstrated in fig. 3, showing the analysis of underivatized phenols on an immobilized polymethylphenyl film.

REFERENCES

1. VERZELE, M., VAN ROELENBOSCH, M., DIRICKX, G., DAVID, F. and SANDRA, P. (1983). Proc. Fifth Int. Symp. Cap. Chrom., Riva del Garda, Italy 26-28 April. Ed. J. Rijks. Publ. Elsevier Amsterdam, pg. 94.
2. OGATA, M. and MIYAKE, Y. (1980). J. Chromatogr. Sci., 18, 594.
3. FRIOCOURT, M.P., BERTHOU, F. and PICART, D. (1982). Toxicological and Environmental Chemistry, vol. 2, 205.
4. BERTHOU, F., personal communication.

ACKNOWLEDGEMENTS

Thanks are due to Dr. F. Berthou for supplying the dibenzothiophene standards.
F.D. thanks the "Instituut tot Aanmoediging van het Wetenschappelijk Onderzoek in Nijverheid en Landbouw - IWONL" for a grant.

COLLABORATIVE STUDY ON THE MASS SPECTROMETRIC QUANTITATIVE
DETERMINATION OF TOLUENE USING ISOTOPE DILUTION TECHNIQUE

Arne Büchert
National Food Institute
19 Mørkhøj Bygade
DK 2860 Søborg

Summary

11 Danish mass spectrometers were applied for the quantitative determination of toluene by the isotope dilution technique. The purpose of this exercise was to compare the virtual performances of the GC/MS-systems and to make the participants familiar with this technique.

In 1982, a section of the Danish Mass Spectrometry Group decided to compare the precision and accuracy of their quadropole and magnetic instruments using the mass spectrometric isotope dilution technique.

This technique has been generally recognized as a very valuable technique for quantitative determination of organic compounds. The method is normally based on the measurement of the intensity of an ion, which is characteristic of the compound being determined, together with the intensity of a similar ion derived from a stable isotope labelled analogue of the compound, which is added to the sample as internal standard. The ratio between the intensities of the two ions can be determined from repeated scans covering the whole mass range or by "selected ion monitoring" (SIM) measuring only the ions of interest.

Ten laboratories with 11 different mass spectrometers took part in the study (table I), which was conducted by Elfinn Larsen, Risø National Laboratory, Roskilde, Helga Flachs, Rigshospitalet, Copenhagen and Gustav Schroll, H.C. Ørsted Institute, Copenhagen. This group was responsible for the preparation and the mailing of samples as well as the collection and the evaluation of the analytical results.

EXPERIMENTAL

All laboratories recieved 5 standard solutions containing 2.04; 5.10; 15.30 and 20.40 mg/l toluene and 11.5 mg/l of d_8-toluene dissolved in cyclohexane. Furthermore, the laboratories got two samples A and B containing toluene and 11.5 mg/l of the international standard dissolved in cyclohexane. The actual concentrations of toluene were unknown to the participants, but the samples were prepared at 3.56 mg/l and 16.42 mg/l toluene respectively. Finally, a sample of pure cyclohexane was included in the program.

The participants were asked to measure the ratio between the intensities of the molecular ions of toluene (m/z = 92) and d_8-toluene (m/z = 100). The two standards marked 2.04 and 20.4 mg/l were to be measured 10 times each and the three other standards marked 5.10; 10.20 and 15.30 mg/l two times each. The samples A and B should be measured 5 times each. All the samples should be used directly without any preconcentration. No specifications were given for injection, type of GC-column, MS-parameters etc.

The 11 participating mass spectrometers are described in table II. Electron impact (about 70 eV) ionization was applied in all cases. Selected ion monitoring was used except for laboratory no 5, where the mass range corresponding to the molecular ions was scanned.

RESULTS AND DISCUSSION

The calibration curve was determined from the measured $I_{92}:I_{100}$-ratio in the 5 individual standards. Ideally, this curve is a straight line. The actual results from the 5 standards are summarized in table III. For each instrument the slope α and the intercept β of the line is calculated by linear regression using all 26 measurements and by linear interpolation using only the 20 measurements from the strongest and from the most diluted standard. However, there is no significant difference between the two sets of α and β as it appears from table III. The precision of the measurements is reflected in the standard deviations $S_{2.04}$ and $S_{20.4}$ calculated from the measurements of the two standards. These values ranging from 0.77 % to 9.50 % are higher than expected. Consequently the linear correlation of the results as expressed by the r and Sx.y values, is for most of the instruments rather poor. It should be emphasized that the values in table III do not show any significant differences between the magnetic and the quadrupole instruments nor between capillary columns and packed columns.

The results of the GC/MS-measurements of the two unknown samples A and B are shown in fig. II and III. The mean values of the concentrations of toluene are determined to be 3.53 mg/l and 16.52 mg/l toluene respectively. This should be compared to the 3.56 mg/l and 16.42 mg/l at which the samples were prepared. Like the results from the standard solutions, the results from A and B does neither show any significant differences between the type of instrument nor the type of GC-column. Even the age of the instrument was not significant.

CONCLUSION

The main purpose of this intercalibration was to compare the performances of the individual GC/MS-systems and to give the participants a possibility to become familiar with the isotope dilution technique. It was not possible to demonstrate any significant instrument-dependent differences between the results, but the exercise has without any doubt been very valuable for the participants.

TABLE I List of participants

Büchert, Arne	National Food Institute, DK 2860 Søborg
Egsgaard, Helge & Larsen, Elfinn	Risø National Laboratory DK 4000 Roskilde
Flachs, Helga	Rigshospitalet DK 2100 Copenhagen
Folke, Jens	Water Quality Institute DK 2970 Hørsholm
Haunsø, Niels	Jydsk Technological Institute DK 8000 Aarhus
Larsen, Kjeld & Lund, P. Anker	Danish National Institute of Occupational Health DK 2900 Hellerup
Møller, Jørgen	Odense University DK 5230 Odense
Nielsen, Peter	The State Chemical Supervision Service DK 2800 Lyngby
Olsen, Carl Erik	Royal Vet. & Agric. University DK 1872 Copenhagen
Olsen, Henrik	Technological Institute DK 2630 Taastrup

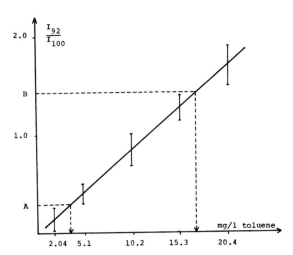

Fig. I Calibration curve used for determination of the toluene content i A and B.

Table II. List of the 11 mass spectrometers participating in the study.

No	Instrument	Installation	Gas Chromatography column	Gas Chromatography injection	GC/MS interface	Data collection and handling
1	HP 5992		capillary	splitless	restrictor	computer
2	HP 5992 B		packed	on-column	jet	computer
3	MS 30	1975	packed	on-column	jet	UV-paper
4	HP 5995 A		capillary	splitless	open split	computer
5	Mat 311	1973	packed	on-column	Biemann	computer
6	Mat 311 A	1976	packed	on-column	jet	computer
7	HP 5992		capillary	splitless	restrictor	computer
8	HP 5992 B	1980	capillary	splitless	restrictor	computer
9	Mat CH 5	1969	packed	on-column	jet	recorder
10	F 4021 T	1980	capillary	splitless	restrictor	computer
11	HP 5985 B	1980	capillary	splitless	restrictor	computer

Table III. Calibration data listing the 95% confidence intervals.

No	1. Linear Regression					2. Linear Interpolation			
	$\alpha_1 \pm t \cdot s_{\alpha_1}$	$\beta_1 \pm t \cdot s_{\beta_1}$	r	$s_{x \cdot y}$	$s_{2.04\%}$	$s_{20.4\%}$	$\alpha_2 \pm t \cdot s_{\alpha_2}$	$\beta_2 \pm t \cdot s_{\beta_2}$
1	0.0862±0.0021	0.0455±0.0288	0.9983	0.497	2.20	3.36	0.0857±0.0023	0.0444±0.0061
2	0.0879±0.0003	0.0015±0.0077	0.9999	0.131	1.17	0.91	0.0877±0.0006	0.0023±0.0021
3	0.1333±0.0057	0.0016±0.0783	0.9949	0.871	6.57	6.21	0.1331±0.0067	0.0149±0.0202
4	0.0910±0.0055	0.0226±0.0763	0.9897	1.236	5.63	8.96	0.0910±0.0066	0.0303±0.0166
5	0.0893±0.0034	0.0475±0.0483	0.9963	0.740	9.50	5.00	0.0889±0.0037	0.0495±0.0189
6	0.0927±0.0008	0.0142±0.0115	0.9998	0.184	2.17	1.23	0.0925±0.0009	0.0152±0.0040
7	0.0857±0.0009	0.0275±0.0122	0.9997	0.212	0.77	1.52	0.0856±0.0010	0.0245±0.0025
8	0.1040±0.0019	0.0713±0.0258	0.9991	0.369	3.98	1.89	0.1044±0.0017	0.0532±0.0090
9	0.0927±0.0008	0.0090±0.0105	0.9998	0.168	1.26	1.05	0.0927±0.0008	0.0058±0.0025
10	0.1028±0.0010	0.0020±0.0140	0.9997	0.202	5.96	1.17	0.1027±0.0013	0.0017±0.0102
11	0.1065±0.0017	0.0031±0.0235	0.9993	0.329	3.61	2.34	0.1064±0.0020	0.0067±0.0076

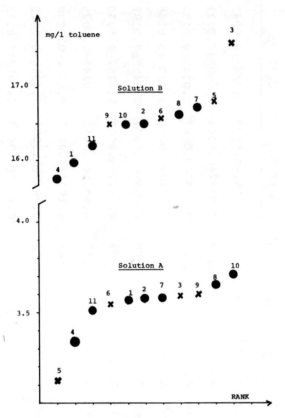

Fig. II The results of the GC/MS measurements of the unknown solutions A and B. Magnetic instruments are marked with a cross.

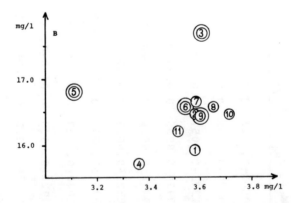

Fig. III Youden plot of the results of the two unknown solutions A and B. (Magnetic instruments are marked with a double-ring).

DETERMINATION OF VOLATILE ORGANIC SUBSTANCES IN WATER BY GC/SIM
RESULTS OF AN INVESTIGATION ON THE RIVERS RHINE AND MAIN

F. KARRENBROCK and K. HABERER
ESWE-Institut für Wasserforschung und Wassertechnologie GmbH
D-6200 Wiesbaden

Summary

Selected Ion Monitoring (SIM) allows a very selective and highly sensitive determination of many organic environmental pollutants. A method for the determination of volatile compounds in surface water and drinking water is described, applying GROB closed-loop-stripping for trace enrichment, capillary gaschromatography for separation and mass spectrometric detection of selected ions for quantification. With this method 23 compounds were monitored over an one year period in surface waters from the River Rhine. High concentrations of chlorinated aromatic hydrocarbons have been detected in samples taken from the River Main downstream of some waste water discharges.

1. INTRODUCTION

The content of organic pollutants is one of the main quality criteria for surface water from the River Rhine, which is extensively used for drinking water preparation. Among these, chlorinated hydrocarbons are of special interest because of their supposed adverse health effect. This is reflected in the fact, that most of 129 pollutants compiled on a recently proposed "Black List" for surface waters are chlorinated substances, mostly volatiles[1].Monitoring of these contaminants in raw water and finished water should therefore be a vital part of the quality assurance program in drinking water treatment plants.
The water works Wiesbaden-Schierstein is situated on the right bank of the River Rhine at km 507, about 7 km downstream the mouth of the River Main, which is known to be far more polluted than the River Rhine. Up to the water works intake area the two rivers have not yet mixed up. As the River Main water flowing along the right bank of the stream is more polluted than the water of the River Rhine, the water for the treatment plant is taken from the middle of the stream. Nevertheless this heavily polluted water flowing along the right bank should be investigated, because it makes a substantial contribution to the total pollutant load of the lower part of the River Rhine. This paper is concerned with the determination of volatile contaminants in surface waters from both Rivers, Rhine and Main.

2. CHOICE OF ANALYTICAL METHOD

For the determination of volatile organic compounds in water GROB closed-loop-stripping (CLSA) is a well established technique [2, 3]. Already in 1975 many organic contaminants have been identified in River Rhine water at the ng/l level using CLSA in combination with GC/MS [4]. For quantitative determinations the combination of CLSA for trace enrichment, high resolution capillary gaschromatography and mass spectrometric detection in the selected ion mode (SIM) is advantageous [5].

3. Experimental

1. Sampling:

   Grab samples were taken from surface waters. Glass bottles (1 l) were filled bubble-free and plugged with a glass stopper. Usually, analysis was carried out immediately after sampling, occasionaly samples were stored at a temperature of 4 °C for 24 h.

2. Stripping:

   The samples were stripped with the closed-loop-stripping apparatus developped by GROB [5] for 1,5 h at 20 °C using a metal bellows pump to recirculate the headspace air in the stripping bottle. The volatile organics were adsorbed on a small filter, containing 5 mg activated carbon.
   After stripping the loaded filter was spiked with 500 ng n-chlordecan (Internal Standard) and extracted with 100 µl carbon disulfide.

3. GC/MS-Analysis:

   The analysis applied is characterized by the following items.
   Instrument: Computerized quadrupol GC/MS-System
   HP 5995 (Hewlett Packard)
   Injector: GROB-Type splitless injector
   GC/MS-Interface: "Open Split"-Interface
   GC-Column: Fused silica capillary column DB-5, 25 m, I.D. 0,31 mm
   Column oven Program: <30 °C at injection, 50 °C for 6 min,
   6 °C/min from 50 °C to 250 °C
   Carrier Gas: helium, 1,5 ml/min
   MS-Conditions: EI, 70 eV,
   selected ion monitoring mode (SIM)
   dwell time 20-50 msec

4. Quantification

The retention times of 23 compounds and the monitored ions used for their quantification are shown in table 1. n-chlorodecane is used as internal standard. For all compounds response factors relative to the peak area of the ion m/e = 91 of chlorodecane were determined. The response factors werde checked frequently and corrected if necessary.

5. Performance characteristics of the method

SIM-Chromatograms of a surface water sample are shown in Fig. 1. The chromatograms clearly demonstrate the good selectivity of the method.

If the mass spectra do not interfere, even coeluting compounds, e.g. benzene and tetrachloromethane, can be determined without difficulties.

Isomers (e.g. of dichlorobenzene) with identical mass spectra were identified by their different retention times, with the exception of m- and p-Xylene, which are not separated.
The mean recoveries for 23 compounds are given in Table 2. The relative Standard Deviation for all Compounds is ≤ 13 %.

To check the reproducibility of the method, six independent analyses of a surface water sample were made. Mean concentrations are given in table 2. In this case, the relative standard deviation was ≤ 20 %.

Using 1 l samples, the detection limit for all compounds was less than 0,05 µg/l.

To prevent the carbon filter from overload, sample size was decreased to 0,25 l for heavy polluted waters.

6. Results and Discussion

From August 1982 to July 1983 samples were taken from the right bank of the River Rhine at km 506 on a week-to-week basis. The results of the total of 62 samples are given in table 3. Relative high mean values were observed for tetrachloromethane and chlorobenzene. For shorter periods levels up to 20 - 40 µg/l were also found for aromatics like benzene, toluene, dichlorobenzenes and chlorotoluenes.

Obviously there is a significant fluctuation in the concentrations of many compounds from week to week, as is demonstrated in Fig.2. In the case of chlorobenzene a seasonal variation is obvious with higher concentrations in winter and springtime.

No simple relations were found between the concentrations of organics and the water flow of the river. The observed fluctuations can probably be attributed to discontinuous discharges of industrial waste waters.

More information about the origin of these compounds was gained by a series of samples taken along the lower part of the River Main. Profiles for three compounds of interest are shown in Fig. 3. The concentrations of chlorinated aromatics were relative high downstream of some chemical factories. These effluents are apparently the major sources of pollution on the lower River Main.

References

1. KRISOR, K.: Umwelt 4, 234-235 (1982)
2. GROB, K.: J. Chromatog. 84, 255 (1973)
3. MELTON, R.G. et al: In: Advances in the Identification and Analysis of Organic Pollutants in Water. Vol. 2, L.H. Keith, Ed. (Ann Arbor Science Publ., 1981), Chapter 36 and Literature cited there
4. STIEGLITZ, L. et al: Vom Wasser 47, 347-377 (1976)
5. KARRENBROCK, F. and HABERER, K.: Vom Wasser 60. 237-254 (1983)
6. GROB, K. and ZÜRCHER, F.: J. Chromatog. 117. 285-294 (1976)

FIG. 1: SINGLE ION CHROMATOGRAMS OF A SURFACE WATER SAMPLE FROM THE RIVER RHINE

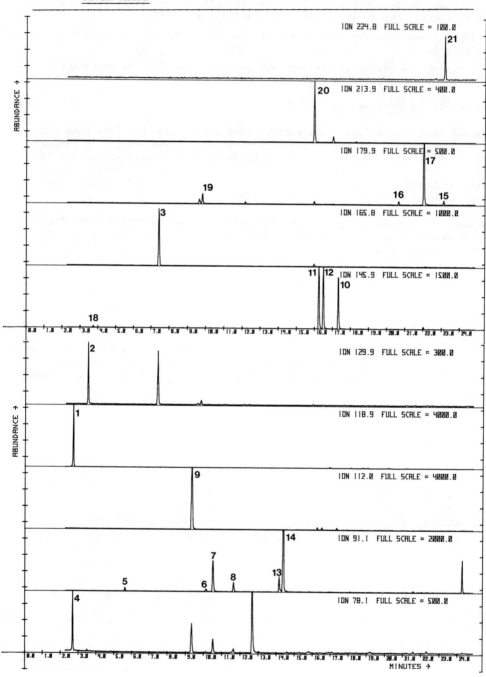

EXPLANATION (FIG. 1)

		µg/l
1.	Tetrachloromethane	4,6
2.	Trichloroethylene	0,32
3.	Tetrachloroethylene	0,74
4.	Benzene	0,07
5.	Toluene	0,02
6.	Ethylbenzene	0,02
7.	m/p-Xylene	0,30
8.	o-Xylene	0,09
9.	Chlorobenzene	2,2
10.	1,2-Dichlorobenzene	0,46
11.	1,3-Dichlorobenzene	0,61
12.	1,4-Dichlorobenzene	0,54
13.	2-Chlorotoluene	0,13
14.	4-Chlorotoluene	0,95
15.	1,2,3-Trichlorobenzene	0,01
16.	1,2,4-Trichlorobenzene	0,01
17.	1,3,5-Trichlorobenzene	0,22
18.	Trifluoromethylbenzene	0,03
19.	4-Chlorotrifluoromethylbenzene	0,07
20.	2,4-Dichlorotrifluoromethylbenzene	0,17
21.	Hexachlorobutadiene	0,03

Tab. 1: RETENTION TIME AND SELECTED IONS FOR 23 PURGEABLE COMPOUNDS

	RT(min)	Ion(m/e)
Tetrachloromethane	2,71	119
Trichloroethylene	3,51	130
Tetrachloroethylene	7,44	166
Benzene	2,70	78
Toluene	5,62	91
Ethylbenzene	10,14	91
m/p-Xylene	10,52	91
o-Xylene	11,64	91
Chlorobenzene	9,35	112
1,2-Dichlorobenzene	17,33	146
1,3-Dichlorobenzene	16,26	146
1,4-Dichlorobenzene	16,51	146
2-Chlorotoluene	14,16	91
4-Chlorotoluene	14,38	91
2,4-Dichlorotoluene	20,21	125
2,6-Dichlorotoluene	20,37	125
1,2,3-Trichlorobenzene	23,12	180
1,2,4-Trichlorobenzene	20,64	180
1,3,5-Trichlorobenzene	22,03	180
Trifluoromethylbenzene	3,75	146
Chlorotrifluoromethylbenzene	9,83	180
Dichlorotrifluoromethylbenzene	17,02	214
Hexachlorobutadiene	23,16	225
1-Chlorodecane (Int.St.)	24,29	91

TAB. 2: RECOVERY DATA AND REPRODUCIBILITY OF THE METHOD

COMPOUND	RECOVERY DATA % RECOVERY		REPRODUCIBILITY CONCENTRATION (µg/l)	
	Mean(N=5)	% SD	Mean(N=6)	% SD
Tetrachloromethane	62	11,6	1,48	11,3
Trichloroethylene	51	10,2	0,56	14,6
Tetrachloroethylene	67	13,2	1,12	16,8
Benzene	62	9,4	0,97	18,6
Toluene	69	8,4	2,30	25,2
Ethylbenzene	69	10,1	0,44	14,3
m/p-Xylene	77	10,4	0,64	11,6
o-Xylene	71	8,3	0,39	12,4
Chlorobenzene	80	11,9	3,42	12,4
1,2-Dichlorobenzene	72	8,3	1,02	9,5
1,3-Dichlorobenzene	77	11,1	1,88	12,7
1,4-Dichlorobenzene	80	11,8	1,70	10,9
2-Dichlorotoluene	69	7,2	0,37	12,1
4-Dichlorotoluene	70	11,7	0,97	10,4
2,4-Dichlorotoluene	73	5,3	0,09	9,7
2,6-Dichlorotoluene	72	6,0	nd	-
1,2,3-Trichlorobenzene	68	7,5	nd	-
1,2,4-Trichlorobenzene	79	4,4	nd	-
1,3,5-Trichlorobenzene	73	10,1	0,06	13,0
Trifluoromethylbenzene	64	12,0	0,03	11,1
Chlorotrifluoromethylbenzene	69	11,0	0,07	10,8
Dichlorotrifluoromethylbenzene	72	7,3	0,06	12,5
Hexachlorobutadiene	70	7,1	0,09	16,3

EXPLANATIONS:

o The recovery was determined from deionized water spiked with 2,5 µg/l of each compound (5 Samples)

o The reproducibility is based on 6 determinations of a surface water sample (River Rhine)

TAB. 3: CONCENTRATIONS (µg/l) OF VOLATILE ORGANIC COMPOUNDS IN SURFACE WATER TAKEN FROM THE RIGHT BANK OF RIVER RHINE NEAR KM 506 (CONTAINING WATER FROM THE RIVER MAIN)

	X	MIN	MAX	50-P	90-P
Tetrachloromethane	4,3	0,45	44,4	1,3	8,4
Trichloroethylene	0,75	0,2	5,1	0,5	1,1
Tetrachloroethylene	1,2	0,26	4,6	0,95	2,1
Benzene	1,3	nd	20,3	0,6	2,2
Toluene	2,3	0,01	16,1	1,4	5,5
Ethylbenzene	0,3	nd	2,0	0,2	0,8
m/p-Xylene	1,3	0,02	7,1	0,8	3,1
o-Xylene	0,23	nd	0,9	0,2	0,4
Chlorobenzene	4,7	0,15	20,9	2,1	13,1
1,2-Dichlorobenzene	1,1	0,23	19,7	0,7	1,4
1,3-Dichlorobenzene	0,9	0,18	2,4	0,8	1,7
1,4-Dichlorobenzene	0,8	0,20	7,2	0,7	1,4
2-Chlorotoluene	0,44	0,02	4,5	0,24	0,8
4-Chlorotoluene	1,15	0,02	43,2	0,28	1,0
2,4-Dichlorotoluene	0,07	nd	0,47	0,04	0,1
2,6-Dichlortoluene	0,02	nd	0,12	0,02	0,04
1,2,3-Trichlorbenzene	0,01	nd	0,09	0,01	0,02
1,2,4-Trichlorobenzene	0,01	nd	0,05	0,01	0,03
1,3,5-Trichlorobenzene	0,05	nd	0,42	0,05	0,08
Trifluoromethylbenzene	0,55	nd	5,1	0,08	1,6
Chlorotrifluoromethylbenzene	0,47	nd	3,9	0,2	0,9
Dichlorotrifluoromethylbenzene	0,12	nd	1,1	0,05	0,25
Hexachlorobutadiene	0,08	nd	0,78	0,05	0,17

EXPLANATIONS:

X: Mean value of 62 samples, taken from August 1982 to July 1983 on a week to week basis

min: Lowest observed value; max: highest observed value

50-P: Median value, 50-Perzentil; 90-P: 9o-Perzentil

FIG. 2: CONCENTRATIONS OF CHLOROBENZENE AND CHLOROTOLUENE DURING A ONE YEAR PERIOD

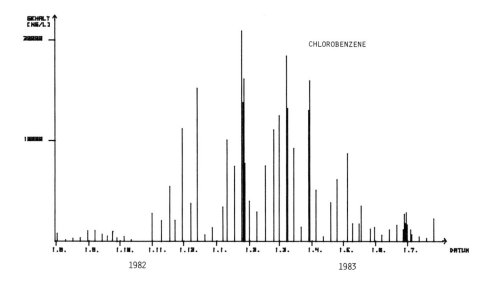

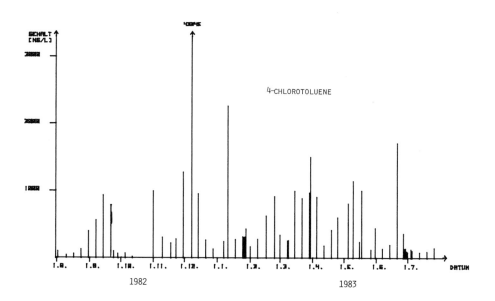

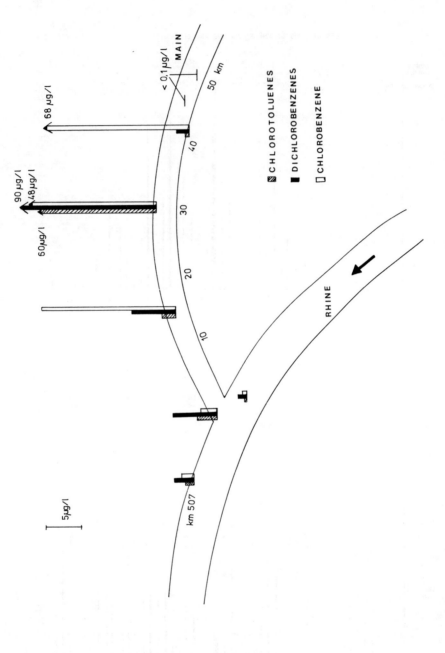

FIG. 3: CONCENTRATION PROFILES FOR SOME CHLOROAROMATICS ALONG THE RIVER MAIN

SESSION III - SPECIFIC ANALYTICAL PROBLEMS

Chairmen: G. PIET and R.C.C. WEGMAN

Analytische Bestimmung organischer Halogenverbindungen

Determination of organic sulphur, organic phosphorus and organic nitrogen

Results from Round Robin Exercises of phenolic compounds within the COST 64b bis project

Instrumental analysis of petroleum hydrocarbons

Screening of pollution from a former municipal waste dump at the bank of a Danish inlet

The impact on the ecology of polychlorinated phenols and other organics dumped at the bank of a small marine inlet

Organic micropollutants from the pulp industry: analysis of lake Päijänne water

Ground water pollution by organic solvents and their microbial degradation products

Comparison of three methods for organic halogen determination in industrial effluents

Organic micropollutants in surface waters of the Glatt Valley, Switzerland

Species and persistence of pollutants in the pond water from an orchard area treated with organophosphorus pesticides

Occurrence and origin of brominated phenols in Barcelona's water supply

The relationship between the concentration of organic matter in natural waters and the production of lipophilic volatile organohalogen compounds during their chlorination

ANALYTISCHE BESTIMMUNG ORGANISCHER HALOGENVERBINDUNGEN

M. SCHNITZLER und W. KOHN
DVGW-Forschungsstelle am Engler-Bunte-Institut, Universität Karlsruhe
Bundesrepublik Deutschland

Summary

The principles of different analytical methods for the measurement of organic halogen compounds in water are described and the results are compared. All methods commonly enclose the following four steps: Enrichment, separation of interfering inorganic substances, mineralization and detection of the mineralized hetero atoms. Numerous determinations of waters of quite different quality have clearly shown, that the measurement of the adsorbable organic halogen (AOX) leads to the most complete enrichment and to values which are very close or identical to the total organic halogen (TOX). Therefore AOX is a suitable parameter for an adequate description of the water quality in the water circle. The completion of the enrichment by adsorption applying a batch technique can be controlled by analysis or calculation. Commonly the enrichment efficiency exceeds 90 % for one adsorption step. Only a part of the compounds which are measured by the AOX-method can be enriched by fluid-fluid-extraction (EOX) or (POX) by purging. It is proposed to measure the organic halogen compounds by the AOX-method only or by a combination of POX and AOX.
Moreover the principle and some results of a new method for the simultaneous determination of organic halogen and sulphur are shown. This procedure enables discrimination of the different halides as well.

Zusammenfassung

Verschiedene Methoden zur summarischen Erfassung organischer Halogenverbindungen in Wässern werden im Prinzip beschrieben und die Meßergebnisse verglichen. Alle Verfahren weisen in der Regel die Teilschritte Anreicherung, Abtrennung anorganischer Störstoffe, Mineralisation und Detektion der mineralisierten Heteroatome auf. Zahlreiche Analysen sehr unterschiedlicher Wässer haben ergeben, daß die Bestimmung des adsorbierbaren organisch gebundenen Halogens (AOX) zur am weitestgehenden Erfassung führt und somit einen Wert liefert, der dem gesamten organisch gebundenen Halogen (TOX) am ehesten entspricht. Daher ist die AOX-Methode für eine aussagefähige Beschreibung der Wasserqualität im Wasserkreislauf geeignet. Die Vollständigkeit der Anreicherung durch Adsorption kann analytisch und auch rechnerisch überprüft werden. In der Regel liegt der Anreicherungsgrad einer Adsorptionsstufe bei über 90 %. Über eine Flüssig-Flüssig-Extraktion (EOX) oder durch Ausblasen (POX) wird nur ein Teil der Verbindungen erfaßt, die auch mit der AOX-Methode angereichert werden. Es wird vorgeschlagen, die organischen Halogenverbindungen mit der AOX-Methode allein oder mit einer Kombination von POX und AOX summarisch zu bestimmen.
Ferner wird das Prinzip einer neuen Methode zur gemeinsamen Bestimmung des organisch gebundenen Halogens und Schwefels vorgestellt, das gleichzeitig die Unterscheidung zwischen den verschiedenen Halogenen ermöglicht.

1. Einleitung

Das Vorkommen organischer Halogenverbindungen in Wässern ist in der Regel nahezu vollständig auf anthropogene Einwirkungen zurückzuführen. Diese Verbindungen können daher als geeignete Leitsubstanzen für eine industriell bedingte Umweltbelastung angesehen werden. Zahlreiche der als Einzelsubstanzen in Wässern analysierten organischen Halogenverbindungen sind als hygienisch bedenklich anzusehen, da sie toxische oder auch mutagene bzw. cancerogene Wirkungen zeigen (1). Auch bei Einsatz aufwendiger Analysensysteme können derzeit etwa nur 10 % des in Oberflächenwässern vorhandenen organisch gebundenen Halogens definierten Einzelsubstanzen zugeordnet werden. Bei den identifizierbaren Verbindungen handelt es sich mengenmäßig vor allem um relativ flüchtige Substanzen (2).

In Abbildung 1 ist die Relation zwischen identifizierten und nichtidentifizierten organischen Halogenverbindungen schematisch dargestellt.

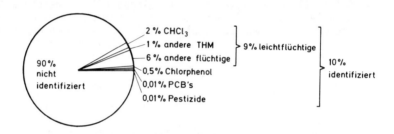

Abbildung 1: Identifizierbarkeit organischer Chlorverbindungen in Wässern (Quelle: OECD)

Da die als Einzelsubstanzen nicht identifizierten restlichen 90 % der organischen Halogenverbindungen nicht ohne weiteres für hygienisch unbedenklich zu erklären sind, ist es zweckmäßig, die Belastung eines Wassers mit organischen Halogenverbindungen über den summarisch bestimmten Gruppenparameter Organische Halogenverbindungen zu beschreiben. Ein solcher Parameter stellt, sofern mit ihm die Gesamtheit dieser Verbindungen erfaßt wird, einmal einen Indikator für die Wasserqualität dar und zum anderen ist er gut geeignet, die Wirksamkeit von Aufbereitungsmaßnahmen zu beurteilen und Aufbereitungsverfahren wie beispielsweise Oxidationsverfahren zu steuern (3).

Für die summarische Bestimmung müssen die organischen Verbindungen in der Regel aus der Wasserprobe angereichert und von den anorganischen Halogeniden abgetrennt werden. Über eine Anreicherung der organischen Wasserinhaltsstoffe durch Ausblasen wird das ausblasbare organisch gebundene Halogen POX (Purgeable Organic Halogen), durch Flüssig-Flüssig-Extraktion das extrahierbare organisch gebundene Halogen EOX (Extractable Organic Halogen) und durch Adsorption das adsorbierbare organisch gebundene Halogen AOX (Adsorbable Organic Halogen) erhalten (4). Diese Anreicherungsverfahren dienen gleichzeitig zur Abtrennung der anorganischen Halogenide. Beim POX ist diese Trennung ohne besondere Maßnahmen sichergestellt, beim EOX ist sie nur bei vollständiger Phasentrennung und Einsatz eines geeigneten Extraktionsmittels gewährleistet. Bei der AOX-Bestimmung kann sie durch besondere Maßnahmen ohne großen Aufwand sichergestellt werden (5).

Bislang liegen keine Ergebnisse vergleichender Untersuchungen an Oberflächen-, Grund- oder Abwässern vor, nach denen einer der drei, letztlich nur operationell definierten Parameter wegen seiner besonderen hygienischen Bedeutung bevorzugt werden sollte. Erstes Ziel summarischer Analysen muß daher sein, die organischen Halogenverbindungen möglichst vollständig zu erfassen, so daß der Meßwert möglichst dem TOX (total OX) entspricht. Diesem Anspruch kann der POX nur bei den wenigen Wässern genügen, in denen überwiegend leichtflüchtige organische Halogenverbindungen vorliegen. Daher werden im folgenden die Methoden der EOX- und AOX-Bestimmung gegenübergestellt.

2. Vergleich von EOX- und AOX-Bestimmung

Bei der EOX-Bestimmung wird zu dem zu untersuchenden Wasser ein organisches Lösungsmittel gegeben, das mit Wasser nur wenig mischbar ist. Die zwei Phasen werden durch Rühren oder Schütteln in innigen Kontakt gebracht. Die organische Phase wird abgetrennt, verbrannt und das dabei gebildete Halogenid ermittelt. Um eine möglichst weitgehende Erfassung der organischen Verbindungen zu erreichen, werden zweistufige Extraktionen, auch unter Verwendung von verschiedenen Lösungsmitteln angewendet (4). Bei der Anwendung verschiedener Lösungsmittel kann allerdings das Ausmaß der Anreicherung über die Ermittlung des Verteilungskoeffizienten nicht rechnerisch überprüft werden.

Bei der AOX-Bestimmung wird heute in der Regel Aktivkohle als Sorbens verwendet (4). Alle angewendeten Methoden stellen im wesentlichen Variationen der am Engler-Bunte-Institut von KÜHN entwickelten Verfahrensweise dar (6, 7).

Zur AOX-Bestimmung wird die Wasserprobe angesäuert und mit Nitrat vernetzt. Dadurch wird die Anreicherung der organischen Verbindungen optimiert und über einen großen Konzentrationsbereich die Chloridadsorption vollständig unterdrückt (5, 6, 7). Nach Zugabe von pulverförmiger Aktivkohle wird die Suspension zur Gleichgewichtseinstellung mindestens 1 h geschüttelt. Die beladene Aktivkohle wird über eine Membran abfiltriert und der Kohlekuchen zur Sicherstellung der Chloridentfernung mit Nitratlösung behandelt (7).

Neben der beschriebenen Schütteltechnik werden auch Minisäulen, die mit körniger Aktivkohle gefüllt sind, zur Anreicherung verwendet (8, 9). Diese Methode hat den Nachteil einer kurzen, invariablen Kontaktzeit. Außerdem kann eine Kanalbildung in der Säule nicht ausgeschlossen werden. Ferner kann die Vollständigkeit der Anreicherung nicht rechnerisch überprüft werden, da kein auswertbares Adsorptionsgleichgewicht erreicht wird (10).

Nach der Anreicherung wird die beladene Aktivkohle im Sauerstoffstrom bei 1000 °C verbrannt (7). Die dabei entstehenden Halogenide mit Ausnahme von Fluorid werden als Chlorid "on line" mikrocoulometrisch erfaßt. Für die Verbrennung und Chloridbestimmung hat sich eine Apparatur bewährt, die von KIWA und EBI entwickelt worden ist. Der Verbrennungsteil ist in Abbildung 2 dargestellt.

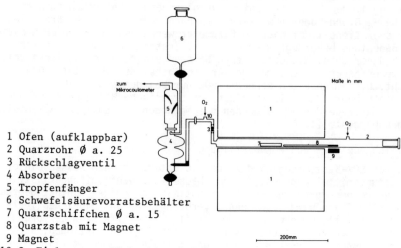

1 Ofen (aufklappbar)
2 Quarzrohr Ø a. 25
3 Rückschlagventil
4 Absorber
5 Tropfenfänger
6 Schwefelsäurevorratsbehälter
7 Quarzschiffchen Ø a. 15
8 Quarzstab mit Magnet
9 Magnet
10 O_2-Einlass zur Rückschlagsicherung

Abbildung 2: Apparatur zur Verbrennung der beladenen Aktivkohle bei der AOX-Bestimmung

Sie besteht im wesentlichen aus einem beheizten Quarzrohr, das über einen Wäscher, der mit konzentrierter Schwefelsäure gefüllt ist, mit der Titrationszelle eines Mikrocoulometers verbunden ist.

Um die Bedeutung der gewählten Anreicherungsmethode für die Wiederfindungsrate in Wässern zu zeigen, wurden verschiedene Wässer mit der EOX- und AOX-Methode vergleichend untersucht. Die AOX-Bestimmung wurde nach der am Engler-Bunte-Institut entwickelten Verfahrensweise durchgeführt, die EOX-Bestimmung nach einem DIN-Entwurf, der eine zweistufige Extraktion mit Pentan und anschließend mit Diisopropyläther beinhaltet (7, 11). Die dabei erhaltenen Ergebnisse sind in Tabelle 1 aufgeführt.

Tabelle 1: Vergleich von EOX und AOX für unterschiedliche Wässer

Probe	EOX µg/l	AOX µg/l	EOX/AOX %
Rhein, Maxau	34	60	57
Main	51	435	12
Rhein, Düsseldorf	30	199	15
Wolfegger Aach	50	626	8
Abwasser A	490	3425	14
Abwasser B	740	3355	22
Abwasser C	1630	9685	17

Aus den Ergebnisse in Tabelle 1 wird ersichtlich, daß bei den untersuchten Wässern über die EOX-Bestimmung maximal nur etwa 60 % der über eine Adsorption an Aktivkohle erfaßten Menge AOX gefunden werden. Bei den nichtflüchtigen Verbindungen in Abwässern aus der Zellstoffindustrie beträgt der EOX-Wert maximal nur 2 % des AOX, wie die Ergebnisse in Tabelle 2 zeigen.

Tabelle 2: Vergleich von EOX und AOX für Zellstoffabwässer nach Ausblasen der flüchtigen Verbindungen

EOX µg/l	AOX µg/l	EOX/AOX %
8	2240	0,4
23	2100	1,1
1558	122500	1,3
360	21500	1,7
1183	59600	2,0

Ausgehend von den beschriebenen Beobachtungen, die frühere Ergebnisse (12) bestätigen, erscheint allein der AOX für eine möglichst vollständige summarische Bestimmung der organischen Halogenverbindungen in Wässern geeignet zu sein. Darüber hinaus können mit der AOX-Methode ungelöste organische oder auch an Feststoffen adsorbierte organische Halogenverbindungen mit erfaßt werden. Über eine Feststoffabtrennung vor dem Anreicherungsschritt durch Filtration oder Zentrifugation können die Feststoffe auch getrennt untersucht werden. Unlösliche anorganische Chloride, von denen es allerdings nur wenige gibt, werden bei der Feststoffuntersuchung mit erfaßt.

3. Vollständigkeit der Erfassung bei der AOX-Bestimmung

Auch wenn über die Anreicherung durch Adsorption an Aktivkohle die organischen Halogenverbindungen am weitestgehenden erfaßt werden, ist damit allein eine vollständige Erfassung noch nicht sichergestellt oder bewiesen. Um schließlich einen richtigen Wert für das gesamte organisch gebundene Halogen TOX zu erhalten, muß bei Wässern, bei denen das Adsorptionsverhalten der gelösten Wasserinhaltsstoffe unbekannt ist, die Vollständigkeit der Anreicherung überprüft werden. Dazu können folgende Methoden angewendet werden (13, 14):

1) Über eine Messung des gelösten organisch gebundenen Kohlenstoffs DOC vor und nach der Anreicherung wird ein Maß für den Anreicherungsgrad bezüglich der gesamten organischen Substanz erhalten. Bei Einhaltung der Randbedingungen (7) wird in einem Adsorptionsschritt in der Regel die organische Substanz zu mehr als 90 %, gerechnet als DOC, erfaßt.

2) Über eine zweite Adsorption im Filtrat nach der ersten Anreicherung kann der Anreicherungsgrad für die organischen Halogenverbindungen analytisch überprüft und schließlich auch das gesamte organisch gebundene Halogen TOX berechnet werden (5).

3) Durch die Anwendung unterschiedlicher Verhältnisse von Kohlemenge zu Probenvolumen ergeben sich verschiedene Beladungen, aus denen sich ebenfalls der TOX berechnen läßt (13).

Voraussetzung für die letzten beiden Methoden sind die Einstellung des Sorptionsgleichgewichts und die Gültigkeit einer linearen bzw. Freundlich-Isotherme zur Beschreibung des Zusammenhangs zwischen Beladung und Konzentration. Außerdem wird vorausgesetzt, daß alle Substanzen in einem meßbaren Anteil erfaßt werden (13). Aus den in Abbildung 3 dargestellten Isothermen für chlorierte Essigsäurederivate ist zu erkennen, daß selbst diese polaren Verbindungen unter den Bedingungen der AOX-Bestimmung (L/m = 2 l/g) merklich adsorbiert werden (13).

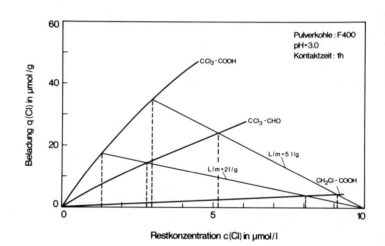

Abbildung 3: Isothermen chlorierter Essigsäurederivate

Die Methode der Mehrfachadsorption ist einfach durchzuführen und führt zu anschaulichen Ergebnissen. In Tabelle 3 sind Ergebnisse von Untersuchungen an Abwässern aus der Chemischen- und Zellstoff-Industrie zusammengestellt (5). Die unterschiedlichen Verdünnungsfaktoren ergeben sich aus den jeweiligen DOC-Werten der untersuchten Wässer.

Die aufgeführten Ergebnisse lassen erkennen, daß der Meßwert nach dem ersten Adsorptionsschritt AOX_1 in allen Fällen über 90 % der Summe AOX_1 + AOX_2 beträgt. Ferner führt nur bei den beiden Zellstoffabwässern, die zuletzt aufgeführt sind, die zweite Anreicherungsstufe zu einer signifikanten Erhöhung der Summe AOX_1 + AOX_2. Eine rechnerische Auswertung der Ergebnisse der Zweifachadsorption unter den oben aufgeführten Annahmen ergibt dabei, daß selbst bei den schlecht adsorbierbaren Wasserinhaltsstoffen der Zellstoffabwässer die Summe AOX_1 + AOX_2 dem TOX gleichgesetzt werden kann; bei den anderen Wässern entspricht bereits der Wert AOX_1 dem TOX (5).

Tabelle 3: Ergebnisse der Überprüfung der Vollständigkeit der Anreicherung durch Mehrfachadsorption an Abwässern aus verschiedenen Industriebetrieben

Branche	Verdünnungsfaktor	AOX_1 µg/l	AOX_2 µg/l	$AOX_1 + AOX_2$ µg/l	$AOX_1/(AOX_1 + AOX_2)$ %
Chemie	1000	76000	< 8000	< 84000	> 90
Chemie	1	390	< 8	< 398	> 98
Chemie	20	2060	< 160	< 2220	> 93
Chemie	50	7100	< 400	< 7500	> 95
Chemie	10	2670	< 80	< 2750	> 97
Zellstoff	20	20000	1600	21600	93
Zellstoff	100	122000	13000	135000	90

Die Ergebnisse bestätigen, daß das Ergebnis der AOX-Bestimmung in den meisten Fällen dem TOX-Wert gleichgesetzt werden kann und daher die AOX-Bestimmung für die vollständige summarische Analyse der organischen Halogenverbindungen gut geeignet ist. Aus den gezeigten Ergebnissen wird außerdem deutlich, daß mit der AOX-Methode auch die polaren organischen Verbindungen erfaßt werden.

4. Erfassung flüchtiger Organohalogenverbindungen

Ein Teil der über die AOX-Bestimmung erfaßbaren Verbindungen läßt sich auch aus der Wasserprobe ausblasen (5, 15). Dazu wird am Engler-Bunte-Institut die in Abbildung 4 dargestellte Apparatur verwendet.

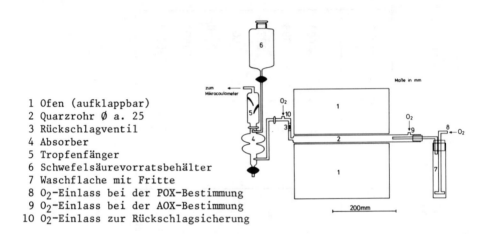

1 Ofen (aufklappbar)
2 Quarzrohr Ø a. 25
3 Rückschlagventil
4 Absorber
5 Tropfenfänger
6 Schwefelsäurevorratsbehälter
7 Waschflache mit Fritte
8 O_2-Einlass bei der POX-Bestimmung
9 O_2-Einlass bei der AOX-Bestimmung
10 O_2-Einlass zur Rückschlagsicherung

Abbildung 4: Apparatur zur Messung des POX

Die Apparatur besteht im wesentlichen aus einer Waschflasche, die über ein beheiztes Quarzrohr mit einem Mikrocoulometer verbunden ist. Sie entspricht in ihren Abmessungen der Pyrolyseapparatur zur Bestimmung des AOX (7, 16). Zur Bestimmung des ausblasbaren organisch gebundenen Halogens POX wird ein Aliquot der Originalprobe bzw. eine Verdünnung in die Waschflasche gefüllt. Durch die Probe in der Waschflasche wird ein Sauerstoffstrom geleitet, der zusammen mit den ausgeblasenen Verbindungen in den Ofen gelangt. Hier werden bei 1000 °C die organischen Verbindungen mineralisiert. Zur Wasserentfernung durchströmen die Reaktionsgase wie bei der AOX-Bestimmung einen Absorber und gelangen in die Titrationszelle eines Mikrocoulometers.

Um Informationen über die Wirksamkeit des Ausblasens und die Art der ausgeblasenen Verbindungen zu erhalten, wurden Grundwässer gaschromatographisch über eine Extraktion mit Pentan und mit der beschriebenen Ausblasmethode vergleichend untersucht. In Tabelle 4 sind die Ergebnisse dieser Untersuchung aufgeführt. Angegeben ist jeweils die Summe der gaschromatographisch bestimmten Einzelsubstanzen (GC), berechnet als Chlor, und der summarisch ermittelte Wert POX (5).

Tabelle 4: Vergleich von POX und der Summe der gaschromatographisch ermittelten Einzelsubstanzen, gerechnet als Chlor - Grundwässer

GC µg/l	POX µg/l	POX/GC -
18	22	1,22
41	39	0,95
88	75	0,85
183	185	1,01
286	245	0,86
290	230	0,79
348	315	0,91

Aus den aufgeführten Ergebnissen für Grundwässer wird deutlich, daß über das Ausblasen die gaschromatographisch analysierbaren Verbindungen mit guter Wiederfindungsrate erfaßt werden. Bei Abwässern trifft dieser Sachverhalt, je nach deren Gehalt an Emulgatoren und Lösungsvermittlern, nicht immer zu (5).

Die ausblasbaren, flüchtigen Verbindungen können auch gemeinsam mit den nichtflüchtigen über eine Adsorption an Aktivkohle erfaßt werden, wie die in Tabelle 5 aufgeführten Ergebnisse belegen.

Tabelle 5: Vergleich der direkten AOX-Bestimmung mit der kombinierten POX-AOX-Bestimmung - Grundwässer

POX µg/l	NVAOX x) µg/l	POX + NVAOX µg/l	AOX xx) µg/l	AOX/(POX+NVAOX) -
29	9	38	43	1,13
60	13	73	79	1,08
70	4	74	74	1,00
100	11	111	108	0,97
146	6	152	156	1,03
230	18	248	250	1,01
315	15	330	323	0,98

x) nach Strippen
xx) direkt und gesamt

Hier wurde in Grundwässern das ausblasbare organisch gebundene Halogen POX und anschließend in der ausgeblasenen Probe das nichtflüchtige adsorbierbare organisch gebundene Halogen NVAOX (non volatile AOX) ermittelt. Außerdem wurde das adsorbierbare organisch gebundene Halogen AOX in der Originalprobe direkt bestimmt. Aus der Übereinstimmung der Werte für das direkt bestimmte adsorbierbare organisch gebundene Halogen AOX mit der Summe von ausblasbarem und nichtflüchtigem adsorbierbaren organisch gebundenem Halogen POX + NVAOX wird deutlich, daß die flüchtigen Verbindungen ebenfalls über eine Anreicherung auf Aktivkohle erfaßt werden können. Wenn die leichtflüchtigen Verbindungen über die AOX-Bestimmung miterfaßt werden sollen, muß die Probe so genommen, gelagert und behandelt werden, daß keine Verluste bei den flüchtigen Verbindungen auftreten.

Sollen suspendierte Feststoffe getrennt auf ihren Gehalt an organischen Halogenverbindungen untersucht werden, ist es zweckmäßig zuerst das ausblasbare organisch gebundene Halogen POX zu ermitteln, anschließend die Feststoffe durch Membranfiltration oder Zentrifugation abzutrennen und schließlich das nichtflüchtige adsorbierbare organisch gebundene Halogen NVAOX zu bestimmen. Die Ergebnisse der getrennten Bestimmung von POX und NVAOX können außerdem bei der Entscheidung für Wasseraufbereitungsmaßnahmen zur Entfernung organischer Halogenverbindungen herangezogen werden. Bei vielen Abwässern überwiegt der nichtflüchtige Anteil. In mit organischen Halogenverbindungen belasteten Grundwässern werden dagegen vor allem flüchtige Organohalogenverbindungen gefunden.

5. Detektion des mineralisierten organisch gebundenen Halogens

Bei der summarischen Bestimmung des organisch gebundenen Halogens stellt die Detektion der mineralisierten organischen Halogene, im wesentlichen Chlorid, kein Problem dar. Zur Detektion der gebildeten Halogenide hat sich heute die Mikrocoulometrie weitgehend durchgesetzt (4). Bei der POX- und AOX-Bestimmung kann das Mikrocoulometer "on line" mit der Pyrolyseapparatur betrieben werden. Ein Nachteil der Mikrocoulometrie besteht darin, daß organisch gebundenes Fluor nicht erfaßt wird und daß eine Unterscheidung zwischen Brom und Chlor nicht möglich ist. Zur Unterscheidung der verschiedenen Halogenide haben ZÜRCHER und LUTZ bei der summarischen Bestimmung des POX einen Ionenchromatographen im "off-line" Betrieb eingesetzt (17).

6. Gemeinsame Bestimmung des organisch gebundenen Halogens und Schwefels

Da bei einer Anreicherung durch Adsorption an Aktivkohle alle organischen Wasserinhaltsstoffe erfaßt werden können, lassen sich nach folgendem Konzept sowohl die organisch gebundenen Halogene als auch der organisch gebundene Schwefel nach SCHNITZLER bestimmen (5, 14, 18, 19).

1) Anreicherung der organischen Wasserinhaltsstoffe auf einer schwefel- und halogen- sowie aschearmen Aktivkohle.
2) Verdrängung der störenden Chloride und Sulfationen durch Nitrat.
3) Verbrennung der beladenen Aktivkohle im Sauerstoffstrom und Absorption der sauren Reaktionsgase in einem basischen, wasserstoffperoxidhaltigem Puffer.
4) Ionenchromatographische Detektion der gebildeten Halogenide und von Sulfat.

Zentrales Problem bei der Verwirklichung dieses Konzepts war die Herstellung einer geeigneten, schwefelarmen Aktivkohle. Über die Verkokung von Zucker konnte schließlich eine Aktivkohle hergestellt werden, die für die gemeinsame Analyse von organisch gebundenen Halogenen und Schwefel geeignet ist. In Abbildung 5 ist unter zwei Standardionenchromatogrammen ein typisches Ergebnis einer gemeinsamen AOX/AOS-Bestimmung zu sehen.

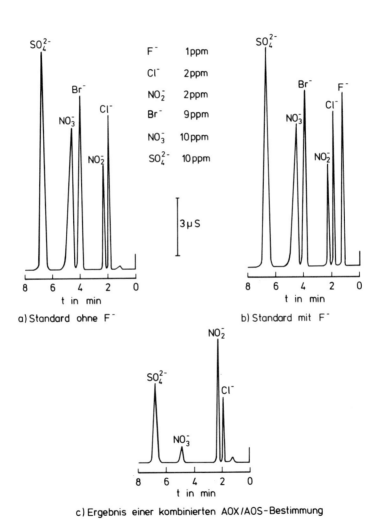

Abbildung 5: Zur ionenchromatographischen Detektion bei der gemeinsamen AOX/AOS-Bestimmung

Ergebnisse der gemeinsamen Bestimmung von AOCl und AOS in Wässern sind in Tabelle 6 aufgeführt (5).

Tabelle 6: Ergebnisse der gemeinsamen Bestimmung von AOCl und AOS

Wasserprobe	AOCl µg/l	AOS µg/l
Blindwert (n = 6)	18,1 ± 3,2	15,3 ± 2,9
Abwasser	343	92
Abwasser	180	30
Rhein, Rheinfelden	47	69
Rhein, Maxau	84	70
Rhein, Düsseldorf	51	93
Uferfiltrat, Düsseldorf	22	38
Ozontes Uferfiltrat	19	25
Grundwasser	1122	< 10
Trinkwasser	34	22

Es wird deutlich, daß bei den untersuchten Wässern die Belastung mit organischen Chlor- und Schwefelverbindungen sehr unterschiedlich ist. Es erscheint daher zweckmäßig zur weitergehenden Beschreibung organischer Wasserinhaltsstoffe sowohl das organisch gebundene Halogen als auch den organisch gebundenen Schwefel zu ermitteln. Mit der zuletzt beschriebenen Methode kann dies in einem Arbeitsgang durchgeführt werden (5).

7. Schlußfolgerungen

Wie die Ausführungen gezeigt haben, bestimmt die angewendete Anreicherungsmethode letztlich das Ausmaß der Erfassung der organischen Halogenverbindungen. Über eine Anreicherung durch Adsorption an Aktivkohle werden flüchtige und nichtflüchtige sowie polare und unpolare Verbindungen erfaßt. Die AOX-Methode führt daher zu einem Wert, der dem gesamten organisch gebundenen Halogen TOX am nächsten kommt und in vielen Fällen diesem gleichgesetzt werden kann. Außerdem werden über die AOX-Methode auf einfache Weise auch die polaren Verbindungen erfaßt, deren Entfernung bei der Trinkwasseraufbereitung Schwierigkeiten bereitet. Um Auswirkungen von Abwassereinleitungen auf Fluß-, Grund- oder Trinkwasserqualität beurteilen zu können, erscheint es daher zweckmäßig, auch bei Abwässern die AOX-Methode anzuwenden, da nur dann eine aussagefähige Beschreibung der Wasserqualität im Wasserkreislauf vorgenommen werden kann.

LITERATUR

(1) -
Water chlorination: Environmental impact and health effects, Vol 4, Book 2. Ann Arbor Science Publishers, Ann Arbor, Michigan, 1983
(2) KÜHN, W.: Organische Chlorverbindungen in Wässern - Analytik und Herkunft in: Aktuelle Probleme der Wasserchemie und Wasseraufbereitung. Veröffentlichungen des Bereichs und des Lehrstuhls für Wasserchemie der Universität Karlsruhe, Heft 20, ZfGW-Verlag, Frankfurt, 1982.
(3) -
Oxidationsverfahren in der Trinkwasseraufbereitung. Hrsg: Kühn, W. und Sontheimer, H., ZfGW-Verlag, Frankfurt, 1979.
(4) WEGMANN, R.C.C.: Determination of organic halogens; a critical review of sum parameters, S. 249 - 263, in: Analysis of Organic Micropollutants in Water (Hrsg.: A Bjørseth und G. Angeletti), D. Reidel Publishing Company, Dordrecht, Boston, London, 1982.
(5) SCHNITZLER, M.; LEVAY, G.; KÜHN, W. und SONTHEIMER, H.: Zur summarischen Erfassung organischer Halogen- und Schwefelverbindungen in Wässern. Vom Wasser 61 (1983), im Druck.
(6) KÜHN, W.: Untersuchungen zur Bestimmung von organischen Chlorverbindungen auf Aktivkohle. Dissertation, Universität Karlsruhe, 1974.
(7) Analytische Erfassung organischer Chlorverbindungen. Veröffentlichungen des Bereichs und des Lehrstuhls für Wasserchemie der Universität Karlsruhe, Heft 15, ZfGW-Verlag, Frankfurt, 1980.
(8) TAKAHASHI, Y.: A review of analysis techniques for organic carbon and organic halide in drinking water. Conference on practical application of adsorption techniques in drinking water. Reston, V.A., USA, 1979.
(9) DRESSMAN, R.C.; NAJAR, B.A. und REDZIKOWSKI, R.: The analysis of organohalides (OX) in water as a group parameter. Paper presented at the Water Quality Technology Conference, American Water Works Association, Philadelphia, P.A., 1979.
(10) SCHNITZLER, M. und PIET, G.J.: Guidelines for the determination of Adsorbable Organic Halogen in waters. Proceedings of Working Party 8, Cost Project 64 B bis, Voorburg, The Netherlands, March 1983 (Hrsg.: Commission of the European Communities, Directorate-General for Science, Research and Development, Joint Research Centre).
(11) DIN: Entwurf - Summarische Wirkungs- und Stoffkenngrößen (Gruppe H), Bestimmung der extrahierbaren organisch gebundenen Halogene (EOX) (H 8), DIN 38409 Teil 8. Beuth Verlag, Berlin, 1982.
(12) IAWR: Rheinbericht 1979/1980.
(13) SONTHEIMER, H. und SCHNITZLER, M.: EOX oder AOX? - Zur Anwendung von Anreicherungsverfahren bei der analytischen Bestimmung von chemischen Gruppenparametern. Vom Wasser 59, 169 - 179, 1982.
(14) SCHNITZLER, M.: Control-Methods for the Enrichment of Organic Water Pollutants by Adsorption onto Activated Carbon in the Determination of Adsorbable Organic Sulfur (AOS). Proceedings of Working Party 8, Cost Project 64 B bis, Voorburg, The Netherlands, March 1983. (Hrsg.: Commission of the European Communities, Directorate - General for Science, Research and Development, Joint Research Centre).
(15) JEKEL, M. und ROBERTS, P.V.: Total Organic Halogen as Parameter for the Characterization of Reclaimed Waters: Measurement, Occurence, Formation and Removal. Environm. Sci. Techn. 14, 970 - 975, 1980.

(16) SANDER, R.: Untersuchungen zum optimalen Einsatz von Chlor bei der Aufbereitung von Oberflächenwässern. Dissertation, Universität Karlsruhe (TH), 1981.
(17) ZORCHER, F. und LUTZ, F.: Simultaneous Determination of Total Volatile Organic Chlorine,- Bromine and Fluorine Compounds in Water by Ion Chromatography. Workshop on Ion-Chromatography, Petten, The Netherlands, June 1980.
(18) SCHNITZLER, M.: Entwicklung einer Methode zur Bestimmung des gelösten organisch gebundenen Schwefels in Wässern. Dissertation, Universität Karlsruhe, 1982.
(19) SCHNITZLER, M. und SONTHEIMER, H.: Eine Methode zur Bestimmung des gelösten organisch gebundenen Schwefels in Wässern (DOS). Vom Wasser 59, 159 - 167, 1982.

DETERMINATION OF ORGANIC SULPHUR, ORGANIC PHOSPHORUS
AND ORGANIC NITROGEN

G. Veenendaal
KIWA Ltd.
The Netherlands Waterworks' Testing and Research Institute Ltd.

Summary

A review is given of methods for the determination of the
group-parameters organic sulphur, phosphorus and nitrogen
in water samples. Special attention is paid to those techniques which include an isolationstep. The methods used
at Engles Bunte Institut and KIWA are given, including
some practical results and possibilities for application.

1. INTRODUCTION

Unlike those of chlorine, organic compounds of sulphur,
phosphorus and nitrogen are very common in nature. This means
that measurements of the total content of organo-sulphur, organo-phosphorus and organo-nitrogen give no direct information
on any contaminant compounds present. The interpretation of
determinations of these compounds is therefore much more difficult than that of organic-chlorine. In this paper we discuss
techniques for the measurements of these parameters. The interpretation of such measurements and their applications will
also be dealt with here.

2. HISTORICAL

Organo-phosphorous and organo-nitrogen in water have been
measured for many years as total organo-phosphorus and
kjeldahl nitrogen. In both methods the measurement is made
directly in the water, the organic content being derived from
the difference between the total and the inorganic contents.
In 1979 Brull and Golden (1) developed a method for the determination of the total sulphur in water by means of micro-coulometry. When the inorganic sulphur (sulphate) can be determined independently, the organic sulphur content is obtained
from the difference.
    These parameters give information on the content of various naturally occurring compounds. As these methods are difference methods, their detection limits are relatively high.
We shall not discuss them further.
    Methods have been developed in recent years in which the
organic compounds are removed from the water by an isolation
technique, after which the organo-sulphur, phosphor or nitrogen contents of the isolate can be determined. These techniques yield methods that are sensitive and selective.
    A method for the determination of organo-sulphur was pu-

blished by Schweer and Fuchs (7) in 1975. In this method the organic compounds are extracted from the water by means of a packed column of granular carbon. The carbon is then eluated with an organic solvent and the sulphur determined in this extract. A similar method was described by Otson (2) in 1979. Research on organo-sulphur is carried out at present mainly at the Engler-Bunte Institute and at KIWA. At the KIWA, work is also being carried out on organo-phosphorus and organo-nitrogen.

3. METHODS

The work at the Engler-Bunte Institute is concerned with the AOS method developed by Schnitzler (5). The determination of AOS involves four steps.:
1. Extraction and concentration by adsorption on carbon powder.
2. Removal of interfering anions.
3. Destruction by combustion whereby organo-sulphur is converted to the sulphate.
4. Micro-coulometric determination of sulphate.

For the extraction of organic compounds from water, one litre of the sample is taken. The pH adjusted to 3 and the nitrate concentration is increased to about 10 m mole/l by means of nitric acid or sodium hydroxide and sodium nitrate. This ensures the optimum adsorption of organic compounds and the adsorption of interfering anoins is almost completely suppressed. The sample is then shaken with 1 gram of powdered carbon. The carbon is filtered off and then shaken up with a nitrate solution (10 m mole/l) to extract the sulphate. After filtering the carbon is burnt at 1000 °C in oxygen in the apparatus shown in figure 1.

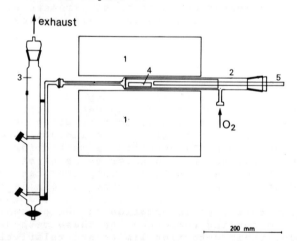

Figure 1   Pyrolysis apparatus for the AOS determination

The pyrolysis apparatus consists of a furnace (1) which heats a silica tube (2) connected to a two stage adsorber (3). The carbon is placed in a quarz boat (4) and pushed into the furnace by means of a quarz rod (5).

Oxygen is passed through the reaction tube and organic sulphur is converted into SO_2 en SO_3. These gases pass into the adsorber which is filled with hydrogen peroxide and are thereby converted into sulphate. The hydrogen peroxide solution is then concentrated by evaporation, potassium permanganate being added as catalyst to remove the excess peroxide. The quantity of sulphate is now determined micro-coulometrically after the catalytic thermal conversion of sulphate into SO_2 (6). The coulometric apparatus is shown in figure 2.

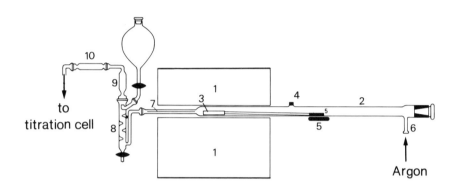

Figure 2   Apparatus for the determination of sulphate

A proportional fraction of the sulphate solution is injected via a septum (4) into a quarz boat (3) which can be moved into and out of the quarz tube (2) by means of a magnet (5).The quarz tube is heated in a furnace to 1000 °C. Argon is passed through the furnace and the sulphate is converted into SO_2 in the hot zone of the tube. The SO_2 gas is carried out off in the argon and passes via a phosphoric acid scrubber (8) and a silver-wool scrubber (10) to the titration cell of the micro-coulometer. The electrolyte in the cell consists of acetic acid (70 %), potassium chlorate and water. The SO_2 reduces the chlorate to chloride. The chloride is titrated micro-coulometrically by the generation of silver ions from a silver anode.
With a water sample of 1 litre, the limit of detection of this method of analysis is about 50 ug/l

Methods have been developed at the KIWA for the analysis of organo-sulphur, phosphorus and nitrogen in organic solvents. The organic compounds are removed from the water by extraction or by adsorption on synthetic resins, followed by elution with an organic solvent.
The determination of organo-nitrogen by this method is illustrated schematically in figure 3.

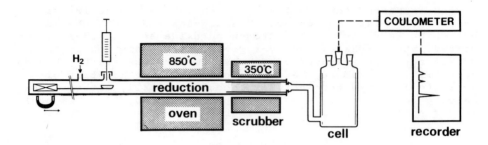

Figure 3 Apparatus for the determination of organic nitrogen

A quantity of the extract (50 ul maximum) is introduced into a small quarz boat by means of a micro-syringe (1). The boat is then slowly pushed into the furnace (2), so slowly that the solvent evaporates before the boat reaches the hot zone. This technique is used because certain extracts (particularly of XAD) contain much material of high molecular weight. This material is of such low volatility that it would never enter the furnace if direct injection were used, but instead would remain in the needle of the syringe or on the walls where injected. In the hot zone of the furnace the boat is at 850 °C. Hydrogen is passed through the furnace and, in the presence of a nickel catalyst, the organic material is converted into CH_4, HCl, H_2S, NH_3, etc. These cracking products are carried along in the gas stream via a lithium hydroxide scrubber (3) to the cell (4). The scrubber removes any acids that may be present, such as HCl and H_2S. The ammonia is determined in the cell by a micro-coulometric acid-base titration.

The limit of detection of this method is about 0.1 mg/l nitrogen in a solvent.

Figure 4 shows schematically a cross-section of the apparatus for the determination of organo-phosphorus and organo-sulpur (3).

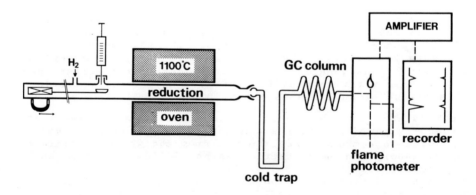

Figure 4 Apparatus for the determination of organic phosphorus and sulphur

The same injection technique with a boat (1) is used as in the case of organo-nitrogen. In the furnace, now at 1100 °C, the organic material is converted into PH_3 (phosphine), H_2S, CH_4. PH_3 and possibly other phosphorous compounds have the peculiarity that they are strongly adsorbed on quarz so that if measurements are made directly at the outlet of the furnace a much smeared-out peak is obtained. To avoid this, the reaction products are caught in a cold-trap (3), a U-tube immersed in liquid air. After 15 minutes, when the greater part (more than 99 %) of the reaction products have left the furnace, the U-tube is heated rapidly to 150 °C. The H_2S, PH_3 and CH_4 are carried along by the gas stream via a short gas-chromatograph column to the detector (4). The latter is a two-channel flame-photometer. One is sensitive to phosphorus and the other to sulphur. The phosphorus channel is affected by sulphur and methane. To overcome this complication the GC-column (5) is introduced. This column, of 60 cm lenght and 5 mm diameter, filled with Porapak T, is sufficient to separate H_2S and methane from the phosphine. For organo-phosphorus the method has a detection limit of 0.1 ng (absolute) and for sulphur 1 ng (absolute). The detection limit in a solvent is 1 mg/l.

4. DETERMINATION IN WATER

Group-parameters are always determined in combination with an isolation technique.

Figure 5 gives a diagrammatic survey of the various isolation techniques used in the investigation into group-parameters (4).

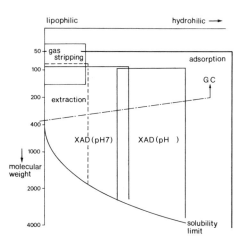

Figure 5   Areas of application of isolation techniques used for the investigation of group-parameters

This figure gives a classification of isolated compounds, according to their chemical and physical properties. The molecular weight is plotted along the y-axis and the degree of lipophilicity of hydrophlicity along the x-axis. The figure gives only a qualitative picture. The numbers shown are intended only as a rough indication.

Extraction with hexane isolates all lipophilic compounds but when concentrating by evaporation, compounds with a molecular weight less than about 80 are lost. With adsorption on XAD at a pH of 7, non-dissociated compounds having at most one polar substitution are isolated. For adsorption on XAD at a pH 2, weakly hydrophilic and acid-reacting compounds having one strong polar substitution or two weak substitutions are isolated. Almost perfect isolation of organic compounds in water is obtained by adsorption on carbon. Only a very small fraction of the organic compounds can be identified with GC/MS. The great majority of the compounds found in natural water have too high a molecular weight and can be determined only by means of other techniques such as HPLC.

Figure 6 gives a survey of the possible combinations of groupparameters and isolation techniques and the corresponding abbreviations.

	EXTRACTION	XAD		ACT. CARBON
		pH = 7	pH = 2	
NITROGEN	EON	X_7ON	X_2ON	
PHOSPHORUS	EOP	X_7OP	X_2OP	
SULPHUR	EOS	X_7OS	X_2OS	AOS

Figure 6   Abbreviations used for the various combinations of isolation techniques and group-parameters

Extraction with hexane isolates primarily the compounds that can be determined by means of GC/MS. These are mainly compounds of non-natural origin.
It is found that EON mainly consists of anilines, pyridines and nitro- compounds.
EOP consists mainly of tri-alkyl phosphates and tri-alkyl thiophosphates.
EOS consists mainly of suphides, disulphides and thiophosphates.

Figure 7 gives a survey of the extractable parameters in Rhine water, in 1979.

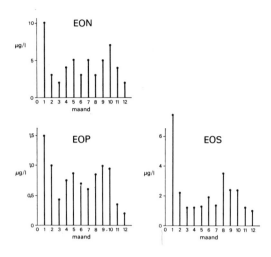

Figure 7   EON, EOP en EOS in Rhine-water in the Netherlands, 1979.

With regard to nitrogen and sulphur, adsorption on XAD isolates mainly the compounds of high molecular weight which are not ameable to gas-chromatography. X_2ON en X_2OS contain more than 90 % of components not determinable by gas-chromatography.

The parameter X_7OP however consists mainly of compounds that are determinable by GC/MS, as a rule non-natural compounds. This is also the case, but although to a lesser degree, for X_2OP. The natural phosphorus compounds, mainly the inositol phosphates, are not isolated by XAD.

Figure 8 shows some results of group-parameter determinations in ground-water.

	Sample no. 1			Sample no. 2			Sample no. 3			Sample no. 4		
	extr.	XAD pH=7	XAD pH=2	extr.	XAD pH=7	XAD pH=2	extr.	XAD pH=7	XAD pH=2	extr.	XAD pH=7	XAD pH=2
ON	0.3	8.5	.	0.2	5.5	.	0.2	9.0	.	0.1	16	.
OP	< 0.3	0.1	0.2	< 0.3	0.3	0.3	< 0.3	0.1	0.1	< 0.3	0.1	0.1
OS	< 1	13	19	< 1	10	15	< 1	14	27	< 1	28	80
DOC	2.5			1.6			3.6			11.5		

Figure 8   Group-parameters in ground-water

Both EON and EOS differ very considerably from XON and XOS respectively while EOP is very low in both cases.
From the results thus far it can be tentatively concluded that the extractable parameters X_7OP and X_2OP consist mainly of contaminant compounds. This group however, includes only a small fraction of all the contaminants in natural water.
The parameters X_7ON, X_2ON, X_7OS and X_2OS consists largely of natural compounds.
As already indicated, using adsorption on carbon gives an estimate of the total content. This method is used only for organo-sulphur.
Some results of AOS determinations are shown in figure 9.

Sample	AOS mg/l	DOC/AOS mole/mole
Waste water A	36,1	10,3
Waste water B	12,6	5,5
Waste water C	0,600	14,6
Rhein Koblenz	0,201	37,2
Rhein Köln	0,196	34,2
Rhein Düsseldorf	0,237	37,4
Bank filtrate, Düsseldorf	0,141	31,1

Figure 9   AOS of waste waters, river-waters and bankfiltrates

From the ratio of AOS to DOC it is found that the content of organo-sulphur in waste waters is higher than that in riverwater of river-bank filtrates.
Up to the present time there have been no comparisons made between determinations of AOS and other organo-sulphur parameters. From comparative experiments with the group-parameter organo-chlorine it is found that isolation by means of XAD is 40-100 % as effective as isolation with carbon.

5. APPLICATIONS

The group-parameters are very useful for:
- Monitoring the quality of river-water and drinking-water.
- Monitoring and control of the water purification process.
- Analysis and monitoring of waste water.
- Special applications, such as the investigation of soil contamination.

The various isolation techniques can be used in all these applications. When information is required about a small fraction of the total, mainly of contaminants, an extraction method can be used. If however the total content is required, then adsorption on carbon is the best method. Compounds are then also determined that are not effectively held by a carbon filter.

A very useful application is the characterization of organic substances in water and the use of the group-parameters in connection with analysis for individual components. The groupparameters give information concerning the content of substances not yet identified. In combination with isolation techniques, information can also be obtained on the physico-chemical properties of these (unknown) compounds, such as the molecular weight.

An advantage of extraction or adsorption on XAD is that, apart from groupparameters, other analyses can be carried out such as GC-MS, HPLC and the Ames-test.

6. FUTURE WORK

Research in this field will be mainly directed towards applications of the group-parameters. At the Engler-Bunte Institute the relation between AOS and DOC in various types of water such as surface water and waste water is under investigation. The behaviour of this parameter in the various steps of process for the purification of drinking water is also being investigated.

At the KIWA investigations are proceeding on the relation between the group-parameters, individual compounds and the Amestest (in the same isolates), together with various isolation techniques. This work is done on a number of different types of water: ground-water, surface-water and water during the purification process.

REFERENCES

1. BRULL, E.E., GOLDEN, G.S. Microcoulometric determination of total inorganic and organic sulphur trace amounts in water. Anal. Chim. Acta; 110 (1979) 1, p. 167-170.
2. OTSON, R., BOTHWELL, P.D., WILLIAMS, O.T. An evaluation of the organics-carbon adsorbable method for use with potable water. JAWWA; 71 (1979) 1, p. 42-45.
3. PUIJKER, L.M., VEENENDAAL, G., JANSSEN, H.M.J. GRIEPING, B. Determination of organo phosphorus and organo sulphur at the sub ng-level for the use in water analyses. Fresenius Z. Anal. Chem. 306 (1981), p. 1-6.
4. NOORDSIJ, A., VAN BEVEREN, J., BRANDT, A. Isolation of organic compounds from water for chemical analysis and toxicological testing. Intern. J. Environ. Anal. Chem. 13 (1983), p. 205-217.
5. SCHNITZLER, M., SONTHEIMER, H. Eine Methode zur Bestimmung des gelöstes organisch gebundenen Schwefels in Wässern (DOS); Vom Wasser 59 (1982), p. 159-167.
6. SCHNITZLER, M., SANDER, R., SONTHEIMER, H. Sulfatbestimmung in Wässern durch thermische Zersetzung und mikrocoulometrische SO_2-analyse in Chlorat-Silber-Titrationszelle. Vom Wasser 58 (1982), p. 59-67.
7. SCHWEER, K.H., FUCHS, F., SONTHEIMER, H. Untersuchungen zur summarischen Bestimmung von organisch gebunden Schwefel in Wässern und auf Aktivkohlen. Vom Wasser 45 (1975), p. 29-43.

RESULTS FROM ROUND ROBIN EXERCISES OF PHENOLIC COMPOUNDS WITHIN THE COST 64b bis PROJECT

Lars Renberg
National Swedish Environment Protection Board
Special Analytical Laboratory
Wallenberg Laboratory
S - 1u6 91 STOCKHOLM, Sweden

SUMMARY

Within the Working Party 8 of the COST 64b bis project, a general opinion has been that intercalibration exercises should be encouraged to improve both accuracy and precision of existing methods. As a result of the discussion concerning determination of specific phenolic compounds a Round Robin Exercise was arranged in which the participants were asked to analyse two water samples, containing significantly different amounts of phenols, with at least two separate analytical methods. When rejecting outliers on a 99 % confidence level the following results were obtained. Using Method A, designed for the determination of polychlorinated phenols the relative standard deviations for each compound ranged 14 - 24 % (mean values 35 - 40 ug/L) for sample I (high phenol levels), the corresponding values for sample II were 18 - 53 % relative standard deviation (mean values 3.5 - 8.8 ug/L). Using Method B, designed also for the determination of non-halogenated phenols, the relative standard deviation for each compound ranged 20 - 36 % (mean values 32 - 87 ug/L) for sample I, the corresponding values for sample II were 19 - 78 % (mean values 3.7 - 9.1 ug/L). It was concluded that the results were quite satisfactory, but further intercalibration exercises should be carried out in the future to improve these results.

INTRODUCTION

During the sessions of Working Party 8 within the COST 64b bis project, there was a pronounced interest for intercalibrating analytical methods, developed for the determination of phenolic compounds. Such intercalibration exercises - if properly carried out - do not only give valuable information of the particular analytical methods but will also encourage previously unexperienced laboratories to apply or develop methods of their own.

The first Round Robin Exercise, arranged by EAWAG-Switzerland, took place during 1982. The sample to be analysed consisted of outgoing waste water from a municipal sewage treatment plant and the participants were asked to determine the presence of chlorinated phenols. The results have been discussed elsewhere (Renberg and Björseth 1983).

There was a general agreement at the meeting in Voorburg, March 2 - 4, 1983, that it would be valuable to carry out a second similar and extended exercise. When designing such a new exercise, the following aspects were taken into consideration.

- Sufficient determinations should be carried out to allow at least basic statistical analyses.
- Determinations should be carried out using at least two methods. The use of a third method, preferably a method familiar to the individual laboratory, should be encouraged.
- At least two samples, containing different concentrations of the individual compounds, should be analysed.

The main object with this paper is to present results of this Round Robin Exercise for phenolic compounds.

EXPERIMENTAL

The two samples chosen for the exercise were obtained from Baltic seawater (brackish water, pH 8.3, salinity 8 ‰) taken from the archipelago off Stockholm and spiked with ethanol solutions of different phenolic compounds at various levels. The laboratories, which previously had expressed their interest to participate, received two water samples, one with relatively high levels of phenolic compounds and one with relatively low levels, respectively. In addition, the participants received descriptions of two recommended methods and protocol forms to be filled in and returned to the organiser. The laboratories were asked to analyse the two water samples with the recommended methods (referred to as Method A and Method B, see appendix) and preferrably also with a method of their own choice. Each determination should be carried out in triplicates.

RESULTS AND DISCUSSION

In order to estimate whether any changes of the phenolic content occurred due to transport and storage of the water samples, spiked samples were analysed immediately after spiking and after one and two months of storage, respectively (4°C, darkness). It was observed that certain losses (10 - 30 %) of the phenols occur during the spiking procedures compared to the theoretical values. However, the levels did not significantly vary with time when comparing the results from the three individual determinations. It was then concluded that variation of the phenolic content due to storage was of minor importance compared to the errors introduced by the analytical methods.

Of the twelve laboratories, which had expressed their interest to participate, a complete set of results were reported by seven laboratories. Partial results were reported by two laboratories. The names of the participating laboratories are given in Table 1. The results are summarised in Table 2.

The following discussion will be concentrated on general aspects. A thorough statistical treatment is underway and these results will be published separately (Björseth et al 1983).

To properly interprete the analytical results of a collaborative test the scientist must give careful consideration to the various sources of error in the data obtained. Any analytical results may be regarded as a complex of three factors:

1. The random error.
2. The inherent systematic error in the procedure.
3. The modification in this systematic error that is a consequence of any particular laboratory's environment equipment and any personal way of using the procedure.

If it is assumed that the statistical material is normally distributed, the error in the determination for one substance can be estimted by the variance for the corresponding whole set of data (s_D^2). However, we are also interested in the variation within the laboratories (s_P^2) which can be regarded as a measurement of precision. The variation between laboratories (s_R^2) can easily be calculated from

$$s_R^2 = s_D^2 - s_P^2$$

The results from this Round Robin Exercise, summarised in Table 2, shows that s_P is always smaller than s_R (usually $s_R \approx 5 - 10$ times s_P), ie that between laboratory error dominate the situation. Apparently, this is the usual situation when dealing with data from a collaborative test (Youden and Steiner, 1975).

The mean values for each single substance and the corresponding standard deviations are given in Table 3. The relative standard deviation can here be regarded as a measurement of "uncertainty", ie lack of precision. It should be stressed that no outliers have here been rejected, hence some relatively large standard deviations may appear. If outliers are rejected, the standard deviations will decrease substantially. As an example, the relative standard deviation for 2,4,6-trichlorophenol will decrease from 34.5 % to 14 % if outliers are rejected on a 99 % confidence level.

Under the assumption that the mean values of all the measurements for each individual compound represent the true content of the water samples, we can calculate the distributions for each determination around the mean value for each laboratory. Figures 1 and 2 show such distributions for Method A and Method B. A narrow range for an individual laboratory will here indicate that a specific laboratory was working with a high precision. If the values are distributed around the mean value, the conclusion can be drawn that the particular laboratory has been working with a high degree of accuracy. Another conclusion to be drawn is that Method A appears to have a better precision than Method B.

It is, however, more difficult to draw general conclusions of "Method C" as this "method" consists of several different methods. The results obtained by these methods seem in some cases more consistent than the results obtained by the other methods. A main impression in that the methods included in "Method C" are modifications of Methods A and B to suite the conditions of the individual laboratory.

CONCLUSIONS

The results of the Round Robin Exercise can be summarised by some conclusive remarks:

1. Both methods appear to be reasonably simple for the trace level determination of phenolic compounds.

2. While Method A seems to have the best precision, it is restricted to polyhalogenated phenols. Method B can in addition be used for the analysis of many non-halogenated phenols, but gives less precision in the analytical determinations.

3. Both methods can be applied to the lower ppb level (and probably much lower) without any difficulties.

4. As alternative analytical methods ("Method C") to the recommended methods no significantly different methods were used by any of the laboratories. The methods suggested merely seem to be variations of Method A and Method B. In certain cases, individual laboratories reported more consistent results using "Method C".

5. The major contribution to the deviation from the mean values consisted of "between-laboratory" discrepancies. The reproducibility (precision) within the individual laboratories was usually high.

Finally, it should be stressed that the accuracy and precision of a particular analytical method must be regarded in respect to what purpose the results are to be used for. Different applications may require different techniques. Typical applications for these methods are toxicological evaluations and monitoring of waste water or recipient waters.

The results from this Round Robin Exercise are encouraging and should promote also other laboratories to test these methods for further evaluation.

ACKNOWLEDGEMENT

The work and the interest of the participating laboratories are greatly acknowledged. Without their help no Round Robin Exercise could have taken place.

REFERENCES

Björseth A, Mjösund A and Renberg L, Manuscript in preparation (1983).

Renberg L and Björseth A, Determination of phenols in water - a brief summary of current knowledge and proposed analytical methods, COST 64b bis, Organic Micropollutants in Water, OMP/38 /83.

Youden W J and Steiner E H, Statistical manual of the AOAC, Association of Official Analytical Chemists (1975).

Appendix

DETERMINATION OF CHLOROPHENOLS AND ALKYL PHENOLS USING GAS CHROMATOGRAPHY WITH ELECTRON CAPTURE DETECTION - ANALYTICAL PROCEDURE

Common steps for Methods A and B (1 - 3)

1. Transfer the water sample (mL) to a test tube with screw cap and teflon packing.

2. Add 1 M sodium hydroxide (1 mL) and iso-octane (2 mL) preferably containing a suitable **Phenolic** internal standard and shake the test tube for at least 2 minutes. After centrifugation, discard the organic phase.

3. Add 1 M sodium bicarbonate (1 mL) and iso-octane (2 mL), preferably containing a suitable **neutral** internal standard.

For the determination of chlorophenols with three chlorine atoms or more proceed accordingly to Method A.

For the determination of low-chlorinated phenols and hydrocarbon-based phenols proceed according to Method B.

Method A

Add acetic anhydride (100 uL) and shake the test tube immediately for at least 2 minutes. After centrifugation remove the organic phase, which is now ready for gas chromatograpahic analysis.

Method B

Add 10 uL of pentafluorobenzoyl chloride and shake the test tube for 5 minutes. After centrifugation remove, by the aid of a Pasteur pipette, as much as possible of the aqueous phase. Add 2 M sodium hydroxide (10 mL) and shake the test tube for 2 minutes. After centrifugation remove the orgaic phase, which is now ready for gas chromatographic analysis.

The following publications can be recommended for further studies.

Acetylation

Chau A S Y, Coburn J A. Determination of pentachlorophenol in natural and waste waters. JAOAC $\underline{57}$ (1974) 389.
Coutts R T, Hargesheimer E E, Pasutto F M. Gas chromatographic analysis of trace phenols by direct acetylation in aqueous solution. J Chromatogr $\underline{179}$ (1979) 291.
Renberg L, Lindström K. C_{18}-reversed phase trace enrichment of chlorinated phenols, guaiacols and catechols. J Chromatogr $\underline{214}$ (1981) 327.

Voss R H, Wearing J T, Wong A. A novel gas chromatographic method for the analysis of chlorinated phenolics in pulp mill effluents. In "Advances in the Identification & Analysis of Organic Pollutants in Water", L H Keith (Ed), Ann Arbor Science (1981).

Benzoylation

Renberg L. Gas chromatographic determination of phenols in water samples as their pentafluorobenzoyl derivatives. Chemosphere 10 (1981a) 767.

Table 1 - List of participating laboratories

Country	Contact	Institution
Denmark	H Lökke/ A Büchert	National Food Institute 19 Moerkhoej Bygade DK-2860 SOEBORG
	A Rudkjaer/ J Folke	Water Quality Institute 11 Agern Alle DK-2970 HORSHOLM
	N Hansen	Levnedsmiddelkontrollen I/S Dyregårdsvej 1 DK-2740 SKOVLUNDE
The Netherlands	G Piet	Rijksinstituut voor der Drinkwatervoorziening (Geb. Damsigt) Nieuwe Havenstraat 6 NL - LEIDSCHENDAM
Norway	K Martensen/ A Kringstad	Central Institute for Industrial Research PB 350 - Blindern N-OSLO 3
Spain	J Rivera/ F Ventura	Instituto QUIMICA - Bio-Organica SCIC c/Jorge Girona Salgado S/N E-BARCELONA 34
Sweden	M Adolfsson-Erici/ L Renberg	National Swedish Environment Protection Board - Special Analytical Lab. - Wallenberg Lab. S - 106 91 STOCKHOLM
	T Xie	Chalmers Univ. of Technology University of Gothenburg Department of Analytical and Marine Chemistry S-412 96 GOTHENBURG
Yugoslavia	V Drevenkar	Institute for Mecial Research and Occupational Health Mose Pijade 158 YU-41000 ZAGREB

Table 2 - RESULTS FROM THE DETERMINATIONS OF SOME PHENOLIC COMPOUNDS IN SURFACE WATER
(concentrations in µg/l, each value is a mean of three determinations)

Compound	A M1	A M2	B M1	B M2	B M3	C M1	C M2	C M3	D M1	D M2	D M3*	E M1	E M2	E M3	E M4	F M1	F M2	F M3	G M2	H M1	I M1	I M2
Sample I																						
m-Cresol	–	26	–	60	–	–	57	63	–	48	28	–	11	–	62	–	–	–	2.7	–	–	42
Thymol	–	35	–	48	–	–	3.7	3	–	29	32	–	7.1	–	129	–	–	–	8.8	–	–	34
2,4,6-Trichlorophenol	40	74	42	42	43	40	42	34	33	28	39	8.1	10	38	49	28	36	31	4.3	4.8	35	38
2,4,5-Trichlorophenol	45	56	41	47	41	35	34	28	44	51	38	9.9	8.1	55	70	32	–	38	3.9	23	44	61
o-Phenylphenol	–	109	–	88	–	–	125	114	–	53	64	–	11	–	–	–	–	–	11	–	–	63
2,3,4,6-Tetrachlorophenol	28	39	33	33	33	22	21	22	28	29	22	5.7	7.7	33	41	31	31	30	4.5	–	34	39
Pentachlorophenol	25	32	46	42	49	26	21	33	43	59	20	4.9	5.8	42	53	35	23	33	5.2	4.9	32	40
Sample II																						
m-Cresol	–	5.7	–	10	–	–	9	8	–	11	4	–	2.1	–	12	–	–	–	0.08	–	–	5.8
Thymol	–	3.0	–	4.2	–	–	0.4	0.3	–	4.2	–	–	0.8	–	13	–	–	–	0.12	–	–	3.1
2,4,6-Trichlorophenol	7.1	12	8.4	9.1	9.7	9	11	8	7.4	7.1	6	16	2.7	10	9.6	5.5	8.6	5.4	0.11	16	6.8	7.4
2,4,5-Trichlorophenol	1.5	3.6	3.4	3.5	3.5	2	3	2	4.3	6.7	11	0.9	0.7	5.9	5	3.2	–	4.7	0.19	2.4	3.7	5
o-Phenylphenol	–	6.1	–	3.9	–	–	7	5.7	–	4.6	4	–	0.6	–	–	–	–	–	0.09	–	–	3.2
2,3,4,6-Tetrachlorophenol	6.7	8.2	8.3	9.8	8.7	8	6	5	7.8	9.9	4	1.3	1.3	12	13	12	8.5	12	0.21	–	9.7	10
Pentachlorophenol	1.5	2.0	2.6	2.8	2.6	5	2	2	3.5	9.3	–	0.2	0.2	2.8	3.3	6.0	–	7	0.12	18	1.9	2.5

*single determination

Table 3

Results from the Round Robin Test compiled as the mean value ($\bar{x}$, µg/l), standard deviation (st d, µg/l) and relative standard deviation (% st d) for the individual compounds

Compound	SAMPLE I ("high" levels)			SAMPLE II ("low" levels)		
	$\bar{x}$	±st d	% st d	$\bar{x}$	±st d	% st d
Method A						
2,4,6-Trichlorophenol	32	11	35	6.6	2.4	36
2,4,5-Trichlorophenol	36	12	33	2.7	1.2	44
2,3,4,6-Tetrachlorophenol	26	9.2	36	7.7	3.1	41
Pentachlorophenol	30	13	43	3.0	2.0	67
Method B						
m-Cresol	41	18	44	7.2	3.1	43
Thymol	26	16	62	2.6	1.6	61
2,4,6-Trichlorophenol	39	19	48	8.2	2.9	36
2,4,5-Trichlorophenol	43	19	44	3.7	2.0	53
o-Phenylphenol	75	39	52	4.3	2.3	53
2,3,4,6-Tetrachlorophenol	28	11	37	7.7	3.0	39
Pentachlorophenol	32	16	51	3.2	3.0	94

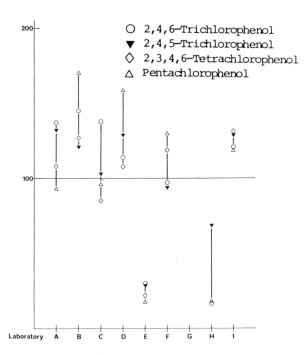

Figure 1 - Method A: Distribution of the mean values for each laboratory around the overall mean values (calculations carried out separately for each single compound)

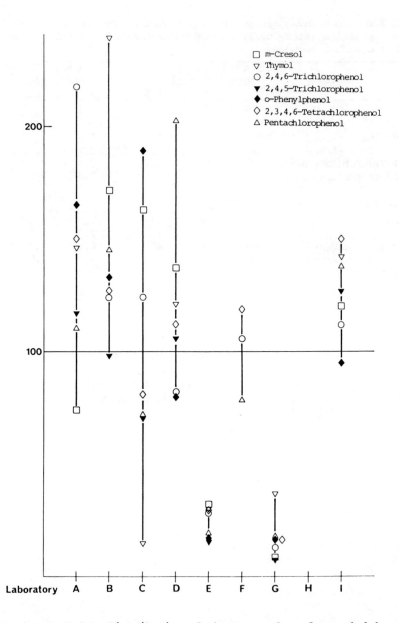

Figure 2 - Method B: Distribution of the mean values for each laboratory around the overall mean values (calculations carried out separately for each single compound)

INSTRUMENTAL ANALYSIS OF PETROLEUM HYDROCARBONS

R.G. Lichtenthaler
Central Institute for Industrial Research
P.O. Box 350 Blindern, Oslo 3, Norway

Summary

Analytical techniques for the analyses of petroleum hydrocarbons in the aquatic environment are overviewed. Due to the complex composition of petroleum and the rapid changes petroleum undergoes after release to the aquatic environment, analyses have to be concentrated on selected compound types, likely to those which are of environmental concern. Methods are described for determinations of total hydrocarbon contents, volatile hydrocarbons, hydrocarbon classes and selected single components. Further, analytical techniques for the identification of oil spills are included.

1. INTRODUCTION

Concern about the fates and effects of petroleum hydrocarbons in the aquatic environment has stimulated to an increased effort of research activity within all aspects of hydrocarbon pollution. An integral part of most studies is the chemical analysis of petroleum hydrocarbons in environmental samples. Different reasons may justify the establishment of hydrocarbon levels by chemical analyses: (i) because of the desire to establish background values prior to industrial activities (drilling, production, refinery etc.), (ii) because the area in question has been polluted by an oil spill, (iii) because the area is subjected to chronic pollution by industrial and urban waste or run-off, natural seeps, athmospheric rainout, or heavy ship traffic and (iiii) because the dosage of hydrocarbons to aquatic organisms is to be monitored in experimental biological studies.
Petroleum introduced to the aquatic environment from whatever sources is subjected to transformation involving physical, chemical and biological processes. These include evaporation, spreading, emulsification, natural dispersion, solution, adsorption, sedimentation, photooxidation, biodegradation and bioaccumulation. As expressed in a report from the Intergovernmental Oceanographic Commission (1) "it is, therefore, no point to analyse for 'oil' as such, since it does not exist any longer as the original oil, when it has reached the aquatic environment. The philosophy in petroleum hydrocarbon analysis should be to ask: What sort of components or group of components are most harmful? When this question has been answered the time has come to analyse individual or certain groups of components".

2. PETROLEUM COMPOSITION

The types of petroleum that end up in the aquatic environment differ significantly. Both crude oils and a wide range of refined petroleum products may have different chemical composition. It is essential to realize that petroleum can contain many tens of thousands of compounds. Crude oils consist principally of hydrocarbons and their derivatives, with appreciable amounts of sulfur and some nitrogen and oxygen. In addition, small concentrations of metals, including nickel and vanadium are complexed with organic polar constituents such as porphyrins. The sulfur, nitrogen and metal constituents are often associated with the higher boiling fractions.

All crude oils contain three general classes of hydrocarbons: alkanes, cycloalkanes and aromatics.

Normal alkanes from methane to beyond C_{60} are often the predominant individual components. Many parallell, homologous branched-chain alkanes are present, including a series of isoprenoid alkanes.

A complex mixture of cycloalkanes is present in crude oil with the alkyl-substituted hydrocarbons being more abundant than their parent hydrocarbons.

Aromatics include complex mixtures of parent compounds like benzene, naphthalene and polyaromatic hydrocarbons as well as of their mono-, di-, tri-, and multiple alkyl homologues. Also included in this class are hydrocarbons often designed as naphtenoaromatics, because of the mixed nature of the compounds, i.e. partly aromatic and partly cyclic saturated with alkyl substitutions.

Various refined petroleum products such as gasoline, diesel, distillate and residual fuels contain all of the classes of hydrocarbons found in crude oils. Additionally, alkenes (olefines) may occur in products produced from cracking stocks.

3. ANALYTICAL METHODS

3.1 General considerations

No single analytical technique can solve all types of problems concerned with the determination of hydrocarbons in the aquatic environment. Hydrocarbon concentrations may range from pure oil on the water surface to extreme low levels dissolved in the water column. Furthermore, the complex composition of petroleum with fractions over a wide boiling range as well as the changes caused by weathering processes, make the application of several analytical techniques necessary.

The choice of one particular or of several analytical methods in petroleum hydrocarbon determinations is very depending on the equipment available and on the nature of subjects to be investigated. One example is the differentiation of petroleum hydrocarbons from biogenic hydrocarbons. This may readily be achieved by gas chromatography showing the n-alkane distributions of biogenic and petroleum derived hydrocarbons. In contrast, gravimetric or infrared analyses are of little value in quantifying these differences.

Prior to the analysis of petroleum hydrocarbons in environmental samples special care has to be taken to ensure proper sample collection and preservation avoiding contamination or bacterial alteration of hydrocarbons. These subjects are described elsewhere in detail (2, 3).

Essential steps for the analysis of the complex mixtures of petroleum in environmental samples are (i) the isolation, enrichment and determination of the total hydrocarbon fraction from the sample matrix, (ii) the separation of hydrocarbons into individual compound classes and (iii) the identification and quantitative determination of individual components. The following chapters describe analytical techniques which are suitable for one or several of these tasks. An overall scheme for the analysis of hydrocarbons in different environmental samples is given in Fig. 1.

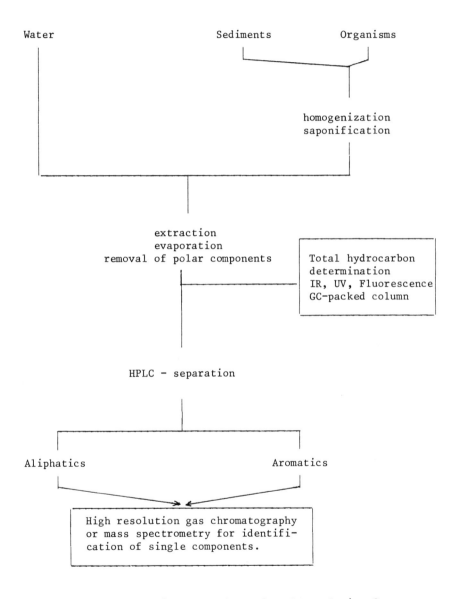

Fig. 1. Flow scheme of analytical procedures for the analysis of petroleum hydrocarbon in environmental samples.

3.2 Analytical methods for volatile hydrocarbons

Volatile hydrocarbons may be measured using different "head-space"-techniques. McAuliffe (4) describes a gas equilibration gas chromatographic technique, which allows the determination of distribution coefficients of hydrocarbons in water/gas systems and hence their quantitative determination. Up to 30 single components may be analysed in water which has been in contact with crude oil or light petroleum products. The sensitivity is approximately 10 ppt (parts per trillion).

Even higher sensitivities, which might be necessary in background determinations of hydrocarbons in open seawaters, may be achieved by gas stripping methods (5, 6) or by vacuum degassing (7) prior to gas chromatographic analyses.

3.3 Analytical methods for total determinations of C_{7+} hydrocarbons

The methods only provide an estimate of the total hydrocarbon levels within a sample extract (gravimetric or gas chromatographic analyses) or in a portion of the extract having a specific photometric response (ultraviolet, infrared or fluorescence spectroscopy). The procedures must also separate hydrocarbons from solvent extractable organic nonhydrocarbon material. This material derives from living or dead organisms. Extraction of environmental samples may be carried out with suitable organic solvents like CH_2Cl_2, CCl_4, n-pentane or other alkane solvents. Removal of nonhydrocarbon material is achieved by filtrating concentrated extracts over deactivated silica or florisil (8).

Gravimetric Method

An aliquot of the solvent extract is evaporated to dryness and the residue weight on a sensitive balance (9). The method gives the total amount of non-volatile hydrocarbons. The method is subject to errors because of possible contamination by dust and because of some of the lower hydrocarbons ($C_7 - C_{15}$) may be lost during solvent evaporation.

Ultraviolet Spectroscopy

The UV-spectra of solvent extracts are measured in the range of 210 and 350 nm. The absorbance at 256 nm is compared with the absorbance at the same wavelength for a sample of petroleum or a pure aromatic hydrocarbon (10). Errors may arise due to extracted substances which absorb UV-light at 256 nm. The spectra provide no indication on the nature of the absorbing molecules in a mixture.

Fluorescence Spectroscopy

Fluorescence spectroscopy is as rapid as ultraviolet absorption and can provide more details about the components in a mixture. It is recommended (11) for the initial screening of seawater extracts for dissolved hydrocarbons.

Fluorescence spectroscopy requires the use of a standardization mixture (containing the same components as the test sample) for calibration. Thus, the interpretation of the results are difficult in environmental monitoring surveys where the nature of the contaminants is not known. However, this technique is rapid and sensitive for checking known oil pollution incidents or for bioassay monitoring in the laboratory where the aromatic hydrocarbon centent pattern can be established.

In synchronous excitation fluorometry the difference between the emission wavelength and the excitation wavelength remains constant during the simultaneous scanning of both wavelengths (12). Considerably more information regarding the nature of the hydrocarbons can be obtained by that procedure over conventional spectrofluorometry, for example, it is possible to calibrate a spectrum according to aromatic ring number.

Infrared Spectroscopy

Infrared measurements are commonly carried out on CCl_4 - extracts after filtration over florisil (13, 14). The IR-absorbance at 2925 and 2960 cm^{-1} is measured and compared to a reference crude oil or to a mixture of aliphatic and aromatic hydrocarbons. Errors in quantification of IR-responses may arise especially in samples rich in aromatics when compared to highly aliphatic crude oils.

Gas Chromatography

When crude oil or petroleum products are analysed by gas chromatography (GC), the chromatograms often show some unresolved material beneath the resolved peaks (often called "envelope" or "unresolved complex mixture"). The total GC-response (sum of resolved peaks and unresolved envelope) may be integrated and expressed as total hydrocarbons by comparing with the same response of a petroleum product with the same boiling range as found in the sample (1, 15).

3.4 Hydrocarbon Class Separation by Liquid Chromatography

The most common technique for separation of petroleum is liquid adsorption chromatography. Also liquid partition and gel permeation chromatography may be applied. Most techniques described in the literature are based on classical liquid chromatography with gravity feed columns packed with either deactivated silica gel or alumina overlying silica gel. Aliphatics are often eluted with n-pentane, and aromatics by stepwise increasing the polarity of the eluting solvent adding benzene or dichloromethane. This separation technique has been successfully applied in a variety of petroleum pollution studies, but the procedure is time-consuming, requires large amounts of precleaned solvents and the columns have to be repacked for each sample.

A significant improvement has been achieved by the application of high pressure liquid chromatography on silica columns for the separation of hydrocarbons in environmental samples into aliphatic and aromatic fractions (8).

3.5 Single Hydrocarbon Component Determinations

As outlined earlier, petroleum consists of a complex mixture of hydrocarbons. Analyses of single hydrocarbon components in the aquatic environment concentrate mainly on those compounds which are relatively easy to determine (n-alkanes and selected isoprenoid hydrocarbons like pristane and phytane) or which are of special environmental concern (aromatics). For these purposes the most powerful analytical techniques are high resolution gas chromatography and gas chromatography/mass spectrometry interfaced with a computer.

Analyses are carried out either on total extracts or on hydrocarbon fractions obtained by liquid chromatography. A wide spectrum of non-polar stationary phases like OV-1, OV-101, OV-17 or Apiezon L and SE-30 as well as the polar free fatty acid phase (FFAP) have been proposed for gas chromatographic hydrocarbon analyses (2).

Chromatograms of petroleum hydrocarbons permit beside their quantitative determination using selected internal standards (1, 8) also the calculation of odd-even n-alkane ratios and n-alkane-isoprenoid ratios, which may be diagnostic in the differentiation of biogenic from petroleum hydrocarbons (16).

Aromatic hydrocarbons may be determined by GC after isolation of an aromatic fraction by liquid chromatography (8). Even more powerful and diagnostic are GC/MS-techniques. Operating in the selected ion monitoring (SIM) mode aromatics are determined without prior fractionation of extracts (1, 17). Also, the distribution of selected aromatic hydrocarbons and their C_1-, C_2- and C_3- alkylhomologs may be used in the source identification of polynuclear aromatic hydrocarbons (combustion processes versus petroleum (17).

3.6 Analytical Methods for Oil Spill Identification

Procedures for identification of oil spills utilize passive rather than active tagging techniques to identify the source responsible for an oil spill. Active tagging refers to the use of known materials, such as a mixture of organo-metallics, which can be added to a cargo of oil and subsequently used to identify the oil in the event of a spill. Passive tagging refers to the measurement of the unique chemical composition of the oil to identify the source. The methods developed use the chemical composition of a spilled oil, as determined by a multi-method analysis, at the time of the spill, for comparison with probable sources located within the immediate area. The analytical data obtained by chromatographic and spectroscopic analyses are referred to as "fingerprint" for a particular oil sample. The reason for a multi-method approach is that no single technique has been established which is unequivocal in all cases. This is due mainly to two reasons: frequent contamination of the spill sample and the changes the spilled oil undergoes when it spreads on water.

Any one of the basic methods might be adequate in a given case but the analytical results obtained utilizing only one technique can be influenced by impurities which frequently find their way into the spilled oil. Additionally, the spilled oil will undergo changes through natural weathering processes such as solution, evaporation, oxidation, and biodegradation. Weathering processes and the effect of environmental contamination can alter the fingerprint obtained by an individual technique to varying degrees with the effect of these two factors being controlled by the nature of the spilled oil and the existing environmental conditions.

An additional important reason to adopt the multimethod philosophy for oil spill identification is that the statistical reliability of the analytical results is enhanced with a multimethod approach to the extent that the selected methods are independent of each other.

The analytical methods involved in the identification procedure are in principal all chromatographic and spectroscopic techniques. Prior to the applications of a particular analytical method it is important to investigate the influence of weathering on the parameters to be measured.

Operational oil spill identification systems have been developed by the US Coastguard (18) and the Nordic countries (19).

The operational identification system of the Nordic countries is based on a two step approach including both screening and identification analyses as shown in Fig. 2. Additionally, back up methods like high performance liquid chromatography and computerized gas chromatography/mass spectrometry are used in extremely difficult cases.

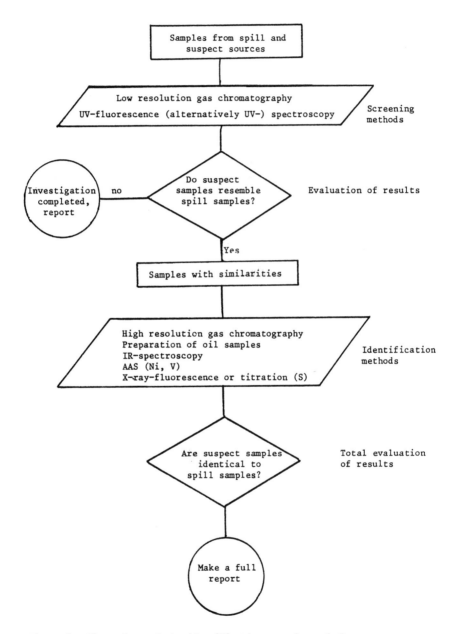

Figure 2. Flow scheme of the identification procedure of the Nordic Countries (19).

4. CONCLUSIONS AND FUTURE OUTLOOK

The possibilities for the analysis of the complex mixture of petroleum hydrocarbons in environmental samples are manifold. Both spectroscopic and chromatographic analytical techniques may be applied in the determination of levels and fates of hydrocarbons in water, sediments and biological materials. The methods vary widely in their detection limits and especially in their diagnostic abilities.

As the complexity of petroleum hydrocarbons increases with molecular weight, the analysis for individual or even classes of hydrocarbons becomes more difficult. High resolution gas chromatography especially when combined with mass spectrometry can, however, partly characterize the high-molecular weight hydrocarbon. So far, these techniques are most powerful and diagnostic, but they are relatively sophisticated and are expensive.

The development of more rapid, but reliable and diagnostic methods for the isolation and analysis of hydrocarbons in environmental samples, should have high priority. High pressure liquid chromatography may be one of the most promising techniques (8).

Studies of the fates and effects of petroleum hydrocarbons may also include the analysis of transformation products formed during the oxidation of hydrocarbons in the environment. These transformation products may be more hazardous to the environment and include compound groups like peroxides, aldehydes, ketones, phenols, acids etc. The analysis of these components in addition to the untransformed hydrocarbons will be much more complex (20).

REFERENCES

1. INTERGOVERNMENTAL OCEANOGRAPHIC COMMISSION. "The Determination of Petroleum Hydrocarbons in Sediments", Manuals and Guides No. 11, Unesco 1982.
2. NATIONAL ACADEMY OF SCIENCES, "Petroleum in the Marine Environment", Proceedings of a Workshop on Input, Fates and Effects of Petroleum in the Marine Environment, Washington D.C., 1975.
3. CLARK, R.C. and BROWN, D.W., "Petroleum: Properties and Analyses in Biotic and Abiotic Systems" in D.C. Malins (Editor) 'Effects of Petroleum on Arctic and Subarctic Marine Environments and Organisms', Vol. I, Academic Press Inc., New York 1977, p. 37 f.
4. McAULIFFE, C.D., "GC of Solutes by Multiple Phase Equilibration", Chem.Technol. 1 (1971), 46.
5. BERG, N. et al., "Comparison of a Stripping Method and a Continous Extraction Method", Proceedings of the "First European Symposium on the Analysis of Organic Micropollutants in Water", Dec. 11-13, 1979, Berlin, edited by the Commission of the European Communities, Brussels, 1981, p. 300.
6. SWINNERTON, I.W. et al., "Gaseous Hydrocarbons in Seawater", Science 156 (1967), 1119.
7. SCHINK, D.R. et al., "Hydrocarbons under the Sea", Offshore Technology Conference, Houston, Texas, April 19-21, 1971, Paper No. OTC 1339, 1 (1971), 130.

8. LICHTENTHALER, R.G. and ORELD, F., "Rapid HPLC-Method for Hydrocarbon Class Separation in Environmental Samples", J. Chromatogr., 1983, in press.
9. American Society for Testing and Materials, Annual Book of ASTM Standards, part 31, ASTM-method D-2278.
10. LEVY, E.M., "The Presence of Petroleum Residues off the East-coast of Nova Scotia, Water Res. 5 (1971), 723.
11. NATIONAL BUREAU OF STANDARDS, "Marine Pollution Monitoring (Petroleum)" Nat.Bur.Stand.Spec.Publ., 409 (1974), 271.
12. WAKEHAM, S.G., "Synchronous Fluorescence Spectroscopy and its Application to Petroleum Derived Hydrocarbons", Environ.Sci.Technol., (1976).
13. Norwegian Standardization Committee, Norwegian Standard NS 4753.
14. BROWN, R.A. et al., "Hydrocarbons in Open Ocean Waters", Scienc, 191, (1976), 847.
15. FARRINGTON, I.W. et al., "Intercalibration of Analyses of Petroleum Hydrocarbons in Marine Lipids", Bull.Environ.Contam.Toxicol., 10 (1973), 129.
16. BLUMER, M. et al., "The W-Falmouth Oil Spill II", Techn. Report 72-19 Woods Hole Oceanographic Institution, 1972.
17. SPORSTØL, S., LICHTENTHALER, R.G. et al., "Source Identification of Aromatic Hydrocarbons in Sediments Using GC/MS", Environ.Sci.Technol. 17, (1983), 282.
18. "Oil Spill Identification System", US Coastguard Research and Development Center, Report No. CG-D-52-77, June 77, Available through NTIS, Springfield, Virginia 22161, USA.
19. LICHTENTHALER, R.G. et al., "Identification of Oil Spills at Sea", Central Institute for Industrial Research, Report No. 77 12 09 - 3, 1982.
20. BARTH, R., TJESSEM, K. and AABERG, A., "Fractionation of Polar Organic Constituents in Environmental Samples", J.Chromatogr. 214, (1981), 83.

SCREENING OF POLLUTION FROM A FORMER MUNICIPAL WASTE DUMP AT THE BANK OF A DANISH INLET

U. LUND[a,b], A. KJÆR SØRENSEN[b] & J.A. FARR[a]
a. VKI, Water Quality Institute
11 Agern Allé, DK-2970 Hørsholm, Denmark
b. Danish Civil Defence, Analytical Chemical Laboratory
2 Universitetsparken, DK-2100 Copenhagen, Denmark

Summary

An investigation of the potential pollution from a former municipal waste dump also containing chemical waste has been performed. The dump is situated at the bank of the inlet Holbæk Fjord, Denmark. The investigation has involved hydrogeological investigations of the dump and surroundings and chemical analyses and ecotoxicological tests of leachate from the dump. The investigations have shown that the water flow from the dump is towards the inlet, and that no ground water resources are threatened. Analyses have shown the presence of approx. 40 compounds in the leachate, of which pentachlorophenol and 2,3,4,6-tetrachlorophenol are thought to involve the highest environmental risk. Acute toxic effects have been shown on approx. 1/3 of the samples of leachate and it has been shown that this effect is removed upon dilution by a factor of 5. It was concluded that the impact of the toxic substances in the leachate on the inlet should be further investigated.

1. INTRODUCTION

The leaching of pollutants from waste dumps has caused rising concern over the last decade, as regards both municipal waste and industrial waste.

A report that toxic chemical wastes had been dumped at a municipal waste disposal site gave rise to an investigation of which kinds of waste had in fact been dumped and later to a screening of the potential pollution from the dump.

The waste disposal site is situated near Holbæk, at the bank of the inlet Holbæk Fjord in the northwestern part of Sjælland, Denmark. The site was in use from 1953 to 1971, the western part of the dump being used until 1964, while the eastern part was used during the later period, mainly for municipal waste. A systematic enquiry to local industries, farms, etc. showed that mercury-stained grain, paint, glue, cleaner, etc. had been dumped. Among these wastes was also a quantity of pentachlorophenol from a wood preservative factory.

The screening involved drillings and collection of hydrogeological data from these drillings and existing data for the area round the waste dump. Water from the drillings was analysed by gas chromatography/mass spectrometry (GC/MS) in order to determine the qualitative composition of the leachate from the dump, and furthermore pentachlorophenol (PCP) and 2,3,4,6-tetrachlorophenol (TeCP) (an impurity in and degradation product from pentachlorophenol) were quantified, and chloride, pH, total organic carbon (TOC) and on some samples selected metals (Hg, Co, Cd, Cr) were determined. The samples were also subjected to an ecotoxicological test for acute toxicity.

The present paper gives the results from the screening investigation, while the detailed investigation of the impact of the pollution on the inlet is reported elsewhere (1). The results from the screening investigation have previously been reported in Danish (2).

2. EXPERIMENTAL

Drillings

24 drillings were made in the area covering the dump site with a concentration of drillings in the middle area, where the probability of finding chemical waste was largest. The situation of the drillings is seen in figures 1 and 2. Two types of drillings were made. Those indicated in the figures with a P were 8" drillings with a 140 mm PVC filter, designed for pumping tests (P-wells). The drillings indicated with an L were 3" piezometers with a 50 mm screen (L-wells). The drillings were 2.5 to 6 m deep, the depth being determined by the thickness of the filling.

Sampling

From the P-wells a 10 l sample was collected in a glass bottle rinsed with hydrochloric acid and dichloromethane after ½ hour of pumping, and another 10 l sample was taken after 3 to 5 hours at the end of the pumping tests. The L-wells were pumped dry and the next day a sample, which was as close as possible to 10 l, was taken in a glass bottle. Most L-wells did not give sufficient water and less than 10 l had to be collected. The samples were left for precipitation of suspended material, pH was measured and the sample was subdivided into 4 samples: 1) a 2 l sample added 15 ml of concentrated hydrochloric acid and 200 ml of dichloromethane in a glass bottle rinsed with hydrochloric acid and dichloromethane for GC/MS, PCP and TeCP determination, 2) a 2.5 l sample in a filled glass bottle for chloride analysis and toxicity tests, 3) a 1 l sample added 10 ml of concentrated nitric acid (Suprapur) in an acid washed glass bottle for metal analysis and 4) a 250 ml sample added 1 ml of concentrated phosphoric acid in a glass bottle for TOC-analysis.

Analyses

For GC/MS-analysis and quantification of PCP and TeCP the samples were extracted as received by stirring over night. The dichloromethane was separated, the volume reduced to 10 ml by rotary evaporation and extracted with 10 ml of 0.5 N sodium hydroxide. The volume of the dichloromethane phase was reduced further by rotary evaporation and then to 0.5 ml by a gentle stream of nitrogen. The extract was transferred to hexane by three times adding 1 ml of hexane and afterwards reducing to 0.5 ml. The aqueous phase was acidified with 4 N hydrochloric acid and extracted with three portions of 10 ml of hexane. The volume of the hexane phase was reduced by rotary evaporation and by a stream of nitrogen to 1 ml. The extract was methylated by reaction with diazomethane for 30 min. at room temperature, whereupon the excess of diazomethane was removed by adding a droplet of acetic acid.

The GC/MS-analyses were performed using a Finnigan gas chromatograph, model 9500, connected to a Finnigan mass spectrometer, model 3100D. The gas chromatograph was equipped with a capillary splitless injection system. Column: 25 m x 0.25 mm i.d. glass capillary column, SE-52, carrier gas: helium 0.8 ml/min. Temperatures: injector: 270°C, column: 60°C for 3 min. then 6°C/min. to 250°C.

Quantification of PCP and TeCP was performed on a Hewlett-Packard gas chromatograph, model 5880A, with an electron capture detector and a capillary splitless injection system. Column: 25 m x 0.25 mm i.d. glass capillary column, SE-52, carrier gas: nitrogen 1.6 ml/min., make-up gas: nitrogen 28 ml/min. Temperatures: injector: 250°C, detector: 300°C, column: 60°C for 0.5 min., then 10°C/min. to 250°C.

Chloride determinations were performed according to (3).

For TOC-analyses inorganic carbon was eliminated from the samples by sparging with nitrogen prior to analysis on a Dohrman DC 80 analyzer.

Mercury, cadmium, cobalt and chromium were analyzed by atomic absorption spectrometry after destruction of the sample with nitric acid. Mercury was measured in a gas cell after reduction with Sn(II), and the three other metals were measured using a graphite furnace. The spectrometers were a Perkin-Elmer model 603 atomic absorption spectrometer equipped with a Mercury/Hydride system, model MHS 20 (Hg), and a Perkin-Elmer model 5000 atomic absorption spectrometer (Cd, Co, Cr).

Ecotoxicological tests

The test was a test for acute toxicity to Artemia salina, the death rate being registered after 24 and 48 hours as a function of the sample concentration. The test was performed according to (4).

3. RESULTS

Hydrogeology

In the area round the waste dump the upper 15 m are mainly clay, under which is a layer of sand and gravel that contain the ground water resources of the area. In the dump all drillings showed filling, which had a thickness between 0.8 and 3.6 m. The filling consisted mainly of sand, clay and top soil with municipal waste, etc.

The piezometric surface in the dump was compared to that

in the underlying aquifer (2). This showed that the main ground water flow was towards the inlet, but that there also was a slight downward gradient from the dump to the aquifer. As the flow in the aquifer was also towards the inlet it could be concluded that no ground water resources in the area were threatened by pollution from the dump.

Since the capacity of most of the drillings was very low, < 0.5 m^3/h, pumping tests were impossible, but the coefficient of permeability could be estimated to 5×10^{-6} m/s (2). Two drillings were placed in a sandy layer below the filling of the dump and on these pumping experiments could be performed. The permeability of this layer was estimated to 1×10^{-3} m/s (2).

The flow from the dump to the inlet was estimated both on the basis of the determined permeability values and on a mass balance for the water. The results indicated a flow between 50.000 and 250.000 m^3/year from the dump to the inlet (2). This is approx. 0.2 to 1 per cent of the total flow of water to the inlet.

Chemistry

GC/MS-analyses and quantification of PCP and TeCP were performed on the samples collected after 3 to 5 hours of pumping, while pH, chloride and TOC were measured on all samples. The differences between values after ½ hour of pumping and after 3 to 5 hours of pumping was, however, small, and in the following only values for samples taken after 3 to 5 hours of pumping are reported.

The pH of all samples was close to neutral, ranging from 6.5 to 7.8. Chloride ranged from 40 to 5960 mg/l, with the highest concentrations in samples close to the inlet. Some exchange of water between the inlet and the dump thus takes place, but the chloride concentration is significantly lower in the leachate than in the inlet. This is in agreement with the hydrogeological evaluation that the main flow is from the dump to the inlet.

The TOC-concentrations in the samples are shown in figure 1. The highest values are in the eastern, and newest, part of the dump. The distribution of the GC/MS-data is similar to the distribution of TOC, with many substances in relatively high concentrations in the eastern part of the dump and few substances in low concentrations in the western part. Approx. 40 compounds were identified by the GC/MS-analyses. The substances most frequently found were PCP, TeCP and PAH (polyaromatic hydrocarbons). Other phenols, both chlorinated and non-chlorinated, were also found, as well as phenoxyalkanoic acids, benzoic acid derivatives, alkyl- and chlorobenzenes and fatty acids. The high frequency of PAH is probably a result of the burning of waste during the use of the dump.

PCP and TeCP were also quantified, and the results from the PCP-analyses are shown in figure 2. TeCP was found in nearly all samples where PCP was found, in concentrations ranging from 0.1 to 80 µg/l.

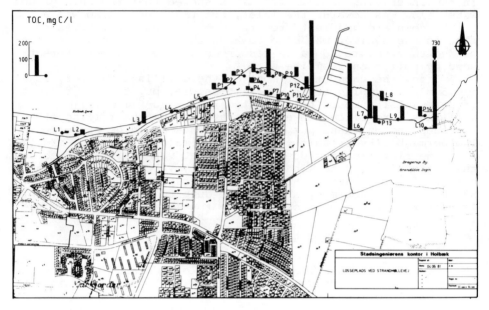

Figure 1. The concentration of total organic carbon (TOC) in water samples.

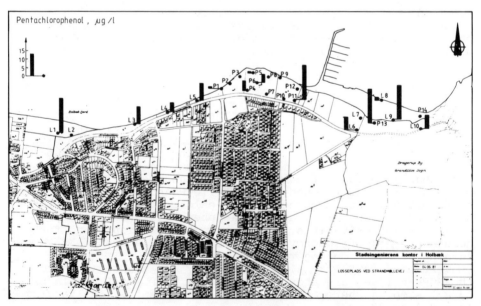

Figure 2. The concentration of pentachlorophenol (PCP) in water samples.

Where no value is shown the concentration was below the detection limit, which for PCP was 0.1 µg/l and for TeCP 0.03 µg/l. PCP and TeCP seem to be rather evenly distributed over the whole dump area.

Heavy metal analysis was performed on eight samples which were selected on the basis of results from the toxicity tests and GC/MS-, PCP- and TeCP-analyses. An even distribution of the samples over the dump area was attempted. The results of these analyses are shown in table I. These values are above background levels in sea water (5), but the values are comparable to maximum permissible values for drinking water (6). An environmental risk from the metals, Hg, Cd, Co and Cr, is thus not anticipated.

Table I.   Concentrations of heavy metals in water samples.

WELL NO.	Hg µg/l	Cd µg/l	Co µg/l	Cr µg/l
P 3	< 0.3	0.09	< 8	7.3
P 11	0.4	3.3	16	9.3
P 13	2.8	0.22	< 8	11
L 1	0.3·	0.58	18	22
L 3	1.0	3.6	31	130
L 5	0.6	1.5	8	27
L 7	< 0.3	0.53	11	17
L 9	< 0.3	0.25	< 8	15

Ecotoxicology

All samples were subjected to a test for acute toxicity with *Artemia salina* as the test organism. The results showed that 8 samples were acutely toxic, while 11 samples showed slight acute toxicity and 5 samples were non-toxic. The test was complicated for some samples in that a heavy precipitation occured during the test, and at the same time the oxygen concentration dropped to between 40 and 70 per cent of saturation. Samples, which gave these problems, were then sparged for 24 hours at 4°C before test. It can, however, not be excluded that this procedure may have affected the test results, e.g. by causing evaporation of toxic substances.

The test results for the samples, which showed acute toxicity, are shown in table II. The dilution factor necessary to remove toxicity is seen to be between 3 and 8.

Table II. 48 hours toxicity to <u>Artemia salina</u> of acutely toxic samples.

SAMPLE	LC 10% 48 HOURS DILUTION FACTOR	LC 50% 48 HOURS DILUTION FACTOR
P 3 +	4.9	1.5
P 9	3.9	1.2
P 11	7.7	3.0
P 13	6.4	2.3
P 14	2.9	1.1
L 6	3.4	1.5
L 7	3.8	1.1
L 10	7.5	3.1

+: 24 hours sparging at $4^{\circ}C$ before test.

4. CONCLUSION

The substances identified by the GC/MS-analyses were evaluated with regard to potential environmental effects. It was concluded that PCP and TeCP involved the greatest risk because 1) they are widely distributed over the whole dump area, 2) they are toxic, and 3) they are not easily degradable.

From the quantitative analyses of PCP and from the estimates of water flow from the dump to the inlet a rough estimate of the load of PCP from the leachate was calculated. Using the lowest estimate of the flow with the mean concentration of PCP in the drillings, and the highest estimate of the flow with the maximum concentration of PCP, respectively, a rough estimate of 1 to 10 g/day of PCP was found.

The tests for acute toxicity showed that dilution by approximately a factor of 5 removed the toxic effects on the test organism, <u>Artemia salina</u>. The test did not include long term effects and acute toxicity to organisms more sensitive than <u>Artemia salina</u> could not be dismissed.

It was thus concluded that further investigations elucidating the impact of the toxic substances in the leachate, primarily PCP, on the environment of the dump were needed. These investigations have been carried out, and are reported in (1).

5. ACKNOWLEDGEMENTS

This work was performed under contract for the Municipality of Holbæk, and we wish to thank the Town Council of Holbæk for permission to publish the results. The publication was supported by the National Council of Technology (file no. 1982-132/328-1).

REFERENCES

1. FOLKE, J., BIRKLUND, J., SØRENSEN, A.K. & LUND. U. (1983). The impact on the ecology of polychlorinated phenols and other organics dumped at the bank of a small marine inlet. Chemosphere, submitted for publication.

2. VANDKVALITETSINSTITUTTET, ATV & CIVILFORSVARETS ANALYTISK-KEMISKE LABORATORIUM. Undersøgelse af den tidligere losseplads ved Strandmøllevej i Holbæk Kommune - Screeningsfasen. Report to Holbæk Kommune, 1981-10-09.

3. DANSK STANDARDISERINGSRÅD. Chlorid (Potentiometrisk metode), DS/R 239 (1978).

4. VANHAECKE, P., PERSOONE, G., CLAUS, C. & SARGELOOS, P. (1981). Proposal for a short term toxicity test with Artemia nauplii. J. Ecotoxicol. Environm. Safety 5, 382.

5. FÜRSTNER, U. & WITTMANN, G.T.W (1979). Metal pollution in the aquatic environment. Springer Verlag.

6. MINISTRY OF ENVIRONMENTAL PROTECTION. Bekendtgørelse om vandkvalitet og tilsyn med vandforsyningsanlæg. Miljøministeriets bekendtgørelse nr. 6, 4. januar 1980.

THE IMPACT ON THE ECOLOGY OF POLYCHLORINATED PHENOLS AND OTHER ORGANICS DUMPED AT THE BANK OF A SMALL MARINE INLET

Jens FOLKE[a*], Jørgen Birklund[a], Arne Kjaer Sørensen[b], Ulla Lund[b]

a: VKI, Water Quality Institute, 11, Agern Allé, DK - 2970 Hørsholm, Denmark

b: Danish Civil Defence, Analytical Chemical Laboratory (CFL), 2, Universitetsparken, DK - 2100 COPENHAGEN Ø

ABSTRACT

Pentachlorophenol (PCP), 2,3,4,6-tetrachlorophenol (TeCP) and other chemicals were leaching into a small marine inlet, the Holbaek Fjord, Denmark, from a dump at a banking zone of 2.5 - 3 km. PCP and TeCP were quantified in the marine ecosystem, and a biological examination of the fauna was conducted. The levels of PCP and TeCP were only slightly increased as compared to the reference area, and no acute toxic effects could be unveiled. However, as the concentrations of organics in the leachate from the dump were quite high, the trends of the ecosystem were to be monitored over the next decade.

INTRODUCTION

Municipal waste was dumped at the bank of a small marine inlet, the Holbaek Fjord, during the period from 1953 to 1971 together with unknown amounts of chemical wastes. Leachate samples from the dump were shown to contain a number of organics, whereof pentachlorophenol (PCP) and 2,3,4,6,- tetrachlorophenol (TeCP) constituted the greatest potential risk in relation to amounts and toxicity. Hydrogeological studies showed that no ground water resources were threatened, i.e. that the toxic impact of the organics would be on the marine ecosystem, because the ground water flow was towards the inlet. Estimations showed that along a banking zone of 2.5 - 3 km, e.g. the PCP-leaching amounted to a maximum of 1 - 10 g per day (18).

Several reviews on the toxicology and environmental impact of the polychlorinated phenols have been published (1, 2, 5, 6, 16, 17, 29, 31, 32). Clearly, TeCP and PCP are both toxic (14, 34, 35) and persistent (3, 7, 8, 9, 11, 22, 38), and have bioaccumulative properties towards aquatic organisms (8, 15, 21, 22, 23, 24, 25, 26, 27, 28, 35). Additionally, they are widespread contaminants (19, 36, 37) and are found in municipal sewage (12, 13) also in Denmark (11), in sea water (8, 36), in sediments (7, 22, 38) and in biological materials (9, 21, 22, 24, 25, 26, 27, 28, 34), including mussels (10).

The present study, which has been reported in Danish (4), has been aimed at (i) estimating the spreading of organics from the dump (18) into the marine inlet, (ii) examining the impact of the organics on the ecosystem of the inlet, and (iii) providing knowledge for the decisions concerning the possible arrangements to limit the pollution from the dump.

PCP and TeCP were chosen as tracer substances for the examination of leaching organics from the dump, because (i) they constituted an environmental risk in relation to amounts and toxicity, (ii) they are easy to detect at low levels in different matrixes, e.g. (31), (iii) they accumulate in biota (vide intra), (iv) they are water soluble at neutral pH, meaning that they would be among the first compounds to be expected in the inlet coming from the dump, and (v) they are relatively persistent (vide intra). PCP and TeCP were quantified in mussels (Mytilus edulis) stocked in growth cages, marine sediments, sea water, and in the secondary effluent of a nearby sewage treatment plant.

The impact of the leaching organics on the marine ecosystem of the Holbaek Fjord was examined by measuring the growth rate and survival ability of the mussels stocked in growth cages. Also, the condition of the natural fauna was examined in terms of species composition, species density, and species diversity.

PCP and TeCP were found in all matrixes analysed at levels slightly higher than the chosen reference area, but no acute effects of the dump on the ecosystem could be unveiled. Thus, it was concluded that the marine ecosystem of the inlet was able to cope with the present burden of exuding chemicals, and that no immediate actions should be taken to limit the present impact. However, the concentrations of pollutants in the leachate from the dump had indicated much more serious effects than seen. This could be due to a slow penetration of organics into the inlet so that future effects could not be excluded. Consequently, it was recommended that the trends of the PCP- and TeCP-concentrations in the natural mussel fauna were monitored over the next decade.

EXPERIMENTAL

Materials

The mussels (Mytilus edulis), 30 - 35 mm in length were obtained from a bay 30 km north of the Holbaek Fjord, the Nykøbing Bugt. The mussel growth cages were constructed at the VKI, one type with the mussels situated 20 cm above the sea bottom for use at low water depth (1 m), and another type with the mussels situated 1.5 m beneath the surface for use at deeper water (2 m).

All solvents and reagents used for the chemical sediment analyses were of analytical grade, and were used without further purification. The other chemical analyses were carried out with "Uvasol"-quality (E. Merck) hexane, diethylether, and toluene, "Nanograde"-quality (Mallinckrodt) isooctane (2,2,4-trimethylpentane), and "Suprapur"-quality (E. Merck) sodium hydroxide. The other chemicals were of analytical grade. Silicone antifoaming agent was a 30 per cent w/w silicone/ water emulsion (BDH), and acetic anhydride (E. Merck) was double distilled before use.

Sampling and analyses

The collected mussels were divided into fourteen populations with an equal length-frequency between 30 - 35 mm. The length of the mussels in each population were determined with an accuracy of 0.1 mm before the mussels were stocked in growth cages, 1982-06-16, at eleven stations covering the inlet of the site of the dump, and two stations at the reference area further away (fig. 3). Furthermore one mussel population was analysed immediately to determine the background levels of PCP and TeCP. Stations 5, 6, 7, 8, 9, 10 and 11 were situated at a depth of water of   2 m (between 60 - 80 cm). The shell length of the mussels on all stations was measured on 1982-07-27 or 1982-08-04 (station 4 was then missing), and a subsample of ten mussels from each station was taken before the rest of the mussels were stocked back in new cages for later sampling (1982-08-30), and shell length measurements.

Marine sediments were taken at the thirteen stations on 1982-06-16 (fig. 3), preferably as 20 cm column-cores, whereof the upper 3 cm was analysed for the PCP content.

Sea water samples were taken on 1982-08-30 at stations 8 and 10 as 10 l samples in glass bottles 20 cm beneath the surface and acidified with 9M sulphuric acid, pH < 2. The sea water level was 12 cm lower than the mean water level and the wind was southerly.

The sewage sludge was taken from the secondary effluent of the biological treatment plant of the area as a 24-hours flow proportional 5 l sample on 1982-08-29/30 and acidified with 9M sulphuric acid, pH < 2. The total effluent of the period was 4229 m^3, which was less than the average (5000 - 7000 m^3).

The mortality of each mussel population was determined by visual inspection. The sampled living mussels were left for 24 hours in the sea water from the inlet before the soft tissue was taken out. The rate of growth of the mussels' soft tissue was determined on the basis of weighing a subsample of each population at the three different dates after drying for four days at 60°C. Also, a subsample of each population was deepfrozen (-20°C) on 1982-08-30 for chemical analysis.

The general condition of the marine ecosystem was analysed from a biological bottom fauna inspection along seven cross sections perpendicular to the bank, five of them at the site of the dump and one on each side. The details of this part of the investigation is described elswhere (4).

TeCP and PCP were isolated from the mussels by steam distillation (15). 20 - 25 g of soft tissue was homogenized and 200 ml of water. 3 ml of 9M sulphuric acid, 2 ml of antifoaming agent and 10 ml of isooctane were added together with the internal standard, 2,4-dibromophenol. 100 ml of water/ isooctane was distilled and recondensed, and 4 ml of 1M sodium hydroxide was added. The organic phase was discarded and the aqueous phase reduced to < 10 ml by rotary evaporation. The sample was then acidified (250 μl conc. H_2SO_4) and adjusted to exactly 10 ml. A 5 ml subsample was extracted with 2 x 1 ml of toluene, and the phenols acetylated as described (11). The TeCP- and PCP-content was quantified as ng/g dry weight (d.w.) soft tissue. The dry weight of the soft tissue was determined gravimetrically upon drying of subsamples at 105°C for 24 hours. The recovery experiments were carried out by spiking pooled mussel soft tissues.

25, 50 or 100 g of sediment was acidified with 4N hydrochloric acid (pH 2 - 3) and soxhlet extracted overnight with 225 ml of acetone/hexane (4 + 1). The extract was reduced to 50 ml, and 50 ml of hexane, 100 ml of water and 5 ml of 4N hydrochloric acid were added. The organic phase was separated, and 2 x 50 ml of hexane was used additionally to extract the aqueous phase. The combined organic phases were extracted with 50 ml of 0.5N sodium hydroxide, and this aqueous phase was then acidified with 4N hydrochloric acid before extraction with 2 x 20 ml of hexane. The combined organic phases were dried with magnesium sulphate, reduced to 5 ml, and derivatized with diazomethane at room temperature for 30 min. The diazomethane in excess was removed by a droplet of acetic acid before 400 μl of internal standard (DDE, 1.02 mg/l) was added prior to the gas chromatographic analysis. The recovery and blank experiments were carried out on a standard soil by spiking. PCP was added in methanol, which was softly evaporated after mixing with the soil. The dry weight of the sediments was determined gravimetrically upon drying of subsamples at $105°C$ until constant weight.

The 10 l sea water sample was extracted with 2 x 100 ml of hexane/ether (2 + 1) at pH = 2 for 2 x 24 hours. The combined extracts were reduced to 2 ml and acetylated (11, 33).

The 5 l sewage sample was extracted and acetylated (11, 33).

Apparatus

The acetylated samples of mussels, marine water and sewage were analysed by gas chromatography using a Hewlett-Packard Model 5840 gas chromatograph equipped with a ^{63}Ni electron-capture detector, an HP 7671A autosampler and a capillary column splitless injection system. The capillary column was a 25 m SE54, 0.3 mm i.d. fused-silica column (Hewlett-Packard). The carrier gas was hydrogen, 2 ml/min and the make-up gas argon/methane (95 = 5), 20 ml/min. Temperatures: Injection port $250°C$, detector $250°C$, and oven $60°C$ for 0.6 min, then $4°C/min$ until $300°C$. Injection volume 1 μl.

The methylated PCP-samples of marine sediments were analysed by gas chromatography on a Hewlett-Packard Model 5880A gas chromatograph equipped with a ^{63}Ni-electron-capture detector and a capillary splitless injection system. The capillary column was a 25 m CP Sil-5, 0.25 mm i.d. fused silica column (Chrompack). The carrier gas was argon/methane (95 + 5), 2 ml/min and the make-up gas argon/methane (95 + 5), 25 ml/min. Temperatures: Injection port $250°C$, detector $300°C$, and oven $60°C$ for 0.5 min., then $10°C$ until $250°C$. Injection volume 1 μl.

RESULTS

The stocked mussel populations showed only insignificant differences in mortality, i.e. after an initial higher rate of mortality, it became very small. The mussels harvested in the Nykøbing Bugt for transplantion contained less soft tissue (d.w.) per unit shell length and less fat per unit soft tissue than after 10 weeks in the Holbaek Fjord. The average rate of growth of the mussels' soft tissue on each station as function of time is shown in fig. 1. There is a positive tendency between the growth rate of the mussels and the depth of the water, and a significant negative correlation between the growth rate of the mussels' soft tissue (d.w.) and their polychlorophenolic content. This probably reflects the competition for foodstuffs between the natural fauna and the stocked mussels (fig.2).

The inspection of the general condition of the ecosystem did not verify any significant differences between the cross-sections studied, i.e. no specific effects relating to the dump could be unveiled.

The possibility of a systematic error in the determination of PCP between the two laboratories participating in the present study, can be excluded on the basis of an intercalibration study of the PCP-content in the sewage sample. Also, the VKI has participated in a European intercalibration of TeCP and PCP in the Rhine water with a satisfactory result (33). The PCP-determination in sediments have been conducted at the CFL, and the biological analyses as well as the rest of the chemical analyses at the VKI.

The levels of TeCP and PCP found in the different matrixes are shown in table I.

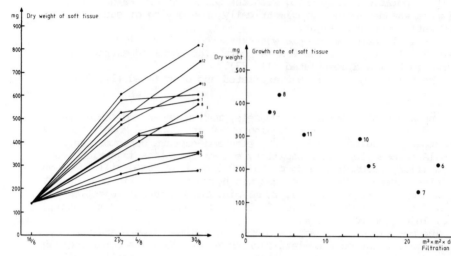

Fig. 1 The growth of the stocked mussels determined on basis of the average increase in soft tissue (d.w.) during the period from 1982-08-30.

Fig. 2 The amount of water filtrated by the natural mussel population per day at each station plotted against the rate of growth of the stocked mussels at stations at low water level (< 1 m). (r = - 0.887).

TABLE I - Levels of TeCP and PCP found in the different matrixes

Matrix	Unit	TeCP	PCP
Leachate (dump) (18)	ng/l	< 30 - 80000	< 100 - 20000
Sewage	ng/l	30	180
Marine water	ng/l	6 - 8	2 - 3
Marine sediment	ng/d d.w. [1)]	_[2)]	10 - 20
Mussels	ng/g d.w.	1.1 - 6.0	1.0 - 4.4

[1)]: d.w. = dry weight
[2)]: analysis not performed

The background concentrations of TeCP and PCP in the mussels from the Nykøbing Bugt were 3.8 ng/g d.w. and 10.0 ng/g d.w., respectively. The concentrations of TeCP and PCP in the stocked mussels (ng/g d.w.) on the thirteen stations after 10 weeks of exposure are shown (fig. 3), to be compared with the results of the PCP-analysis of leachate (ng/l) at different positions of the dump (fig.4) (18).

PCP was quantified only in two sediment samples, station 4 (10 ng/g d.w.) and station 5 (20 ng/g d.w.). Stations 2 and 12 were not analysed, and the rest of the sediment samples contained PCP in amounts less than the detection limit (10 ng/g d.w.).

There are no significant differences between the TeCP- and PCP-content of the two sea water samples.

DISCUSSION

Biology

The initially higher rate of mortality of the stocked mussels on all stations can be explained by the damages caused by the transplantation. At the end of the experimental period a very low mortality was seen, with no significant differences between the stations.

The growth rate of the stocked mussels is related to the content of algae in the water, i.e. the amount of available foodstuffs. The primary production is smaller at low levels of water, and the competition for foodstuffs between the stocked mussels and the natural mussel fauna is greater. This can be illustrated by a calculation of the amount of water that the natural mussel population filtrates per day (m^3/m^2/day) at each station. A plot of this number against the growth rate of the stocked mussel population at each station (fig. 2), renders probable that the food competition and not the polychlorophenolic content of the mussels is the primary variable determining the growth rate of the mussels at each station. This conclusion is supported by the reported toxic levels of TeCP and PCP in laboratory experiments, e.g. (5), which are approximately a factor of 1000 higher than the present levels of the Holbaek Fjord.

Chemistry

The data on TeCP- and PCP-levels in sediments, plants and animals are few, especially in marine systems.

In mussels at the Weser estuary a PCP-level of 3 - 10 ng/g soft tissue d.w. has been reported (10). In connection with a study of the Swedish paper and pulp industry ten different chlorophenols and chloroguaiacols, among them TeCP and PCP, were quantified in mussels (Mytilus edulis) and in different species of fish (muscle and liver) in fresh water, brackish waters, and marine waters (20). Only PCP of the two is reported found in the mussels, and only at trace levels, but including the reference station. Significant contents of TeCP and PCP were reported in fish liver (TeCP up to 80 ng/g fresh weight (f.w.) and PCP, 20 - 1600 ng/g f.w.), and lower levels in fish muscles (TeCP, 0.7 - 7 ng/g f.w. and PCP, 40 - 150 ng/g f.w.). Even higher levels of TeCP and PCP were found two weeks after an accidental discharge of a wood-impregnating solution (34).

In natural water, the concentration of TeCP and PCP has been reported a the Weser estuary at levels of 41 ng/l and 140 - 500 ng/l, respectively (7, 8). Surface water of the Netherlands contained an average of 70 ng/l TeCP and 410 ng/l PCP (27), and in the Tokyo area an average of 180 ng/l was found (19). Higher values have been reported as the result of the accidental discharge (34). It was not possible to detect the background levels of TeCP and PCP in the natural waters during the Swedish study (20), but in the German Bight, the TeCP- and the PCP-levels of sea water were reported at 0,08 - 0,14 ng/l and 0.4 - 1.3 ng/l, respectively (36), a somewhat lower level than in the Holbaek Fjord.

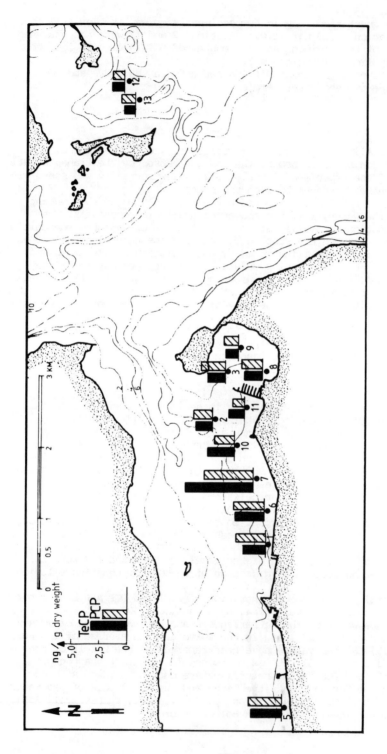

Fig. 3 The positions of the thirteen stations in the Holbaek Fjord and the reference area, and the concentrations of TeCP and PCP in the transplanted mussels stocked in growth cages at these positions.

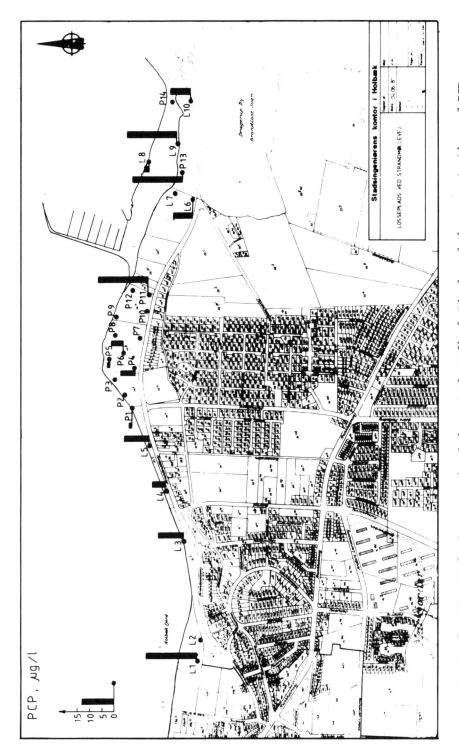

Fig. 4 *The site of the former municpal dump at the Holbaek Fjord, and the concentrations of PCP in the leachate at the specified positions.*

The level of PCP in the sediments of the Holbaek Fjord is close to the detection limit of the employed method (~10 ng/g d.w.). Consequently, quantifications could only be made at two stations, 4 and 5. Considering the geographical distance between these two stations and the results of the mussel analyses, it seems, however, reasonable to assume that the general level of PCP in the area is 1 - 10 ng/g d.w. of the sediment. This level is then comparable with other observations at the Weser estuary and the Netherlands, 0.1 - 60 ng/g d.w. (36, 38), but in these instances the observations were combined with considerably higher PCP-concentrations (vide intra) in the natural waters. Sediment analyses in Sweden showed a PCP-level of 30 - 185 ng/g wet weight in a zone close to a paper and pulp industry (20).

The level of PCP in the sewage (180 ng/l) is comparable to other observations in Denmark (11). TeCP and PCP have also been reported in the secondary effluent of the sewage treatment plant of Dübendorf at higher levels, 900 ng/l and 1000 ng/l, respectively (12, 13).

Biochemistry

Laboratory and in situ studies show a fast accumulation and excretion of chlorophenols in aquatic organisms (15, 23, 34). A constant level is obtained in mussels during exposure within 2 - 3 days (10), and PCP is excreted again when the mussels are transferred to pure water with an estimated half time of the same order (10, 15). The transplantation experiment of the Holbaek Fjord took place over a period of more than 10 weeks, and it is therefore reasonable to assume that the levels of TeCP and PCP found in the mussels after transplantation reflects the exposure levels in the Holbaek Fjord rather than the (unexpected) high background level found before transplantation in the Nykøbing Bugt. At present, the results obtained at the Nykøbing Bugt are being investigated.

The bioaccumulation factors that can be calculated from the present in situ experiment are tabulated for comparison with some values calculated from octanol/water partition coefficients and other experimental data (table II). There are no disagreements between the obtained values and other experimental data, but the calculated values based on the octanol/water distribution of the compounds seem to be somewhat overestimated.

TABLE II - Bioaccumulation of PCP and TeCP in mussels from the marine inlet compared with the results from the laboratory experiments and theoretical calculations

Bioaccumulation coefficients	PCP	TeCP
Calculated values (10)	3620	610
Laboratory experiments (10, 12, 23)	30 - 390	
Marine inlet, station 8	200	45
station 10	100	75
stations 1 - 11*	170	60

* Average of mussels, stations 1 - 11, and water samples stations 8 and 10.

Impact of chemicals

The impact of chemicals from the dump can be estimated from the fate of PCP in the marine inlet. The sources of PCP are (i) from the dump, 1 - 10 g/day (18), (ii) from the sewage, 1.0 - 1.3 g/day^1, (iii) the precipitation, 0.2 - 0.4 g/day^2, and other diffuse sources (e.g. streams like the Tuse Å). The sea water of the inlet contains an estimate of 75 - 110 g PCP^3, and the bottom fauna ~1 g^4. It is not possible to estimate the sediment content on the present basis. Assuming a steady-state situation for the inlet system, the transfusion of PCP into the inlet is counterbalanced by the degradation processes (biodegradation and photochemical oxidation mainly), the sedimentation, and the net transport out of the inlet.

The present study has not proven that chemicals are leaching from the dump, but this has been rendered probable by the significantly higher levels of TeCP and PCP in the stocked mussels outside the dump as compared to the reference area (fig. 3), and by the homogenicity of the TeCP- and the PCP-content in the mussel populations at all stations covering the site of the dump. If TeCP and PCP had not been leaching from all over the wide banking zone, it would have been expected that the highest concentrations of TeCP and PCP in the mussels were seen at station 5, near the sewage effluent, as this then would be the main source of PCP. The higher concentrations of TeCP and PCP in the leachate from the east-end of the dump as compared to the rest of the dump, are not, as could be expected, reflected by a similar higher level of TeCP and PCP in the mussel population at the stations 3, 8 and 9. This can be due to the fact that this part of the dump is the newest and that leaching has not fully started yet.

CONCLUSION

The chemical analysis of the leachate from the dump showed that PCP and TeCP were the two organics to be concerned about (18). The biological, chemical and biochemical analyses have shown that the present impact of the chemicals on this ecosystem is not alarming.

The present impact of the dump on the marine ecosystem based on the PCP burden is a factor fo 1000 lower thant the estimated effect level. It is recommended, however, that the situation is monitored until it is evident that the leaching of organics from the dump has maximized, and is declining.

The monitoring program has been set up in 1984 to ensure that the impact of the dump on the ecology of the Holbaek Fjord does not increase significantly. This program includes (1) the quantification of TeCP and PCP in the leachate of the drillings P9 and L8 (fig. 4), and (ii) the quantification of TeCP and PCP in the natural mussel fauna sampled at the stations 5, 7, 8, 13 and at the Nykøbing Bugt. The future frequency and

^{1}: 5000-7000 m^3 sewage of 0.2 µg PCP/l
^{2}: 0.55 m precipitation/year of 13.8 m^2 (area of the inlet) of 10 - 20 ng PCP/l
^{3}: 35.8 x 10^6 m^3 sea water of 2 - 3 ng PCP/l
^{4}: 475.000 kg dry weight biomass of 2.3 ng PCP/g d.w. mussel soft tissue.

the extent of the monitoring program will be decided upon the results of 1984.

ACKNOWLEDGEMENTS

This study was contracted work for the Municipality of Holbaek, and the authors acknowledge the permission from the Town Council of Holbaek to publish the work. The publication was supported by the National Council of Technology (file no. 1982 - 132/238 - 1), and was first published in Chemosphere.

REFERENCES

1) Ambient Water Quality Criteria for Pentachlorophenol. US Environmental Protection Agency, EPA 440/5-80-065, pp A1-C49 (1980).
(2) Ambient Water Quality Criteria for Chlorinated Phenols. US Environmental Protection Agency, EPA 440/5-80-032, pp A1-C124 (1980).
(3) Baker, M.D., C.I. Mayfield & W.E. Inniss. Degradation of chlorophenols in soil, sediment and water at low temperature, Water Res. 14: 1765-1771 (1980).
(4) Birklund, J., Folke, J., Sørensen, A.K. and Lund, U. Undersøgelser i Holbaek Fjord af spredningen og den biologiske efffekt af udsivende stoffer fra den tidligere losseplads ved Strandmøllevej i Hoelbaek kommune. Vandkvalitetsinstituttet, ATV (VKI), Report No. 81.795, 1-193 (1982).
(5) Buikema, A.L., McGinniss, M.J. and Cairns, J. Phenolics in aquatic ecosystems: A selected review of recent literature. Marine Environmental Research, 2(2): 87-181 (1979).
(6) Crosby, D.G. Environmental chemistry: An overview. Environm. Toxicol. Chem., 1: 1-8 (1982).
(7) Eder, G. and Weber, K. Chlorinated phenols in sediments and suspended matter of the Weser estuary. Chemosphere, 9: 111-118 (1980).
(8) Ernst, W and Weber, K. The fate of pentachlorophenol in the Weser estuary and the German Bight. Veröff. Inst. Meeresforsch. Bremerh. 17: 45-53 (1978).
(9) Ernst, W. and Weber, K. Chlorinated phenols in selected estuarine bottom fauna. Chemosphere, 11: 867-872 (1978).
(10) Ernst, W. Factors affecting the evaluation of chemicals in laboratory experiments using marine organisms. Ecotoxicol. Environm. Safety, 3: 90-98 (1979).
(11) Folke J. and Lund, U. The occurrence of low and high chlorinated phenols in municipal sewage before and after passing biological purifying plants. in: Rijks, J. (ed.), Proceedings of the Fifth International Symposium on Capillary Chromatography, Rival del Garda (Italy), April 26-28. Elsevier Scientific Publishing Company, Amsterdam (Holland), 230-241 (1983).
(12) Giger, W. and Schaffner, C. Determination of phenolic water pollutants by glass capillary gas chromatography, in: Keith, L.H. (ed.), Advances in the identification and analysis of organic pollutants in water, Vol. 1, Ann Arbor Science, Ann Arbor, Michigan (USA), 141-154 (1981).
(13) Giger, W., Vasilic, Z. and Schaffner, C. Polychlorinated phenols in sewage effluents and rivers. COST 64b bis. Report of the workshop of the working party 8, Specific analytical problems, Dübendorf (Schweiz), May 18-19, Commission of the European Communities, OMP/33/82: 51-53 (1982).

(14) Gupta, S., Verma, S.R. and Saxena, P.K. Toxicity of phenolic compounds in relation to the size of a fresh water fish, Notopterus notopterus (pallas). Exotoxicol. Environm. Safety, 6: 433-438 (1982).
(15) Jensen, S. and Renberg, L. Accumulation of some organo-chlorine compounds from water and their subsequent elimination. Scientific report to the Oslo Commission Working Group on Degradability, Lowestoft/4, 1-14 (1974).
(16) Kozak, V.P., Simsiman, G.V., Chesters, G., Stensby, D. and Harkin, J. Reviews of the environmental effluents of pollutants: X. Chlorophenols. US Environmental Protection Agency, EPA-600/1-79-012, 1-491 (1979).
(17) Kreijl, C.F. van and Slooff, W. Biological effects of phenolic compounds. COST 64b bis. Report of the workshop of the working party 8, Specific analytical problems, Hague (Holland) March 2-4, Commission of the European Communities, OMP/37/83: 199-207 (1983).
(18) Lund, U., Sørensen, A.K. and Farr, J.A. Screening of pollution from a former municipal waste dump at the bank of a Danish inlet, in: Proceedings of the Third European Symposium on Analysis of Organic Micropollutants in Water, Oslo (Norway), September 19-21, Commission of the European Communities (1983).
(19) Matsumato, G. Comparative study on organic constituents in polluted and unpolluted inland aquatic environments - III. Phenols and aromatic acids in polluted and unpolluted waters. Water Res., 16: 551-557 (1982).
(20) Miljövänlig tillverkning av blekt massa. Final report of the project: Bleached pulp production with minimum environmental impact (1979-81). Industrins Processkonsult AB IPK, Stockholm (Sweden), 1-192 (1981).
(21) Miyazaki, T., Kaneko, S., Horii, S. and Yamagishi, T. Identification of polyhalogenated anisoles and phenols in oysters collected from Tokyo Bay. Bull. Environm. Contam. Toxicol., 26: 557-584 (1981).
(22) Murray, H.E., Ray, L.E. and Giam., C.S. Analysis of marine sediment, water and biota for selected organic pollutants. Chemosphere, 10: 1327-1334 (1981).
(23) Niimi, A.J. and McFadden, C.A. Uptake of sodium pentachlorophenate (NaPCP) from water by rainbow trout, Salmo gairdneri, exposed to concentrations in the ng/l range, Bull. Environm. Contam. Toxicol., 28: 11-19 (1982).
(24) Paasivirta, J. and Linko, R. Environmental toxins in Finnish wildlife. A study on time trends of residue contents in fish during 1973-1978, Chemosphere, 9: 643-661 (1980).
(25) Paasivirta, J., Särkka, J., Leskijärri, T. and Roos. Aa. Transportation and enrichment of chlorinated phenolic compounds in different aquatic food chains, Chemosphere, 9: 441-456 (1980).
(26) Paasivirta, J. Enrichment of chlorobleaching residues in food chain, Proceedings from the 17th Scandinavian Symposium on Water Research, Porsgrunn, Nordforsk Miljövardsserrier, publ. 1, 187-195 (1981).
(27) Paasivirta, J., Särkka, J., Aho, M., Surmo-Aho, K., Tarhanen, J. and Ross., Aa. Recent trends of biocides in pikes of the lake Päijänne, Chemosphere 10(4): 405-414 (1981).
(28) Paasivirta, J., Särkka, J., Surmo-Aho, K., Humpi, T., Kuokanen, T. and Martinen, M. Food chain enrichment of organochlorine compounds and mercury in clean and polluted lakes of Finland. Chemosphere, 12(2): 239-252 (1983).

(29) Rao, K.R. (ed.) Pentachlorophenol: Chemistry, pharmacology and environmental toxicology. Plenum Press, New York, USA (1-402 (1978).
(30) Renberg, L. and Lindström, K. C_{18} reversed-phase trace enrichment of chlorinated phenols, guaiacols and catechols in water, J. Chromatogr., 214: 327-334 (1981).
(31) Renberg, L. Gas Chromatographic determination of chlorophenols in environmental samples. National Swedish Environment Protection Board, Report SNV PM 1410, Stockholm (Sweden), 1-135 (1981).
(32) Renberg, L. Phenolic ompounds - Analytical methods in relation to environmental aspects, COST 64b bis, Report of the workshop of the working party 8, Specific analytical problems, Dübendorf (Schweiz), May 18-19. Commission of the European Communities, OMP/33/82: 36-49 (1982).
(33) Renberg, L. and Björseth, A. Determination of phenols in water. A brief summary of current knowledge and proposed analytical methods. COST 64b bis, Report of the working party 8, Specific analytical problems. Commission of the European Communities, OMP/38/83/: 1-7 (1983).
(34) Renberg, L., Martell, E., Sundström, G. and Adolfsson-Erici, M. Levels of chlorophenols in natural waters and fish after an accidental discharge of a wood-impregnating solution, Ambio, 12(2): 121-123, (1983).
(35) Salkinoja-Salone, M., Saxelin, M.-L., Pere, J., Jaakkola, T., Saarikoski, J., Hakulinen, R. and Koistinen, O. Analysis of toxicity and biodegradability of organochlorine compounds released into the environment in bleaching effluents of kraft pulping, in: Keith, L.H. (ed.), Advances in the identification and analysis of organic pollutants in water, Vol. 2, Ann Arbor Science, Ann Arbor, Michigan (USA), 1131-1164 (1981).
(36) Weber, K. and Ernst, W. Levels and pattern of chlorophenols in water of the Weser estuary and German Bight, Chemosphere, 11: 873-879 (1978).
(37) Wegman, R.C.C. and Hofstee, A.M.W. Chlorophenols in surface waters of the Netherlands (1976-1977). Water Res., 13: 651-657 (1979).
(38) Wegman, R.C.C. and van den Brook, H.H. Chlorophenols in river sediment in the Netherlands. Water Res., 17: 227-230 (1983).

ORGANIC MICROPOLLUTANTS FROM THE PULP INDUSTRY;
ANALYSES OF LAKE PÄIJÄNNE WATER

E. Toivanen, I. Kuningas and S. Laine
Helsinki City Waterworks
Water Examination Bureau

Summary

Lake Päijänne has since 1982 been the raw water supply for the Helsinki metropolitan area. There are pulp mills along the upper reachers of the lake´s drainage area. In order to investigate the appearance and course of organic micropollutants derived from the pulp industry, TOC, lignosulphonates, TOCl, trihalomethanes and some chlorinated phenols were analysed at several sampling stations in the Lake Päijänne area. Also the removal of these compouds in water treatment was studied in pilot plant tests. The highest value: of all analysed parameters has been observed immediately downstream of the pulp mills. All compound concentrations decreased with increase in distance from the pulp mills. At the Asikkalanselkä intake station only traces of organic chlorocompounds and lignosulphonates were detected. Water treatment and especially ozonization proved to be effective in reducing the content of organic compounds derived from the pulp industry and present in the raw waters used in the pilot plant runs. The Helsinki Waterworks will continue regular studies on the organic micropollutants in Päijänne water. The purpose is to secure the present good quality of the raw water and to investigate how organics behave in water treatment.

1. INTRODUCTION

Lake Päijänne has since 1982 been the raw water source for the Helsinki metropolitan area. The water is conducted from the Asikkalanselkä intake station to the water treatment plants of the City of Helsinki in a rock tunnel 120 km in length.
Lake Päijänne water contains humic substances. Also industrial and domestic wastewaters are discharged into the northern and central parts of the lake. Pulp mills along the upper reaches of the lake's drainage area cause severe pollution problems because the pulp bleaching effluents are not treated. However, the gross water pollution caused by oxygen-consuming organic substances has been considerable reduced in the 1970´s and water quality has greatly improved in the southern part of Lake Päijänne because of improvements in industrial processes and in wastewater purification systems. As a consequence water quality is not far from its natural state at Asikkalanselkä, the raw water intake point.
A new modern sulphate pulp mill is beeing built in Äänekoski and will start production in 1985. It will replace smaller existing sulphate and sulphite mills and increase about 50 % their present capacity. The consequences with regard to water quality in the Päijänne area are of great

concern at present. An intensive study program was started in 1982 in order to investigate the appearance and course of organic micropollutants derived from the present pulp industry. Also, their removal in water treatment was studied in pilot plant tests.

2. STUDY METHODS

In order to study the appearance and course of organic micropollutants water samples were collected from several sampling stations in the Lake Päijänne area. Sampling was started 8 km south of the pulp mills in Äänekoski and continued along the watercourse down to Asikkalanselkä where the raw water intake station is located.

Ten cubic meters of water was transported in tanks from the sampling stations to the pilot plant in Helsinki for test runs on the removal of organic micropollutants in water treatment. As the full-scale treatment at the Helsinki Waterworks, the pilot plant treatment consisted of alum coagulation, sand filtration, ozonization and chloramine addition.

The quality of the water obtained was investigated with special attention to organic compounds. The level of total organics in the water samples was determined as TOC. Also $KMnO_4$-consumption and UV-absorbance were measured. To classify the organics, TOCl and lignosulphonates were determined. Trihalomethanes and chlorinated phenolic compounds were analyzed of the compounds formed in pulp bleaching.

In TOCl-measurements activated carbon filtration was used as the isolation step. The chlorine content of carbon was determined either by microcoulometry (nonpurgeable organic chlorine) or by neutron activation.

Trihalomethanes were analyzed by means of gas chromatography and an electron capture detector after pentane extraction. The chlorophenols were determined as their ethyl derivatives by means of gas chromatography-mass spectrometry after extraction and concentration. The separation of various compounds was performed by means of a silica capillary column and selected ion monitoring (SIM) was used as the analyzing technique.

3. RESULTS AND DISCUSSION

Waste discharge from the pulp industry causes higher organics content in water. Especially the concentrations of organic chlorine compounds are very high downstream from the pulp mills. These compounds are formed in pulp bleaching and they are foreign to the ecologic environment. Only a part of them have been identified and there is not enough knowledge on their potential toxicity and health hazard due to long-term, low-level exposure.

Organic compounds in the raw water increase treatment costs and the need for disinfection. All organics are not removed even though more effective purification methods are used. Also, new organic compounds, e.g. chlorocompounds, can be formed in water treatment. Drinking water quality is always lower when a raw water with a high organic content is used.

Some of the detailed results of the study are shown in Figures 2 - 6.

The highest values of all analyzed parameters were observed immediately downstream of the pulp mills. All compound concentrations decreased with an increase in distance from the pulp mills.

At the Asikkalanselkä intake station only traces of organic chlorocompounds and lignosulphonates were detected. The concentration levels were about the same as in the natural waters upstream of the pulp mills.

Organic compounds are quite effectively removed in water treatment.
Especially ozonization has proved to be effective in removing organic compounds such as lignosulphonates, chlorophenols and other chlorinated organic compounds derived from the pulp industry and present in the raw waters used in the pilot plant runs.

The treatment of the raw water presently taken from Asikkalanselkä in Lake Päijänne has caused no problems at the Helsinki Waterworks, and high-quality drinking water is produced.

The Helsinki Waterworks will continue regular studies on the organic micropollutants in Päijänne water. The aim is to secure the present good quality of the raw water, observe in time any factors threatening the quality and control adverse changes in it. Studies on the significance of organic compounds in water treatment will also be continued. The purpose is to investigate how organics behave in water treatment, determine safe concentration limits in raw water and the effect of organics removal on the cost of water treatment.

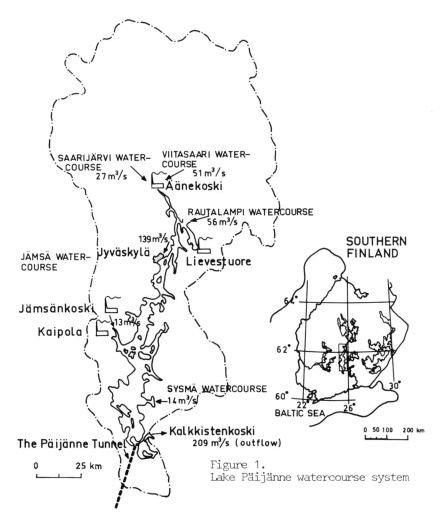

Figure 1.
Lake Päijänne watercourse system

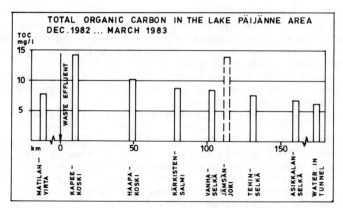

Figure 2. Total organic carbon (TOC) content in the water sample 8 km downstream from the pulps mills in Äänekoski was almost twice as high as the TOC-value in the lake upstream. TOC in Asikkalanselkä water was about the same level as the upstream values. In the raw water of the Helsinki treatment plants TOC has been even lower due to ground water dilution in the tunnel.

The decrease in $KMnO_4$-consumption and UV-absorbance were similar to TOC.

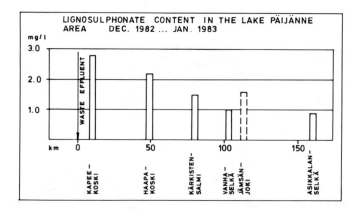

Figure 3. In pulp production some constituents of the wood dissolve; lignine compounds are the most important of these. Lignines are stable in water and their structure resemble natural humus. Their amount is often measured as lignosulphonate, which gives the total lignine value for sulphite mill effluents, but only relative values for sulphate mill effluents. Lignines in raw water raise water treatment costs and can cause taste and odor in the product water, especially if chlorine is used in the treatment process. There is some lignine in Asikkalanselkä raw water, but the concentrations observed at present are easily removed with ozonization and do not cause any problems in water treatment.

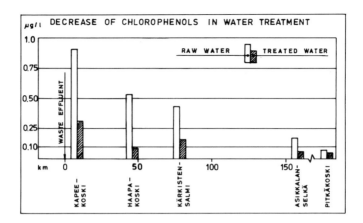

Figure 4. Chlorinated phenolic compounds are formed in pulp bleaching. They are toxic and persistent organic compounds, some of which are proved to accumulate in the food chain. Chlorophenols are not totally removed in normal water treatment. They can even be formed if free chlorine is used in water treatment. In pilot plant experiments ozonization proved to be most effective in removing chlorophenols present in the raw water.

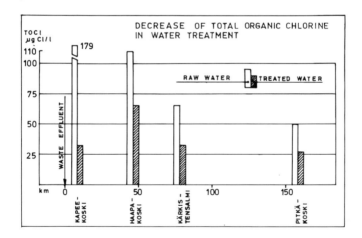

Figure 5. The total amount of various organic chlorine compounds formed in pulp bleaching can be measured as total organic chlorine (TOCl). In this case practically all nonvolatile organic chlorine was measured.
These compounds cannot be totally removed in normal water treatment.

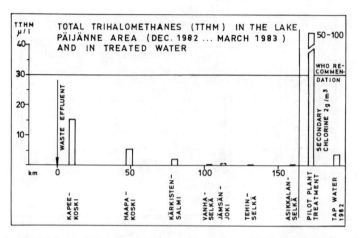

Figure 6. Trihalomethanes (THM) are formed in pulp bleaching as well as in water treatment when free chlorine is used. These substances, harmful to health, form the major part of the volatile organic halogen compounds. THM-concentrations in the northern part of Lake Päijänne are higher in winter when the watercourse is covered with ice. Trihalomethanes derived from pulp effluents have not been detected in Asikkalanselkä. The maximum acceptable THM-level recommended by WHO in drinking water is 30 ug/l. The low THM-level in drinking water in 1982 is due to low chlorine dosage in chloramine disinfection. THM-concentrations vary between 50-100 ug/l in treated water if ozonization cannot be used and over 2 g/m^3 of free chlorine is needed for disinfection.

GROUND WATER POLLUTION BY ORGANIC SOLVENTS AND THEIR
MICROBIAL DEGRADATION PRODUCTS

D. BOTTA, L. CASTELLANI PIRRI, E. MANTICA
Dept. of Industrial Chemistry and Chemical Engineering
Politecnico di Milano, Piazza Leonardo da Vinci 32
20133 Milano (Italy)

Summary

Serious contamination of the ground water at a depth of about 30 m beneath a paint factory was found to have occurred as a result of the dispersion of organic solvents from a number of underground storage tanks. Samples of water in static and in dynamic conditions were taken from a well excavated to a depth of about 61 m close to the probable source of the pollution. Extraction with methylene chloride on a 100 ml water sample at various pH values yielded three fractions. The basic fraction was found to contain no significant traces of organic compounds, whilst the neutral fraction revealed considerable quantities (several tens of ppm) of various organic solvents commonly used in this kind of industry. These were hydrocarbons, prevalently aromatic, alcohols, glycol ethers, ketones, esters, alkyl halides. Other oxygenated compounds were also found to be present; these were alcoholic or ketonic, aliphatic or aromatic in nature, and their presence could hardly be ascribed to dispersions from the industry responsible for the pollution. Aliphatic and aromatic acids and alkyl phenols, which could not be explained by the products used in the factory, were also found in the acidic fraction. The presence of all these substances was attributed to oxidative microbial degradation of the hydrocarbon components present in the soil and water, which hypothesis was supported by the presence in the water samples of a large number of microorganisms capable of oxidizing aromatic hydrocarbons.

1. INTRODUCTION

Ground waters have natural protection against chemical and biological pollution,which was for a long time considered to be unlikely to occur. Nonetheless,in recent years the municipal water supply of Milan, which derives exclusively from ground waters, has been found to be subject to extensive pollution by chlorinated hydrocarbons and chromates, and at the same time to a constant increase in water hardness. Together with these very widespread contaminations, cases have been found of localized underground pollution from a number of chemical plants, giving the local authorities cause for serious concern. This paper outlines the results of a study of one such case, involving pollution of the aquifer by a paint factory. This kind of industry presents a high pollution hazard because of the wide variety of its raw materials and end products, as also for the large quantities involved. Particularly serious problems arise from storage in large underground tanks (Fig.1a) of thousands of tons of solvents and diluents. The solvents thus stored in the case examined here are listed in Table I, and include hydrocarbons (main ly aromatic), alcohols, ether glycols, ketones, esters, chlorinated solvents. Periodic leakage checks on the inaccessible underground storage tanks revealed certain leakages which represent a hazard to the aquifer because the products leaking out slowly percolate through the unsaturated zone of the subsoil until they reach the water table of the aquifer. Further accidental spillages take place during filling and unloading of tanks, also contributing to pollution. Movement of chemical products dispersed into the ground, which mainly consists of several mixed strata or alternations of sand and gravel of various grain sizes, as can be seen in Fig. 1b, depends on the nature and compactness of the soil. In this case it consists of easily-permeable alluvial deposits.

After the discovery of these dispersions of solvents and diluents, it was decided to sink an inspection well to verify the conditions of the ground water (Fig. 1c). The well is close to the storage tanks, has a diameter of 350 mm and a depth of 61 m, with two perforations between 28 and 40 m and between 49 and 58 m. The well has two purposes: to permit direct sampling at a depth of about 30 m, and to begin operations aimed at cleansing the polluted zone by pumping off considerable quantities of water (20 l/s).

2. EXPERIMENTAL

2.1 Sampling and Preservation of Samples

Samples were taken under static conditions (pump not operating), a suitable sampler being employed to take up water from the well at a depth of 30 m, the samples then being transferred to one-litre dark glass bottles sealed with ground-glass stoppers. Other samples were taken in dynamic conditions from a tap fed by the submerged pump. Samples were maintained at 4°C for some hours until further treatment was carried out.

2.2 Separation of Contaminants from aqueous Matrix

2.2.1 Extraction with methylene chloride at various pH values

Extraction of a 100 ml sample of water was carried out using methylene chloride at acid and basic pH as shown in Fig. 3. Three fractions were obtained: neutral, basic and acidic. These, dried on anhydrous sodium sulphate, were concentrated at 1 ml in a Kuderna-Danish evaporator. The three concentrated solutions were then maintained in the refrigerator at -20°C until analysis by gas chromatography or GC/MS.

2.2.2 Purge and trap

A 25 ml water sample was submitted to purge and trap treatment in the conditions shown in Fig. 4. After dry nitrogen flow for 30 min the trap containing adsorbed pollutants was eluted with 1 ml of high purity carbon disulphide (C.Erba for spectrophotometry). A known quantity of internal standard (1-chlorohexane) was added to the liquid obtained and the solution brought to volume (1 ml) was used for the quantitative determination of aromatic hydrocarbons.

2.3 Analyses

2.3.1 Gas chromatography

A Carlo Erba Mod. 2150 gas chromatograph was used for quantitative analysis of aromatic hydrocarbons under the conditions reported in Table II. Quantitative measurements were carried out by the internal-standard method using a Spectra-Physics Mod. 4100 computing integrator.

2.3.2. Gas chromatography / mass spectrometry

A Hewlett-Packard Mod. 5985 B GC/MS/DS system was used for qualitative analyses and for identification of the components present in the three extracts. Working conditions were as indicated in Table III.

3. RESULTS AND DISCUSSION

Gas chromatographic analysis of the neutral, acidic and basic fractions reveals the presence of a lot of pollutants in the former two (chromatograms in Figs. 5 and 6) whilst the basic fraction showed only one meaningful peak, assigned on the basis of GC/MS findings to 2,4-diamino-6-phenyl--1,3,5-triazine (benzoguanamine), used as a raw material in the preparation of resinous products. Identification of the various components present was carried out by GC/MS using, whenever possible, comparison with the system's NBS library, the numerous spectra not recognized by the computer being interpreted manually. When standards were available, identity was confirmed by comparison with retention times and mass spectra of pure components. Tables IV and V give a list of the most abundant or the most significant pollutants found in the sample.

Some of the compounds identified can easily be recognized as primary pollutants directly dispersed by the paint factory during storage of raw materials and products, or during the various stages of production. Such for example are the aromatic hydrocarbons [benzene, toluene, ethylbenzene, xylenes, components of the solvent naphtha (C_9 aromatic hydrocarbons of molecular weight 120 and 118; C_{10} of m.wt. 134 and 132; C_{11} of m.wt. 148 and 146, and C_{12} of m.wt. 162), naphthalene and methylnaphthalene, biphenyl, acenaphthene, fluorene, phenanthrene, anthracene], the alcohols (isobutanol, n-butanol, cyclohexanol, 2-ethylhexanol), the ketones (acetone, methylethyl ketone, methylisobutylketone, mesityl oxide, diacetonalcohol), the ethers of ethylene glycol (cellosolve, isopropylcellosolve, n-butylcellosolve), the esters (isobutyl acetate, n-butyl acetate, cellosolve acetate), the alkyl halogenides (1,2-dichloropropane, thrichloroethylene, tetrachloroethylene).

Some of the components of the acidic fraction are also present because of dispersion, into the soil or into industrial waste waters, of the water used to rinse drums containing mixtures of fatty acids (sunflower-seed, soy bean, linseed, castor oil fatty acids) used in production processes. These are linear-chain acids such as capronic, enanthic, caprylic, pelargonic, caproic, lauric, myristic, palmitic, oleic and stearic acid.

Other contaminants identified, however, cannot be explained by the raw materials used by the paint industry. These are to be found both in the neutral fraction (aliphatic alcohols and ketones, aromatic alcohols and ketones, and high-pK phenols found both in the neutral and in the acidic fraction), and in the acidic fraction (short-chain and branched-chain aliphatic acids, aromatic acids, phenols). These being prevalently oxygenated compounds, it may be supposed that these are produced by microbial oxidation of the products originally dispersed (aliphatic, alicyclic, terpenic, and mainly aromatic hydrocarbons).

As indicated in the literature (see for example 1-4) many varieties of microorganisms to be found in the soil or in waters are capable of degrading completely numerous organic substances belonging to various chemical classes, which can be used as sources of carbon and energy for their growth (biochemical mineralization processes). This capability evolved over extremely long periods of time thanks to contact with organic compounds of natural origin and to the synthesis of enzyme systems necessary for such degradation. Side by side with mineralization processes, ever-growing interest is being dedicated by the ecologists to the biomedical processes of co-oxidation or co-metabolism in which partial oxidation of certain organic molecules takes place. These molecules cannot be used by the microorganisms as a growth substrate, so if they are to survive they also need another compound which acts as a substrate and undergoes complete degradation. Discharge into the environment of synthetic organic substances having a chemical structure similar to that of compounds that have long been present in nature gives rise to highly complex self-purification mechanisms. When, however, the substances discharged have structures differing from those which the microorganisms are able to recognize, we find phenomena of accumulation in the environment (persistent non-biodegradable pollutants) until a metabolic capacity is evolved which is compatible with these new structures. Even in the case of partial degradation (co-oxidation) intermediates may be formed whose accumulation may represent a danger to human health or to plant and animal life.

A class of substances which has taken on special importance because it is found everywhere in the environment is that of the hydrocarbons. These, in fact, in the form of complex mixtures, are used in large quantities in many fields of human activity. Because their use is so widespread, large quantities are inevitably dispersed during extraction, refining, transport and final destination of petroleum derivatives. Dispersed into the environment, these hydrocarbons are exposed to degradation by microorganisms. This usually starts with oxidation, transforming the original hydrophobic molecules into biologically more active, water-soluble substances. In aliphatic hydrocarbons the terminal methyl is oxidized to $-CH_2OH$, $-CHO$ and $-COOH$ groups. The fatty acid thus obtained may be further metabolized, for example by beta oxidation and decarboxylation, producing a ketone with one carbon atom less than the parent acid. In aromatic hydrocarbons microbial attack under aerobic conditions starts with oxygen fixation on two adjacent carbon atoms on the aromatic nucleus, forming a dihydrodiol and then, by dehydrogenation, an ortho phenol, the starting point of the processes of fission of the ring, giving rise to the mineralization of the compound. Possible substituents (methyls or alkyls whose chain may be of any length and branching degree) may be modified before or after the fission of the ring according to their chemical structure and to the type of microorganism involved. A methyl may be converted to a carboxyl or remain unaltered until the nucleus opens, and an alkyl group may be degraded earlier or later according to the chain length.

These transformations take place prevalently by aerobic processes, which are the most important ones in the degradation of organic compounds dispersed in the environment. Also to be remembered, however, are the effects of the anaerobic processes, which follow a different pattern and lead to different transformation products, and of other degradation processes which are less well understood, but produce the same result.

The products identified in the water sample analyzed are in part products of the microbial oxidation of aliphatic hydrocarbons (alcohols, ketones, fatty acids with short branched chain), terpenic hydrocarbons (aromatic alcohols and ketones, phenols, aromatic acids). The presence of microorganisms able to utilize aromatic hydrocarbons as their sole source of carbon and energy was confirmed by a microbiological analysis of the water sample. Fig. 7 reports some examples of compounds initially present and of partially oxidized compounds deriving from these.

4. CONCLUSIONS

The current practice of burying tanks for storage of solvents for use in the paint industry so as to reduce fire hazards is to be considered ecologically unsuitable in the light of the results obtained. Possible dispersion of the products thus stored, due to corrosion phenomena which are frequent in the ground, is discovered too late, when pollutants have already been discharged into the soil in large quantities and have penetrated to considerable depths. Given enough time these may also enter the aquifer, either maintaining their identity or being transformed into other compounds whose toxicity and persistence may give rise to further environmental hazards, lowering the quality of the ground waters and reducing their range of utilization.

ACKNOWLEDGEMENTS

This work was supported by the Municipality of the City of Milan, and we would like to thank the Councillor in charge of Ecology and Environmental Hygiene, Tino Casali, for the award of a research contract. We also thank Dr. Filippo Boldrino of the Industrial Section of the Office of Hygiene for information regarding the case of pollution studied, and Prof.Vittorio Treccani and his staff for the microbiological analysis of the water sample.

BIBLIOGRAPHY

1. Treccani V., Progr. Ind. Microbiol. $\underline{4}$, 1-33 (1964)
2. Treccani V., Z.Allgem.Mikrobiol. $\underline{5}$(4), 332-41 (1965)
3. Gibson D.T. in "Aquatic pollutants: transformation and biological effects" Hutzinger O., Van Lelyveld I.H. and Zoeteman B.C.J. editors, Pergamon Press, Oxford, 1978, pp. 187-204.
4. Galli E., Baggi G., Treccani V., Pecenick G., Atti del Seminario sull'Ecotossicologia 1979, Consiglio Nazionale delle Ricerche, Roma, 1980, pp. 75-107.

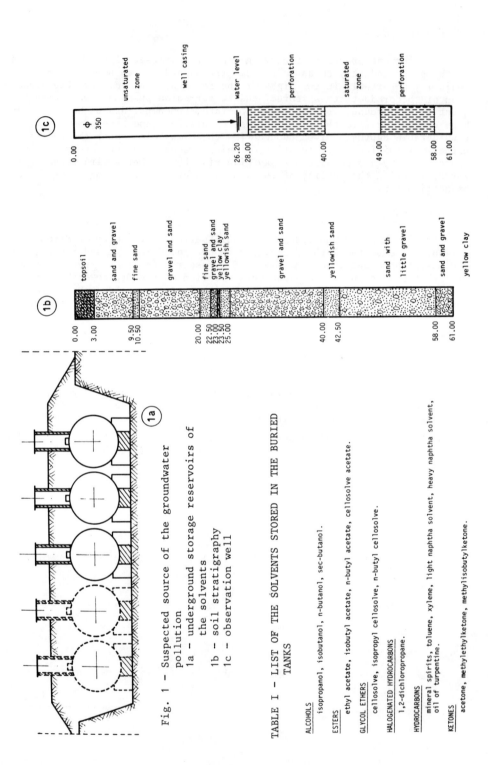

Fig. 1 - Suspected source of the groundwater pollution
1a - underground storage reservoirs of the solvents
1b - soil stratigraphy
1c - observation well

TABLE I - LIST OF THE SOLVENTS STORED IN THE BURIED TANKS

ALCOHOLS
  isopropanol, isobutanol, n-butanol, sec-butanol.
ESTERS
  ethyl acetate, isobutyl acetate, n-butyl acetate, cellosolve acetate.
GLYCOL ETHERS
  cellosolve, isopropyl cellosolve, n-butyl cellosolve.
HALOGENATED HYDROCARBONS
  1,2-dichloropropane.
HYDROCARBONS
  mineral spirits, toluene, xylene, light naphtha solvent, heavy naphtha solvent, oil of turpentine.
KETONES
  acetone, methylethylketone, methylisobutylketone.

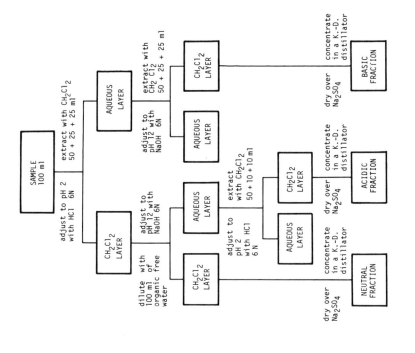

Fig. 3 - Procedure of extraction of a 100 ml portion of the water sample with methylene chloride at different pH values

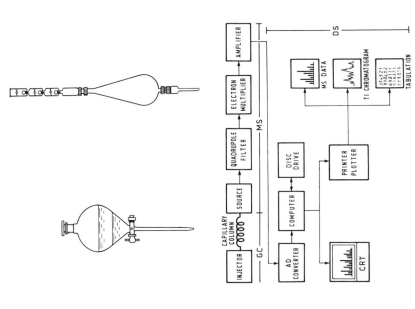

Fig. 2 - Extraction, concentration and GC/MS analysis of the neutral and acidic fraction

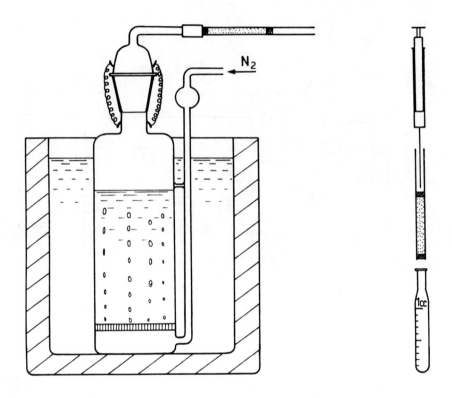

Water sample :	25 ml diluted at 500 ml with organic free water.
Purge gas :	Nitrogen (GC), 50 ml/min, 30 min.
Bath temperature :	70°C .
Adsorbent :	Chromosorb 102 (80-100 mesh), 150 mg, between layers of silanized glass wool.
Tubular trap:	Glass tube, 100 x 6 O.D. x 4 I.D. mm.
Eluent :	Carbon disulfide (Erba for spectrophotometry), 1 ml.

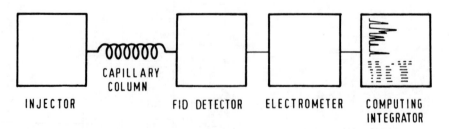

Fig. 4 - Purge and trap, elution and GC analysis

TABLE II - QUANTITATIVE ANALYSIS OF AROMATIC HYCROCARBONS

Gas Chromatograph : Carlo Erba Model 2150

Column : Pyrex glass capillary column, persilanized coated with silicone rubber JXR 0.3 mm I.D., length 15 m, film thickness 2.7 μm

Injection: splitless. Tinj = 275°. Quantity injected : 2 μl of CS_2 solution

Carrier gas : Hydrogen (GC) at 3 ml/min

Oven : room temperature at injection
to 40°C ballistically after 45 s
40°C isothermal for 2 minutes
40 ⟶ 250°C at 3°C/min

Detector : flame ionization (FID). T_{det} = 275°C
H_2 flow : 30 ml/min
Air flow : 300 ml/min

Computing integrator : Spectra-Physics Mod. 4100

Compound	Concentration ppb
TOLUENE	4269
ETHYLBENZENE	1617
m- + p-XYLENE	5385
o-XYLENE	4001
ISOPROPYLBENZENE	1581
n-PROPYLBENZENE	445
1-METHYL-3-ETHYLBENZENE	2074
1-METHYL-4-ETHYLBENZENE	1174
1,3,5-TRIMETHYLBENZENE	952
1-METHYL-2-ETHYLBENZENE	4234
1,2,4-TRIMETHYLBENZENE	1237
1,2,3-TRIMETHYLBENZENE	2420
INDAN	54
1,2,3,4-TETRAMETHYLBENZENE	13
TETRALIN	2590
AROMATIC HYDROCARBONS C_{10}	
NOT IDENTIFIED	
TOTAL	32046

TABLE III - CG-MS ANALYSIS CONDITIONS

GC/MS/DS System Hewlett-Packard mod. 5985 B

Gas Chromatograph: HP 5840 A modified for capillary column use.

Column : Fused silica capillary column coated with SE-54 silicone rubber. 0.32 mm I.D., length 50 m, film thickness 0.17 μm.

Carrier gas: Helium (GC) at 2 ml/min.

Injection: Splitless. T_{inj} = 250°C. Sample injected: 1 μl CH_2Cl_2 solution.

Oven temperature: 32°C at injection
32°C isothermal for 5 minutes
32 ⟶ 250°C at 3°C/min

Interface GC/MS: The column was directly coupled to the ion source. Transfer line temperature = = 250°C

MS : Hyperbolic quadrupole mass filter.

Source: Dual source EI-CI. T_{source} = 200°C. Electron energy = 70 eV

Detector: Electron multiplier at 2200 V

Data system: HP 21MX Series E, with dual disc drive HP 7906
Video terminal HP 2648 A and thermal printer HP 9876 A.

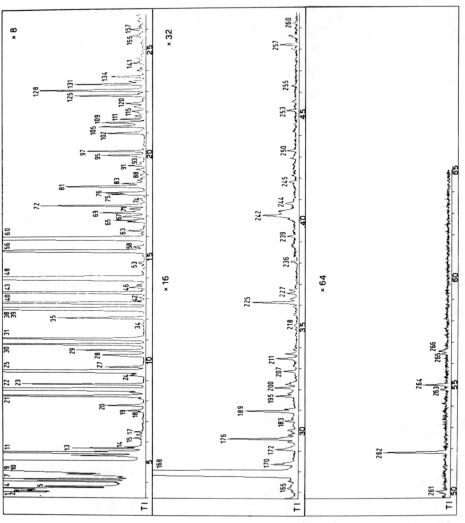

Fig. 5 TI chromatogram of the neutral fraction

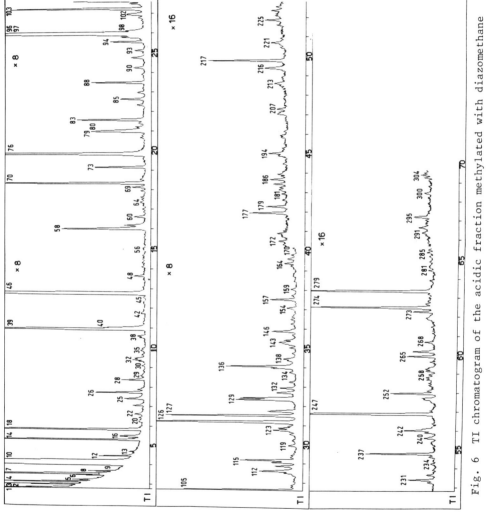

Fig. 6 TI chromatogram of the acidic fraction methylated with diazomethane

TABLE IV – LIST OF COMPOUNDS IDENTIFIED IN THE NEUTRAL FRACTION

PEAK No.	COMPOUND
1	2-BUTANONE
3	ISOBUTANOL
4	BENZENE
5	n-BUTANOL
6	2-PENTANONE
7	TRICHLOROETHYLENE
8	1,2-DICHLOROPROPANE
9	3-METHYL-2-BUTANOL
10	4-METHYL-2-PENTANONE
11	2-ETHOXYETHANOL
12	TOLUENE
13	4-METHYL-2-PENTANOL
14	2-ISOPROPOXYETHANOL
15	ISOBUTYL ACETATE
16	2,4-DIMETHYL-3-PENTANONE
17	4-METHYL-3-PENTEN-2-ONE
18	TETRACHLOROETHYLENE
19	2,4-DIMETHYL-3-PENTANOL
20	2-METHYL-3-HEXANONE
21	4-HYDROXY-4-METHYL-2-PENTANONE
22	ETHYLBENZENE
23	m-/p-XYLENE
24	4-HEPTANONE

PEAK No.	COMPOUND
24	CYCLOHEXANOL
25	o-XYLENE
26	2-HEPTANONE
28	4-HEPTANOL
29	2-BUTOXYETHANOL
30	2-ETHOXYETHYL ACETATE
31	ISOPROPYLBENZENE
32	2-METHYL-4-HEPTANONE
33	TERPENE M.W. 136
34	2,5-HEXANEDIONE
35	5-METHYL-3-HEPTANONE
38	n-PROPYLBENZENE
39	1-ETHYL-3-METHYLBENZENE
40	1-ETHYL-4-METHYLBENZENE
43	1,3,5-TRIMETHYLBENZENE
45	1-ETHYL-2-METHYLBENZENE
48	α-METHYLSTYRENE
49	1,2,4-TRIMETHYLBENZENE
52	t-BUTYLBENZENE
53	BENZOFURAN
55	ISOBUTYLBENZENE
	sec-BUTYLBENZENE
	1,4-CINEOLE

TABLE IV – continued

PEAK No.	COMPOUND
56	1,2,3-TRIMETHYLBENZENE
57	1-ISOPROPYL-3-METHYLBENZENE
58	1-ISOPROPYL-4-METHYLBENZENE
60	INDAN
61	BENZYL ALCOHOL
62	1-ISOPROPYL-2-METHYLBENZENE
63	INDENE
64	1,3-DIETHYLBENZENE
65	1-n-PROPYL-3-METHYLBENZENE
66	1-n-PROPYL-4-METHYLBENZENE
67	1,4-DIETHYLBENZENE
68	n-BUTYLBENZENE
69	1,3-DIMETHYL-5-ETHYLBENZENE
70	o-CRESOL
71	α-METHYLBENZYL ALCOHOL
72	ACETOPHENONE
74	1-n-PROPYL-2-METHYLBENZENE
75	TERPENOID M.W. 152
76	1,4-DIMETHYL-2-ETHYLBENZENE
77	1,3-DIMETHYL-4-ETHYLBENZENE
78	m-CRESOL
79	p-CRESOL
81	1,2-DIMETHYL-4-ETHYLBENZENE

PEAK No.	COMPOUND
82	α,α-DIMETHYLBENZYL ALCOHOL
83	FENCHONE
85	1,3-DIMETHYL-2-ETHYLBENZENE
88	3-METHYLBENZOFURAN
89	2-METHYLBENZOFURAN
91	1,2-DIMETHYL-3-ETHYLBENZENE
92	5-METHYLBENZOFURAN
93	2,6-DIMETHYLPHENOL
94	FENCHYL ALCOHOL
95	1,2,4,5-TETRAMETHYLBENZENE
97	1,2,3,5-TETRAMETHYLBENZENE
100	PHENYLACETONE
101	2-ETHYLPHENOL
102	2-METHYLACETOPHENONE
105	CAMPHOR
106	2-METHYLBENZYL ALCOHOL
107	3-METHYLINDENE
108	1-PHENYL-1-PROPANOL
109	AROMATIC COMPOUND M.W.132
111	1,2,3,4-TETRAMETHYLBENZENE
112	1-METHYLINDENE
113	ALKYLBENZENE C_{11} M.W.148
114	2,4-DIMETHYLPHENOL
115	2,5-DIMETHYLPHENOL

TABLE IV – continued

PEAK No.	COMPOUND	PEAK No.	COMPOUND
117	α,4-DIMETHYLBENZYL ALCOHOL	176	AROMATIC COMPOUND M.W. 146
119	PROPIOPHENON	178	4-INDANOL
120	ALKYLBENZENE C$_{11}$ M.W. 148	183	AROMATIC ALCOHOL M.W. 150
121	BORNEOL	189	2(3H)-BENZOFURANONE
125	3-METHYLACETOPHENONE	195	AROMATIC KETONE M.W. 148
128	NAPHTALENE	200	1-TETRALONE
129	x,y-DIMETHYLBENZYL ALCOHOL	202	BIPHENYL
131	4-METHYLACETOPHENONE	207	AROMATIC COMPOUND M.W. 146
133	AROMATIC ALCOHOL M.W. 150	211	AROMATIC COMPOUND M.W. 146
134	α-TERPINEOL	212	AROMATIC COMPOUND M.W. 146
137	ALKYLBENZENE C$_{11}$ M.W. 148	224	AROMATIC KETONE M.W. 148
136	2,3-DIMETHYLPHENOL	225	AROMATIC KETONE M.W. 148
139	3,5-DIMETHYLPHENOL	226	AROMATIC KETONE M.W. 148
141	TERPENOID M.W. 152	227	AROMATIC KETONE M.W.148
142	x,y,z-TRIMETHYLPHENOL	228	ACENAPHTHENE
143	AROMATIC ALCOHOL M.W. 150	236	DIBENZOFURAN
144	2-ISOPROPYLPHENOL	242	DIETHYLPHTHALATE
154	1-INDANOL	244	FLUORENE
155	ALKYL PHENOL M.W. 136	257	PHENANTHRENE
157	ALKYL PHENOL M.W. 136	258	ANTHRACENE
161	2-INDANOL	262	DIISOBUTYL PHTALATE
168	1-INDANONE	264	DI-n-BUTYL PHTALATE
171	2-METHYLNAPHTHALENE		
173	1-METHYLNAPHTHALENE		

TABLE V – LIST OF COMPOUNDS IDENTIFIED IN THE ACIDIC FRACTION METHYLATED WITH DIAZOMETHANE

PEAK No.	COMPOUND	PEAK No.	COMPOUND
2	METHYL PROPANOATE	83	METHYL BENZOATE
7	METHYL ISOBUTYRATE	84	2,6-DIMETHYLPHENOL
11	METHYL n-BUTANOATE	88	METHYL n-OCTANOATE
18	METHYL ISOPENTANOATE	89	ALIPHATIC ESTER M.W. 172
26	METHYL 2-METHYLBUTYRATE	90	2-ETHYLPHENOL
26	METHYL n-PENTANOATE	91	2,4-DIMETHYLPHENOL
31	METHYL 2-ETHYLBUTYRATE		2,5-DIMETHYLPHENOL
35	METHYL 2-METHYLVALERATE	95	3-ETHYLPHENOL
39	METHYL ISOHEXANOATE		3,4-DIMETHYLPHENOL
45	ALIPHATIC ESTER M.W. 144	96	METHYL o-TOLUATE
46	METHYL HEXANOATE	97	METHYL PHENYLACETATE
48	ALIPHATIC ESTER M.W. 144	98	2,3-DIMETHYLPHENOL
55	ALIPHATIC ESTER M.W. 144	101	ALIPHATIC ESTER M.W. 172
56	ALIPHATIC ESTER M.W. 144	102	3,5-DIMETHYLPHENOL
58	PHENOL		ALIPHATIC ESTER M.W. 172
60	METHYL 2-ETHYLVALERATE	103	METHYL m-TOLUATE
62	o-METHYLANISOLE	104	2-ISOPROPYLPHENOL
65	m-/p-METHYLANISOLE	105	METHYL p-TOLUATE
69	METHYL HEPTANOATE	110	2-n-PROPYLPHENOL
70	ALIPHATIC ESTER	111	METHYL n-NONANOATE
73	METHYL 2-ETHYLHEXANOATE	112	3-ISOPROPYLPHENOL
76	o-CRESOL	113	4-ISOPROPYLPHENOL
79	m-CRESOL		METHYL 2,6-DIMETHYLBENZOATE
80	p-CRESOL	114	ALIPHATIC ESTER

TABLE V - continued

PEAK No.	COMPOUND
115	AROMATIC ESTER M.W. 164
116	ALKYL PHENOL M.W. 136
117	ALIPHATIC ESTER M.W. 186
118	ALKYL PHENOL M.W. 136
119	ALKYL PHENOL M.W. 136
120	ALKYL PHENOL M.W. 136
122	AROMATIC ESTER M.W. 164
	ALKYL PHENOL M.W. 136
123	AROMATIC ESTER M.W. 164
124	AROMATIC ESTER M.W. 164
125	ALKYL PHENOL M.W. 136
126	METHYL 2,5-DIMETHYLBENZOATE
127	METHYL 2,4-DIMETHYLBENZOATE
128	AROMATIC ESTER M.W. 164
129	METHYL 2,3-DIMETHYLBENZOATE
130	METHYL 3,5-DIMETHYLBENZOATE
131	4-INDANOL
132	AROMATIC ESTER M.W. 178
133	METHYL n-DECANOATE
135	METHYL TRIMETHYLBENZOATE
136	METHYL 3,4-DIMETHYLBENZOATE
137	AROMATIC ESTER M.W. 178
139	AROMATIC ESTER M.W. 178
140	AROMATIC ESTER M.W. 178

PEAK No.	COMPOUND
141	AROMATIC ESTER M.W. 178
142	AROMATIC ESTER M.W. 178
143	AROMATIC ESTER M.W. 178
144	AROMATIC ESTER M.W. 178
145	AROMATIC ESTER M.W. 178
146	METHYL TRIMETHYLBENZOATE
149	AROMATIC ESTER M.W. 176
150	AROMATIC ESTER M.W. 178
151	AROMATIC ESTER M.W. 178
152	AROMATIC ESTER M.W. 178
154	AROMATIC ESTER M.W. 178
157	METHYL TRIMETHYLBENZOATE
158	AROMATIC ESTER M.W. 178
159	AROMATIC ESTER M.W. 178
	AROMATIC ESTER M.W. 192
162	AROMATIC ESTER M.W. 176
163	AROMATIC ESTER M.W. 192
164	AROMATIC ESTER M.W. 178
166	AROMATIC ESTER M.W. 178
168	AROMATIC ESTER M.W. 192
170	METHYL TRIMETHYLBENZOATE
173	AROMATIC ESTER M.W. 192
174	AROMATIC ESTER M.W. 176
175	AROMATIC ESTER M.W. 192
177	METHYL LAURATE

TABLE V - continued

PEAK No.	COMPOUND
181	ALIPHATIC ESTER
183	AROMATIC ESTER M.W. 192
184	AROMATIC ESTER M.W. 192
188	AROMATIC ESTER M.W. 192
191	AROMATIC ESTER M.W. 192
193	AROMATIC ESTER M.W. 190
196	AROMATIC ESTER M.W. 190
198	AROMATIC ESTER M.W. 190
199	METHYL n-TRIDECANOATE
200	AROMATIC ESTER M.W. 192
211	ALIPHATIC ESTER M.W. 242
213	AROMATIC ESTER M.W. 192
217	METHYL n-TETRADECANOATE
224	ALIPHATIC ESTER M.W. 270
226	ALIPHATIC ESTER M.W. 256
228	ALIPHATIC ESTER M.W. 256
231	METHYL n-PENTADECANOATE

PEAK No.	COMPOUND
237	UNKNOWN PHTHALATE
240	ALIPHATIC ESTER
242	METHYL 9-HEXADECENOATE (c)
245	UNKNOWN PHTHALATE
247	METHYL PALMITATE
252	PHTHALATE M.W. 278
256	ALIPHATIC ESTER
259	ALIPHATIC ESTER M.W. 284
260	METHYL 2-METHYLHEXADECANOATE
261	ALIPHATIC ESTER M.W. 284
266	METHYL HEPTADECANOATE
273	METHYL 9,12-OCTADECADIENOATE
274	METHYL 9-OCTADECENOATE (c)
279	METHYL STEARATE
285	METHYL 2-METHYLOCTADECANOATE
291	ALIPHATIC ESTER M.W. 312
303	ALIPHATIC ESTER M.W. 312

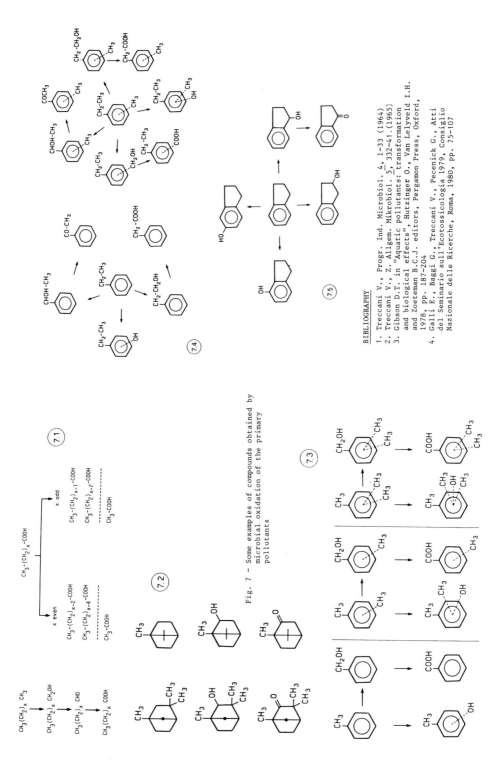

Fig. 7 - Some examples of compounds obtained by microbial oxidation of the primary pollutants

BIBLIOGRAPHY

1. Treccani V., Progr. Ind. Microbiol. 4, 1-33 (1964)
2. Treccani V., Z. Allgem. Mikrobiol. 5, 332-41.(1965)
3. Gibson D.T. in "Aquatic pollutants: transformation and biological effects", Hutzinger O., Van Lelyveld I.H. and Zoeteman B.C.J. editors, Pergamon Press, Oxford, 1978, pp. 187-204
4. Galli E., Baggi G., Treccani V., Pecenick G., Atti del Seminario sull'Ecotossicologia 1979, Consiglio Nazionale delle Ricerche, Roma, 1980, pp. 75-107

COMPARISON OF THREE METHODS FOR ORGANIC HALOGEN DETERMINATION
IN INDUSTRIAL EFFLUENTS

G.E. CARLBERG and A. KRINGSTAD
Central Institute for Industrial Research
P.B. 350 Blindern, Oslo 3, Norway

Summary

Three different methods for organic halogen (OX) determination in effluents from bleacheries in the pulp and paper industry have been compared. One of the methods determines extractable OX, the second determines purgable and carbon adsorbable OX, while the third method determines the high and low molecular weight OX after a preseparation.

1. INTRODUCTION

In Norway the State Pollution Authorities has set a limit on the discharge of organochlorine compounds (OX) from the pulp and paper industry (10 kg organochlorine per ton bleached pulp). The effluents from this industry contain a broad spectrum of chlorinated organic substances, including aliphatic and aromatic hydrocarbons, aldehydes, ketones, acids, phenols, thiophenes and higher molecular weight lignin degradation products.
    The aim of the present investigation was to find a suitable analytical method for determination of OX in the industrial effluents. This was done through a comparison of three different methods.

2. EXPERIMENTAL

    The bleacheries in the factories being investigated all have two separate effluents, acidic and alkaline effluents.
    The acidic effluent contains the spent bleach liquors from the acidic bleaching stages (chlorine and chlorinedioxide stages) and the alkaline effluent contains the spent liquors from the alkaline extraction stages. The effluents were kept at 4 °C until analysis. All effluents were treated with sulphite to get rid of chlorine and chlorinedioxide before the analysis.

The extraction method (1)
    The pH of the sample (200 ml) was adjusted to pH $\sim$ 2 (conc. sulphuric acid) and the sample was extracted twice with butylacetate (15 ml) for 1 h. The two butylacetate extracts were washed twice with distilled water (5 ml) at pH 2 and dried over sodiumsulphate. The content of organic chlorine in the extracts was determined after neutron activation analysis.

The carbon adsorption method
    The Dohrmann DX-20 total organic halide analyser unit was used for this determination.
    The pH of the sample was adjusted to pH $\sim$ 2 (conc. nitric acid). The sample was diluted 300 times and 100 ml of the diluted sample was passed through two glass columns in serial, each containing 40 mg carbon

at a rate of 3 ml/min under nitrogen pressure. The columns were washed with 3 ml of a 5000 ppm nitrate solution. Thereafter the carbon from each column was combusted separately and the organic halogen determined microcoulometricly. From some of the samples the amount of purgeable organic halogen was determined. 10 ml of sample was purged at 45 °C for 10 minutes with carbondioxide.

The ultrafiltration/XAD-4 adsorption method (2)

The effluent sample (50 ml) was divided in a high molecular weight fraction (M > 1000) and a low molecular weight fraction (M < 1000) using an Amicon 404 ultrafiltration cell with a filter with a nominal cut-off level of molecular weight 1000 (Amicon YM2 filter).

The pH of the filtrate was adjusted to pH 2 with conc. sulphuric acid and sodium nitrate was added to a final concentration of 0.5 mol/l. The filtrate was passed through a XAD-4 column (15 x 200 mm) at a rate of 1 ml/min. The retained substances were eluted with acetone.

The organic halogen in the low and high molecular weight fractions was determined by potentiometric titration after combustion in a Schöninger flask.

3. RESULTS AND DISCUSSION

Typical differences between the three methods in determining organic halogen in bleachery effluents are shown in the table. The actual concentrations ranged from 50 to 400 mg OX in the effluents. The concentrations in the table are however given as per cent of the highest value.

TABLE

Relative organic halogen concentrations in bleachery effluents determined by three different methods. Actual concentrations 50 to 400 mg/l. The concentrations are given as per cent of the highest value.

Sample	Extraction EOCl	Dohrmann DX-20 POX	AOX	Ultrafiltration/ XAD-4 adsorption OX		
				M > 1000	M < 1000	Total
Alkaline effluent	18		100	62	18	80
Acidic effluent	38	0.3	100			72

In this investigation the acidic effluent was not ultrafiltrated before the XAD-4 adsorption. The reason being that previous investigations have found acidic bleachery effluents to contain mainly low molecular weight compounds. Higher concentrations (up to 10 per cent) would have been obtained if the ultrafiltration had been performed before the resin adsorption. This clearly shows the influence of incomplete resin adsorption on the concentration determination.

One advantage with the ultrafiltration/XAD-4 adsorption method is that it determines the molecular weight distribution of the organicchlorine compounds in the sample. In the alkaline effluent for instance about 20 per cent of the organochlorine was bound to low molecular weight compounds. Another advantage with this method is that only part of the fractions obtained are used in the detection step. The rest of the fractions can therefore be used for other types of analysis.

The capacity of the carbon columns in the carbon adsorption method is rather limited (up to 200 µg OX) and the samples therefore had to be diluted before the analysis. Two carbon columns in serial are used in this method and the first column always contained 84 per cent or more of the OX determined.

The crucial step in the carbon adsorption method is the packing of the carbon columns and it takes some experience before constant results are obtained.

Compared to the carbon adsorption method the extraction method could only account for 18 and 38 per cent of the OCl in the two samples. The extraction method determines only those chlorinated compounds which are extractable and that means mainly low molecular weight compounds. It is typical for this method that the relatively highest concentration was found in the acidic effluents which mainly contains low molecular weight compounds.

Butylacetate was used as solvent in the extraction method. This increases the extraction efficiency of the more polar compounds in the samples compared to extraction with non polar solvents. The butylacetate will however contain some water and thereby inorganic chloride after the extraction. The extract must therefore be washed with water to get rid of the inorganic chloride and during this washing some of the organic chlorine will also be lost. One advantage with the neutron activation method is that it is possible to determine the concentration of each halogen separately.

The acidic effluent was found to contain 0.3 per cent of purgeable organic chlorine. This is rather low when considering that this effluent contains a large portion of low molecular weight compounds. The results therefore indicates that the low molecular weight compounds in this effluent is fairly polar and water soluble and therefore little amendable to purging.

The carbon adsorption method gave the highest concentrations of the three methods being compared and it was also the fastest to perform. The ultrafiltration/XAD-4 adsorption method is somewhat more time consuming. The method does, however, give additional information on the molecular weight distribution in the sample and the fractions obtained can be used for further analysis. The extraction method is simple to perform. The method uses neutron activation analysis as detection method but this technique is not readily available. Combustion of the extract followed by other detection methods may be used instead of the neutron activation.

Contrary to the other two methods a diluted sample had to be used in the case of the carbon adsorption method. This is due to the small capacity of the carbon columns used.

The detection limit of the carbon adsorption and the extraction methods is about 10 µg organic halogen per liter. The detection limit for the ultrafiltration/XAD-4 adsorption method is about 1 mg organic halogen per liter. Work, is however, in progress to reduce the detection limit for this method down 40 µg organic halogen per liter.

The relative standard deviation is less than 5 per cent for all the three methods investigated.

The suitability of the carbon adsorption and the ultrafiltration/XAD-4 adsorption methods for analysing OX in industrial effluents is further investigated in a joint project between the Swedish Forest Products Research Laboratory and our institute.

4. ACKNOWLEDGMENT

This work was financially supported by the Norwegian State Pollution Authorities.

The authors would like to thank K. Johnsen, The Norwegian Pulp and Paper Institute for performing the OX determination of the fractions from the ultrafiltration/XAD-4 adsorption method.

5. REFERENCES

1. A. Bjørseth, G.E. Carlberg and M. Møller. Determination of halogenated organic compounds and mutagenicity testing of spent bleach liquors. Sci. Total Environ. 11 (1979) 197.

2. L. Sjöström, T. Rådeström and K. Lindström. Determination of total organic chlorine in spent bleach liquors. Svens Papperstidn. 3 (1982) R7.

ORGANIC MICROPOLLUTANTS IN SURFACE WATERS OF THE GLATT VALLEY, SWITZERLAND

M. AHEL*, W. GIGER, E. MOLNAR-KUBICA and C. SCHAFFNER
Swiss Federal Institute for Water Resources and Water Pollution Control (EAWAG), 8600 Dübendorf, Switzerland

Summary

Pentachlorophenol, 2,3,4,6-tetrachlorophenol, 4-nonylphenol, 4-nonylphenolmonoethoxylate, 4-nonylphenoldiethoxylate, tetrachloroethylene and 1,4-dichlorobenzene were quantitatively determined in surface waters of the Glatt Valley, Switzerland. Monthly samples were collected during one year from four locations along the Glatt River, from two small creeks and from the effluents of two municipal wastewater treatment plants. The longitudinal concentration and load profiles found in the Glatt River indicated that all of these micropollutants are mainly introduced into the river via treated wastewater effluents. The highly chlorinated phenols remain unchanged in the river, while the other pollutants are affected by various physico-chemical and biological processes. Depth profiles of tetrachloroethylene in the Greifensee, the lake which feeds the Glatt, showed that this compound was eliminated from the epilimnion by transfer to the atmosphere and by discharge into the Glatt River.

1. INTRODUCTION

Among the organic chemicals which are introduced into the aquatic environment via effluents from sewage treatment plants are compounds which are strongly resistant to biodegradation (e.g. chlorinated compounds). During biological wastewater treatment, other chemicals are transformed to more persistent products which then are discharged into the receiving waters. Alkylphenolethoxylates with short ethoxylate chains belong to this latter group [1].

In an earlier study, we have determined volatile micropollutants in the Glatt River [2], a contaminated Swiss river. This contribution is a preliminary report on an investigation of volatile and phenolic water pollutants in surface waters of the Glatt Valley. Longitudinal concentration and load profiles of the Glatt River, two small creeks, treated sewage effluents and depth profiles in the Greifensee were studied to assess the sources and the fate of these pollutants.

*On leave of absence from the Center for Marine Research, "Rudjer Bosković" Institute, Zagreb, Yugoslavia

2. STUDY AREA AND SAMPLING PROGRAM

The Glatt Valley is located Northeast of Zürich and is densely populated in its upper part (240'000 inhabitants, including parts of the city of Zürich). Figure 1 shows a map of the Glatt Valley including the sewage treatment plants and the sampling locations used for this study. The Glatt River is the major river in the Glatt Valley. It is the outflow of the Greifensee and a tributary to the Rhine River. The Glatt River has an average discharge rate increasing from 3 to 9 m^3/s and it receives water from ten mechanical-biological treatment plants of municipal wastewater. In the lower part of the river 15 to 20% of the water are treated wastewaters. Chimlibach and Chriesbach are two small creeks (0.03 to 0.4 m^3/s) receiving relatively high amounts of treated municipal wastewaters. They are tributaries to the Glatt River. The Greifensee is a highly eutrophic lake with a maximum depth of 32 m. It is stratified from spring to fall and mixed during winter. The Greifensee receives effluents from several treatment plants of municipal wastewaters.

During one year (February 1982 to January 1983), grab samples were collected monthly from four locations along a longitudinal transect of the Glatt River (stations A, B, C and D), from the Chimlibach (station E), from the Chriesbach (station F) and from the effluents of the treatment plants Zürich-Glatt (station H) and Niederglatt (station I). A series of depth profiles of the water column at the deepest point of the Greifensee were sampled from October 1982 to August 1983.

3. ANALYTICAL METHODS

Pentachlorophenol and 2,3,4,6-tetrachlorophenol were determined by a method based on the procedure described by Renberg and Lindström [3]. The lipophilic phenols were extracted by percolating 0.5 ℓ of the acidified water sample (pH 2) through a SepPak C_{18} cartridge (Waters). The adsorbed phenols were eluted with 1.5 mℓ of acetone and acetylated by adding 50 µℓ of acetic anhydride. The excess anhydride was then destroyed by adding 3 mℓ of 0.1 M aqueous K_2CO_3, and the acetylated phenols were extracted with 2 mℓ of pentane. The pentane extract was analyzed by high-resolution gas chromatography (HRGC) with glass capillary columns using electron capture detection. 2,4,6-Tribromophenol was used as internal standard. The detection limit was approximately 20 ng/ℓ.

Tetrachloroethylene and 1,4-dichlorobenzene were concentrated from the water samples by the closed-loop gaseous stripping/adsorption/elution procedure developed by Grob and Zürcher [4]. The water samples (typically 1 ℓ) were stripped for 90 min at 30°C, and the organic compounds were trapped by adsorption on a filter of 1.5 mg of activated charcoal. The filter was then extracted with 20 µℓ of carbon disulfide (CS_2) and the extract analyzed by HRGC. In some Greifensee samples tetrachloroethylene was determined by direct aqueous injection into glass capillary gas chromatography according to Grob and Habich [5].

Nonylphenol, nonylphenolmono- and diethoxylates were extracted from the water samples in a closed-loop apparatus for continuous steam-distillation and solvent-extraction of the distillate [6]. After adding 30 g of sodium chloride, water samples of 2 ℓ were kept under reflux for three hours. 1 - 2 mℓ cyclohexane were used as extracting organic solvent. The extracts were dried over sodium sulfate. After adding a known amount of 2,4,6-trimethylphenol as an internal standard to the extract, aliquots (up to 100 µl) were analyzed by normal-phase high-performance liquid chromatography (HPLC).

HRGC: Carlo Erba gas chromatographs equipped with a Grob-type splitless injector were used. Gas chromatographic separations were performed on persilylated and immobilized glass capillary columns, kindly supplied by K. and G. Grob [7].

HPLC: Solvent delivery was performed by Waters pumps and programmer. For detection a Perkin Elmer LC-55 detector was applied. The following chromatographic conditions were used:
Column: Lichrosorb-NH_2, 250 x 4.6 mm; eluent A: hexane, eluent B: hexane/isopropanol (1/1), elution gradient: from 2 to 30% B in 25 min at a flow of 2 ml/min; detection wavelength: 277 nm.

Examples of resulting chromatograms are presented in another contribution to this volume [8] and detailed procedures will be reported elsewhere [9].

4. RESULTS AND DISCUSSION

Figure 2 shows concentration (µg/l) and load (g/h) profiles for pentachlorophenol and 2,3,4,6-tetrachlorophenol along a longitudinal transect of the Glatt River. Average values from twelve determinations are given. Pentachlorophenol concentrations were increasing steadily from 0.04 µg/l at the outflow of the Greifensee (station A) to 0.17 µg/l at station C and stayed approximately constant to station D. The load profile was increasing along the whole transect. Similar features were observed for 2,3,4,6-tetrachlorophenol but at lower levels and less pronounced than for pentachlorophenol. Results of the individual determinations of pentachlorophenol plotted versus time and versus the flow rate of the river are depicted in Fig. 3A and 3B, respectively. The measured concentration ranges are illustrated in both plots. Maximum concentrations at stations C and D were around 0.3 µg/l and were observed at low flow rates. Figure 3B shows that the pentachlorophenol concentrations at the stations B and C depended upon the discharge rates of the Glatt River. The same result was found for station D but is not included in Figure 3B because of graphical clarity. Such a concentration-flow dependence indicates the effect of dilution at high flow rates of the river. In contrast, the concentrations of pentachlorophenol at station A was not dependent upon the flow rate at the Greifensee outflow.

Table I gives average concentrations and observed ranges for pentachlorophenol at the Glatt station C, in the Chimlibach (station E), in the Chriesbach (station F) and in the effluents from two mechanical-biological treatment plants of municipal wastewaters (stations H and I). High levels of approximately 0.2 µg/l were found in the two small creeks which have high contributions from treated sewage effluents. The maximum concentration observed in the Chriesbach was 0.47 µg/l. Both sewage effluents averaged around 0.6 µg pentachlorophenol per litre water with maximum concentrations of 1.5 µg/l.

The results for pentachlorophenol and 2,3,4,6-tetrachlorophenol indicate that these water pollutants are discharged into the Glatt River mainly from sewage treatment plant and that no significant elimination occurs in the river.

Figure 4 shows the average concentrations from four determinations of 4-nonylphenol (NP), 4-nonylphenolmonoethoxylate (NP1E) and 4-nonylphenoldiethoxylate (NP2E). These compounds are considered as persistent intermediates formed by biological degradation from nonionic surfactants of the 4-nonylphenolpolyethoxylate type [1]. The concentration profiles of NP1E and NP2E showed drastic increases from approximately 0.5 µg/ℓ at station A to 12 and 15 µg/ℓ respectively at station C. These maxima were then followed by sharp decreases to 5 and 6 µg/ℓ at station D. In contrast, NP showed only a small increase to an upper level between 1 and 2 µg/ℓ, not followed by a significant decrease. The maxima in the profiles of NP1E and NP2E are attributed to elimination processes occurring in the river. At present, it is not possible to determine the nature of these processes. Physicochemical processes (sorption, photochemical degradation, volatilisation) and biological transformation could all be involved to cause the observed changes. As can be seen from Table I, NP1E and NP2E are by far the most abundant organic micropollutants which were determined in this investigation.

In Fig. 5 longitudinal concentration profiles for the two volatile compounds tetrachloroethylene and 1,4-dichlorobenzene are shown. The maximum at station C can be explained by the major input of these substances through treated sewage effluents in the upper part of the river followed by elimination in the lower part of the river. In this case transfer to the atmosphere (volatilisation) must be of primary importance because of the large Henry coefficients, the low adsorptivities and the resistance of these chemicals to biodegradation.

The surprisingly high value for tetrachloroethylene at the outflow of the Greifensee was due to particularly high concentrations during winter 1981/2 which then dropped towards the summer, followed by a new, but smaller increase in the winter 1982/3. It was suspected that these concentration changes were caused by the hydrodynamics of the Greifensee which is stratified from spring to fall and is well mixed during winter. We therefore determined tetrachloroethylene concentrations in a series of depth profiles of the water column, taken at the deepest point of the lake. The results of these analyses together with water temperature data are shown in Fig. 6. On the upper abscissum the values determined at the outflow of the Greifensee (Glatt station A) from winter 1981/2 to summer 1982 are also plotted. The first depth profile taken in October 1982 clearly indicates that tetrachloroethylene was depleted in the upper part of the lake above the thermocline (epilimnion). In the deeper part (hypolimnion), tetrachloroethylene remained at a level of 0.5 µg/ℓ. During winter 1982/3 the tetrachloroethylene concentrations in the lake were reduced to approximately 0.2 µg/ℓ stayed constant throughout the water column. Similar behavior of tetrachloroethylene has been observed in the Zürichsee [10] but at much lower concentration levels.

The depth profile of August 1983 showed concentrations around 0.18 µg/ℓ in the hypolimnion, while in the uppermost metres of the water column tetrachloroethylene was left at only 0.05 µg/ℓ. It can be concluded that tetrachloroethylene was removed from the lake by volatilisation and by export via discharge into the Glatt River. Most probably we have observed the declining contamination of the Greifensee after an unknown pollution event.

5. CONCLUSIONS

1. Treated municipal wastewater effluents are the major sources of pentachlorophenol, nonylphenol, nonylphenolmono- and diethoxylate, tetrachloroethylene and 1,4-dichlorobenzene.

2. In the lower part of the Glatt River the concentrations of the different organic micropollutants were affected as follows:

    - pentachlorophenol and nonylphenol remained constant,
    - tetrachloroethylene and 1,4-dichlorobenzene were lowered by mass transfer to the atmosphere,
    - nonylphenolmono- and diethoxylates were decreased by not yet determined elimination processes.

3. An unknown input of tetrachloroethylene had caused elevated concentrations in the Greifensee which were observed in the outflow (Glatt station A) for the first time in February 1982. During stagnation of the Greifensee, tetrachloroethylene remained constant in the hypolimnion while it was depleted in the epilimnion by transfer to the atmosphere and by discharge into the Glatt River.

ACKNOWLEDGMENT

This work was supported in part by the Swiss Department of Commerce (Project COST 64b). We thank K. and G. Grob for supplying glass capillaries for gas chromatography.

REFERENCES

1. Stephanou, E., Giger, W., Environ. Sci. Technol., $\underline{16}$, 800-805 (1982) and
   Giger, W., Stephanou, E., Schaffner, C., Chemosphere, $\underline{10}$, 1253-1263 (1981).
2. Zürcher, F., Giger, W., Vom Wasser, $\underline{47}$, 37-55 (1976).
3. Renberg, L., Lindström, K., J. Chromatogr. $\underline{214}$, 327-334 (1981).
4. Grob, K., Zürcher, F., J. Chromatogr., $\underline{117}$, 285-294 (1976).
5. Grob, K., Habich, A., HRC & CC $\underline{6}$, 11-15 (1983).
6. Veith, G.D., Kiwus, L.M., Bull. Environ. Contam. Toxicol., $\underline{17}$, 631-636 (1977).
7. Grob, K., Grob, G., Blum, W., Walther, W., J. Chromatogr., $\underline{244}$, 197-208 (1982).
8. Giger, W., Ahel, M., Schaffner, C., this volume, p. 91.
9. Ahel, M., Giger, W., in preparation.
10. Schwarzenbach, R.P., Molnar-Kubica, E., Giger, W., Wakeham, S.G., Environ. Sci. Technol., $\underline{13}$, 1367-1373 (1979).

Samples\Compound	Receiving Waters			Secondary Sewage Effluents	
	Glatt (C)	Chimli-bach (E)	Chries-bach (F)	Zürich-Glatt (H)	Nieder-glatt (I)
Pentachloro-phenol	0.17 (0.087-0.32)	0.22 (0.06-0.36)	0.24 (0.13-0.47)	0.62 (0.34-1.5)	0.62 (0.21-1.5)
4-Nonylphenol	1.8 (1.0-2.8)			7.2 (5.2-13.6)	5.8 (2.1-8.9)
4-Nonylphenol-monoethoxylate	12.7 (6.8-15.2)			40 (14-63)	26 (13-42)
4-Nonylphenol-diethoxylate	15.7 (10.3-16.7)			46 (23-72)	29 (7.7-50)
Tetrachloro-ethylene	0.59 (0.22-1.3)	0.055 (0.010-0.18)	0.88 (0.25-4.5)	1.0 (0.31-6.4)	0.16 (0.03-1.1)
1,4-Dichloro-benzene	0.23 (0.081-0.44)	0.44 (0.072-1.9)	0.46 (0.19-1.0)	1.1 (0.63-2.2)	1.1 (0.71-1.6)

Table I Concentrations of phenolic and volatile water pollutants in receiving waters and secondary sewage effluents.
Average concentrations in µg/ℓ
( ): observed ranges.

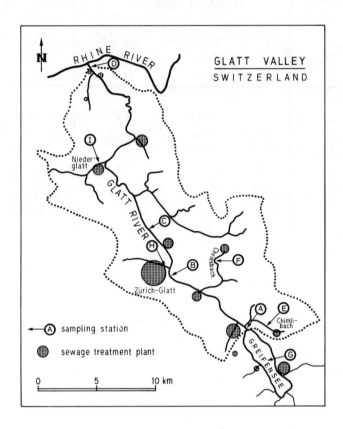

Fig. 1 Map of the Glatt Valley, Switzerland

Sampling stations:

A: Glatt at Fällanden, 0 km
B: Glatt at Hagenholz, 9 km
C: Glatt at Rümlang, 15 km
D: Glatt at Rheinsfelden, 35 km
E: Chimlibach
F: Chriesbach
G: Greifensee
H: Secondary effluent from the sewage treatment plant Zürich-Glatt
I: Secondary effluent from the sewage treatment plant Niederglatt

The diameters of the black dots correspond to the sizes of the sewage treatment plants.

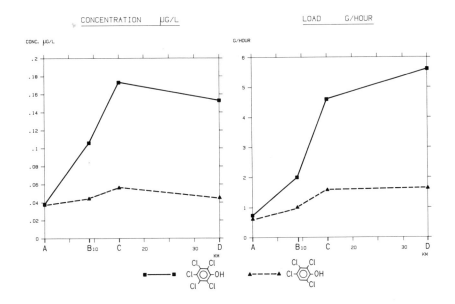

Fig. 2   Concentration and load profiles of pentachlorophenol and 2,3,4,6-tetrachlorophenol in the Glatt River. Yearly averages of twelve monthly determinations.

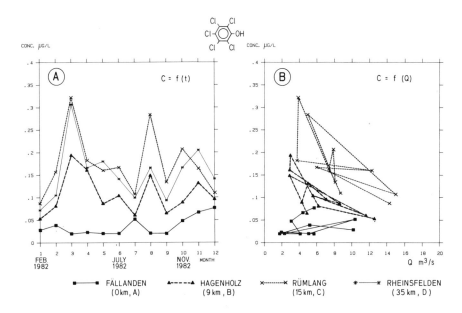

Fig. 3   Pentachlorophenol concentration in the Glatt River as a function of time (A) and flow rate of the river (B).

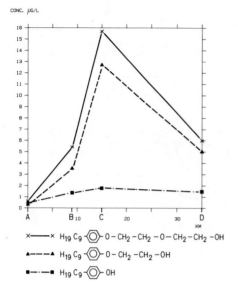
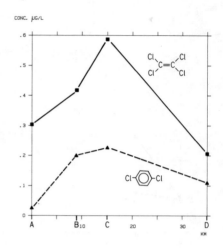

Fig. 4 Concentration profiles of 4-nonylphenol and 4-nonylphenolethoxylates in the Glatt River. Averages of four determinations.

Fig. 5 Concentration profiles of tetrachloroethylene and 1,4-dichlorobenzene in the Glatt River. Yearly averages of twelve monthly determinations.

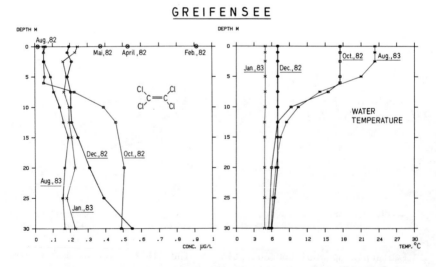

Fig. 6 Depth profiles of tetrachloroethylene and 1,4-dichlorobenzene concentrations and water temperature in the Greifensee.

– 288 –

SPECIES AND PERSISTENCE OF POLLUTANTS IN THE POND WATER FROM AN ORCHARD AREA TREATED WITH ORGANOPHOSPHORUS PESTICIDES

V. Drevenkar, Z. Fröbe, B. Štengl and B. Tkalčević

Institute for Medical Research and Occupational Health,
Zagreb, Yugoslavia

Summary

Dimethyl- and diethyl-thiophosphates and -dithiophosphates were pursued before and after seasonal application of organophosphorus pesticides in the water from three ponds located in an apple orchard. The treatment of fruit-trees with pesticides led to local soil pollution causing the formation of degradation products which are proposed to be rinsed to the pond water. After three treatments during the summer months the level of analysed species in the pond water increased 4 - 166 times. The compounds were quite resistant to further degradation and could be detected in trace amounts as long as one year after pesticides application.

1. INTRODUCTION

One of the principal advantages claimed for organophosphorus pesticides is their ability to undergo degradation and/or chemical alteration shortly after application. However, diesters formed by initial hydrolysis of one functional group are not completely harmless and are usually far more stable than the parent triesters.

The reiterated poisonings of fish masses in three small ponds located in an apple orchard, ascribed to the presence of pesticides, initiated the investigation of the effect of seasonal application of organophosphorus pesticides on the purity of pond water. The water samples were collected two months and immediately before (April and July) and after spraying of fruit-trees with pesticides (September). The samples were analysed for the degradation products dimethyl- and diethyl-thiophosphates /(DMTP)$^-$ and (DETP)$^-$/ and -dithiophosphates /(DMDTP)$^-$ and (DEDTP)$^-$/. These species were accumulated from the pond water by adsorption of their ion-associates with tetraphenylarsonium cation, $(Ph_4As)^+$, on activated carbon microcolumn and elution with dichloromethane (1).

2. EXPERIMENTAL

2.1 Determination of dialkylphosphorus anions in water

A volume of 50 cm^3 or 1 dm^3 of a water sample (8<pH<9), containing (DMTP)$^-$, (DETP)$^-$, (DMDTP)$^-$ and (DEDTP)$^-$ and (Ph$_4$As)$^+$OH$^-$ in the anion to cation molar ratio not lower than 1:2, was passed through an activated carbon microcolumn (5x5 mm). The column was eluted with 3 cm^3 of methanol followed by 7 cm^3 of dichloromethane. Depending on the concentrations of analysed organophosphates an aliquot of eluate or the whole eluate concentrated to 1 cm^3 under a stream of nitrogen was methylated with diazomethane before gas chromatographic analysis.

2.2 Persistence of phosalone in pond water

An amount of 34.808 mg of phosalone was added to 3 dm^3 of Pond 1 water. Changes in the concentration of phosalone degradation products (DETP)$^-$ and (DEDTP)$^-$ in the aerated water were determined by analysis of 50 cm^3 aliquots of water solution during nine days.

2.3 Persistence of dialkylphosphorus anions in pond water

The concentrations of (DMTP)$^-$, (DETP)$^-$, (DMDTP)$^-$ and (DEDTP)$^-$ in a 20 dm^3 aerated Pond 2 water sample with related silt added were measured during a period of 35 days by analysis of 1 dm^3 of water every seven days.

2.4 Gas chromatographic analysis

Varian Aerograph Series 2800 gas chromatograph with Alkali Flame Ionization Detector (Rb$_2$SO$_4$) was used. A glass 1.8 m x 2 mm i.d. gas chromatographic column was packed with 4% SE-30 + 6% OV-210 on Gas Chrom Q, 0.16 - 0.20 mm, starting with a 10 cm length of Chromosorb W/NAW, 0.16 - 0.20 mm, coated with 10% Carbowax 20M. Operating temperatures and gas flows: column 85 °C, injector 200 °C, detector 210 °C; nitrogen carrier gas 30 cm^{3}min^{-1}, air 235 ± 10 cm^{3}min^{-1}, hydrogen 35 ± 3 cm^{3}min^{-1}.

The identity of dialkylphosphorus anions analysed in pond water samples was confirmed by a capillary gas chromatographic-mass spectrometric analysis (GC-MS) performed by Mr. Christian Schaffner, EAWAG, Dübendorf, Switzerland.

3. RESULTS

The results of pond water analysis are presented in Table I. The evaluation was made by standards i.e. aqueous solutions with correspondent dialkylphosphorus anions treated in the same way as the pond water samples. The concentrations in the standards were adjusted so that peak heights in the gas chromatograms of samples and standards were nearly equal. The experimental error was thus minimized as the accumulation recoveries of the analysed anions from deionized and pond water were practically identical.

Traces of $(DMTP)^-$, $(DETP)^-$, $(DMDTP)^-$ and $(DEDTP)^-$ were detected in the samples from all three ponds collected in April and July before seasonal spraying. One day after the first pesticide application no great changes in concentration in Pond 1 water were observed. However, the levels of dialkylphosphorus anions measured two months later, i.e. after two more treatments at the end of July and August, increased 4-166 times in all ponds reaching the maximum values of 3.612 µg $(DMTP)^- dm^{-3}$, 25.322 µg $(DMDTP)^- dm^{-3}$, 2.779 µg $(DETP)^- dm^{-3}$ and 12.880 µg $(DEDTP)^- dm^{-3}$. The identity of methylated species was confirmed by GC-MS analysis. For identification of components with retention times corresponding to the compounds analysed the mass chromatograms were reconstructed of ions prominent in their scanned spectra. The molecular ion and two of the most intensive characteristic fragment ions of each compound were chosen.

The experiments carried out with model systems, i.e. air circulated Pond 2 water with related silt added, showed that after 35 days there still remained 10, 18, 23 and 46% of the initially present $(DETP)^-$, $(DMTP)^-$, $(DMDTP)^-$ and $(DEDTP)^-$, respectively (Fig. 1). On the other hand, after addition of 11.623 mg dm^{-3} of organophosphorus pesticide phosalone to the pond water the relatively small concentrations of the formed $(DETP)^-$ (0.048 µg cm^{-3}) and $(DEDTP)^-$ (0.919 µg cm^{-3}) were found to be nearly constant during nine days of measurement.

4. CONCLUSIONS

The results obtained in the experiment with phosalone support the presumption that after application in the apple orchard organophosphorus pesticides are predominantly degraded before reaching the pond water by rinsing of the soil. For a definite proof the parent compounds have to be analysed simultaneously with their degradation products in the future experiments.

After summer spraying $(DMTP)^-$, $(DETP)^-$, $(DMDTP)^-$ and $(DEDTP)^-$ levels significantly increased in all ponds. The

Table I - Concentrations of dialkylphosphorus anions in pond water from an orchard treated with organophosphorus pesticides

Time of sampling	Pond	Concentration, µg dm^{-3}			
		$(DMTP)^-$	$(DMDTP)^-$	$(DETP)^-$	$(DEDTP)^-$
Before the first treatment					
APRIL	1	0.029	NM	0.077	0.254
	2	0.023	NM	0.113	0.125
	3	0.072	NM	0.188	0.247
JULY	1	0.318	0.624	0.298	0.196
	2	0.038	0.417	0.151	0.078
	3	0.057	0.825	0.538	0.202
One day after the first treatment					
JULY	1	0.329	0.547	0.260	0.225
15 days after the last treatment					
SEPTEMBER	1	2.076	9.568	1.090	2.452
	2	1.220	11.044	2.779	12.880
	3	3.612	25.322	0.837	0.271

NM = not measured

highest increase was in Pond 2 (18-166 times). The amount and type of degradation products obviously depend on the characteristics of the applied pesticide and location of the pond in the orchard territory as well as on the type of the soil.

A relatively slow rate of disappearance of dialkylphosphorus anions from the pond water indicates that these compounds are quite resistant to further degradation. They persist up to several months after spraying and may be detected in trace amounts as long as one year after pesticide application. Therefrom a slow but steady long-termed increase of the pollution level of the ponds can be anticipated.

Reference

1. V. Drevenkar, Z. Fröbe, B. Štengl and B. Tkalčević, Improved accumulation of organophosphates from aqueous media by formation of ion-associates with tetraphenylarsonium cation. In "Analysis of Organic Micropollutants in Water", A. Bjørseth and G. Angeletti (eds.), D. Reidel Publishing Comp., Dordrecht, Boston, London (1982)

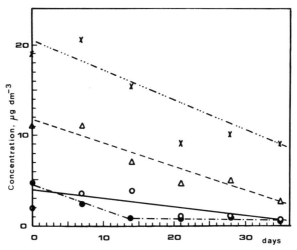

Fig. 1 - Persistence of dialkylphosphorus anions in air circulated water of Pond 2 with related silt added

○——— (DMTP)$^-$
●–·–· (DETP)$^-$
△– – – (DMDTP)$^-$
✕–··– (DEDTP)$^-$

OCCURRENCE AND ORIGIN OF BROMINATED PHENOLS IN BARCELONA's WATER SUPPLY

J. RIVERA* and F. VENTURA**
* Institut de Quimica Bio-Orgànica. J. Girona Salgado, s/n.
  Barcelona-34. SPAIN
** Sociedad General de Aguas de Barcelona. Passeig Sant Joan, 39
   Barcelona-9. SPAIN

Summary

Systematic monitoring of levels of THM's in Llobregat river and treated water has proved the special importance of brominated THM's occuring after chlorination for disinfection purposes. Their origin has been studied in connection to the high contents of bromide salts. In this work we investigate the occurrence of halogenated phenols, special interest being taken in brominated derivatives, as Llobregat river has to bear industrial spills of phenols and phenolic compounds.

INTRODUCTION

The presence of phenol and chlorinated phenols in drinking waters has been shown to affect both taste and odour and to cause negative effects on health (1). The priority pollutants list issued by E.P.A. includes eleven of these compounds as suspected carcinogenics. Many authors have published their results on chlorinated phenols but little attention has been focussed on the presence of bromophenols.

Rook (2), has demonstrated the presence of volatile brominated organic compounds when bromide is present, due to chlorine's action according the following reaction:

$$HOCl + Br^- \longrightarrow HOBr + Cl^- \quad (2)$$

In the same way, brominated phenols might be formed during the chlorination in water works.

Sweetman et al (3) reported the production of bromophenols resulting from the chlorination of distilled water containing bromide and phenol. Kuehl et al (4) have identified them in fish.

Llobregat river that currently supplies water to Barcelona city and its surroundings (fig. 1) -3.2 millions on inhabitants- is a extremely heavy polluted river (Table I) bearings high contents of bromide coming from salt mines in the upper course of the river. Levels of bromide range from 1 mg/l to 3,5 mg/l during 1979-83. Brominated THM's found in Barcelona's tap water (5) in this period, were the most predominant species reaching levels as high as 714 ug/l. Phenol and phenolic compounds are usually found in raw water in ppb range but occasionally spills may increase the phenol content up to ppm. (6 mg/l detected in April 1982).

Many analytical methods for the determination of phenolic compounds in waters using HPLC (6-9) or GC techniques (10-18) have been published. The procedure used to isolate them includes liquid-liquid extraction (6,10-15, 17), concentration by anion-exchange resins (18) or by C_{18} cartridges (19).

Many phenolic compounds are amenable to gas chromatographic analyses either as free phenols (16) or derivatized as acetates (10,11,15) pentafluorobenzylates (13,21) heptafluorobutyrates (17) or pentafluorobenzoylates (12).

In this work, analyses have been carried using liquid-liquid extraction and derivatization of phenols with pentafluorobenzoyl chloride according to the method described by Renberg (12). Gas chromatographic peaks have been identified by their retention time and their spectra by GC-MS. Also mass fragmentograms with significant peaks were obtained.

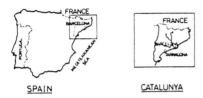

SPAIN   CATALUNYA

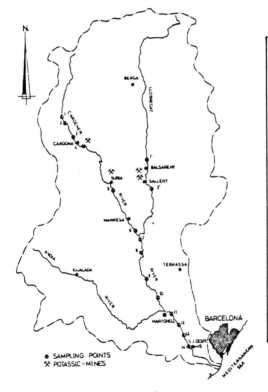

PARAMETERS MG/L	27-7-83	12-9-83
COLOUR (MGPT/L)	35	25
TURBIDITY (MGSiO$_2$/L)	297	410
CONDUCTIVITY (MHO/CM)	2651	2538
T.O.C	12,2	8,3
T.A.C.	185	228
CHLORIDE	730	667
BROMIDE	1,6	1
ANMONIA	1,3	0,8
NITRATES	5,3	-
PHOSPHATES	1,0	-
SULPHATES	205	-
DISSOLVED O$_2$	2,2	5,6
DRY RESIDUE	1901	-
PH	7,85	8,20
PHENOL	-	0,012
HARDNESS	544	-
FE	-	0,37
MN	-	0,35
ZN	038	0,01
CR	0,02	0
CD	0	0
-MEANS "NOT AVAILABLE"		

Fig. 1: Geographical situation

TABLE I: Water quality of Llobregat river

EXPERIMENTAL

Reagents and materials

Solvent used-isooctane-(Fluka, Buchs, Switzerland) was a puriss quality. Potassium hidroxide and sodium bicarbonate were analytical grade (Panreac, Spain). Derivatizing agent, pentafluorobenzoyl chloride (99%) was Fluka. The standards phenol (C.Erba, Italy) (99%), 2 and 4 bromophenols, (95%

and 99%), 2,4 and 2,6 dibromophenols (95% and 97%) and 2,4,6 tribromophenol (98%) were obtained from Fluka. These standards were used whitout further purification.

Gas chromatographic conditions

Konik 2000 gas chromatograph equipped with a ^{63}Ni electron capture detector (Tracor 560) was used. Column was 25 m x 0,25 mm SE-30 fused silica. Gas chromatographic conditions were as follows: H_2 carrier gas 0.6 atm. N_2 as make-up 67 ml/min, injection temperature 250C, detector temperature 330ºC. Temperature programme 65ºC - 300ºC at 8ºC/min. Splitless injection.

Gas chromatography-Mass spectrometry

MS9-VG updated with a Konik 2000 gas chromatograph and VG 11-250 data system was used. Chromatographic conditions as described previously and mass spectrometer operated in EI mode at 4 sec/scan time and 1000 resolution.

Procedure

Studies were performed using raw water from Llobregat river, in order to evaluate the formation of bromophenols varying: a) Cl_2 dose/phenol relationship. b) bromide ion contents of raw water and c) Analysis of real samples along the river course.

a) Cl_2 dose/phenol relationship

Phenol (6 mg/l) was added to a raw water containing bromide (1,6 mg/l not added) and chlorinated at different doses. Contact time was established to 24 hr. The Cl_2 dose-phenol ratio was 1:1, 2:1, 4:1, 7:1, 10:1.

b) Variation of bromide contents

To a raw water containing 1 mg/l of bromide, phenol (6 mg/l) was added and chlorinated at 12 mg/l.To the same raw water, KBr was added to ensure a bromide concentration of 3.5 mg/l. The same amount of phenol and Cl_2 dose were used. Contact time was in both cases 24 hr.

c) Different sampling points

Real samples along the river course were analyzed. Neither phenol nor bromide was added. Samples were chlorinated at 12 mg/l, common chlorine dose at Barcelona's water works.

RESULTS AND DISCUSSION

a) Chlorine dose plays an important role. If chlorine dose/phenol ratio is less than 4:1, halogenated phenols (chlorinated, brominated and mixed) are formed. This is in agreement with Burttschell et al (22). The maximum amount of brominated phenols are obtained at this ratio. When the relationship is changed to values higher than 4:1 ring cleavage of phenol seems to occur. (See figure 2).

b) Increase in bromide contents tends to increase the bromination of the phenolic ring to 2,4,6-tribromophenol, with formation of mono and dibrominated derivatives. A dramatic decrease of chlorinated phenols is observed (See figure 3).

c) Analysis of real samples showed that in normal conditions, brominated phenols are not present in chlorinated raw water from Llobregat river. This is due to several reasons, mainly to low phenol level, and relationship chlorine dose/phenol is higher than 4. This situation is dramatically changed when high spills of phenol are produced (as high as 6 mg/l had been detected). (See fig. 4).

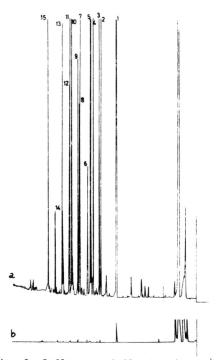

Compounds identified

1 phenol
2 2 Cl phenol
3 4 Cl phenol
4 4 Br phenol
5 4 Br phenol
6 2,4 diCl phenol
7 2,6 diBr phenol
8 unknown
9 2,4 diBr phenol
10 possible Br diCl phenol
11 possible Br diCl phenol
12 unknown
13 possible Cl diBr phenol
14 2,4,6 TriBr phenol
15 unknown

Fig. 2: Influence of Cl_2 dose/phenol relationship
a) ratio 4:1   b) ratio 10:1
Gas chromatographic conditions described in the text

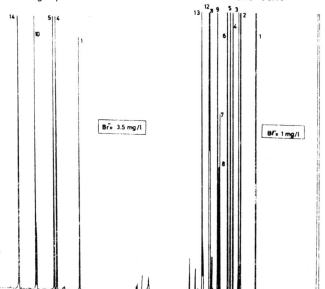

Fig. 3: Variation of bromide contents in raw water (Phenol added 6 mg/l, chlorine dose 12 mg/l) Compounds identification as described above.

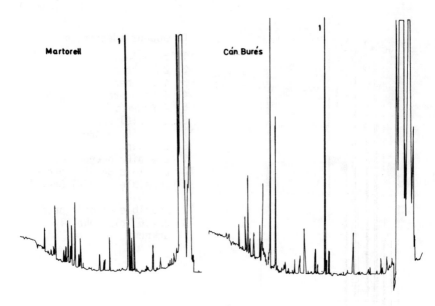

Fig. 4: Analysis of real samples. Left: Sample nº 11. Right: Sample nº 7. Geographical situation shown at fig. 1.

d) Identification of brominated phenols were confirmed by GC-MS. The peak base of pentafluorobenzoylhalogenated phenols is 195 due to $(C_6F_5-CO)^+$ with the molecular peak very small. Mass spectra of some phenols found in the samples of figure 3 are provided. Molecular peaks is magnified in some cases to confirm the unknown compound.(See figure 5).

CONCLUSIONS

Studies performed on Llobregat river raw water showed the formation of bromophenols resulting from occasionally spills of phenol and related compounds and the presence of bromide coming from daily discharges of salt mines in the upper course of the river.

When phenol was added to a raw water, and Cl_2 dose/phenol relationships was ranged from 1:1 to 4:1, we could observe formation of brominated, chlorinated and mixed phenols , with a maximum amount at this ratio. Ring cleavage of phenol was detected at a higher values, as 7:1 and 10:1. The ratio at which phenol cleavage starts to appear will depend on nature of raw water and chlorine consuming substances.

Increase of bromide concentration tends to produce more brominated phenols specially 2,4,6 tribromophenol.

These are previous studies. Current work is done in order to identify all the compounds of the samples to quantify them and to cover a full year of monitoring.

ACKNOWLEDGEMENTS

This work has been partially sponsored by CIRIT (Comissió Interdepartamental de Recerca, Innovació Tecnològica. Generalitat de Catalunya).Direcció General d'Ensenyament Universitari. Conselleria d'Ensenyament. Generalitat de Catalunya.

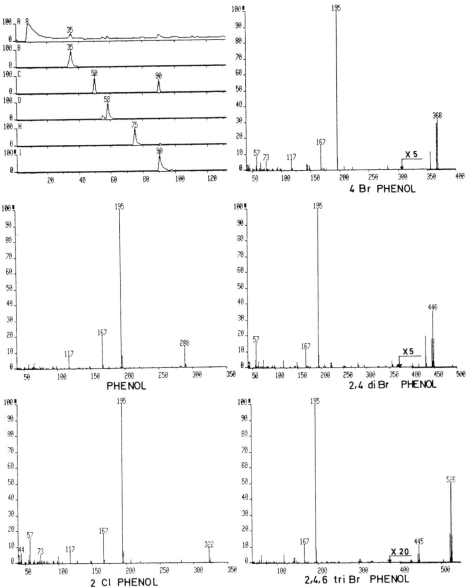

Fig. 5: Mass spectra of some pentafluorobenzoylhalogenated phenols found in the samples of figure 3.

BIBLIOGRAPHY

1- Kozak, V.P., Simsiman, G.V., Chesters, G., Stensby, D. and Harkin, J. Reviews of the environmental effects of pollutants: XI Chlorophenols. Oak Ridge Nat. Lab. ORNL/EIS-128 June 1979 (EPA/600/10).
2- Keith, L.H. and Teillard, W.A. Environ Sci & Techn. 13 (1979) 1469-1471.
3- .Sweetman, J.A. and Simmons, M.S. Water Res. 14 (1980) 287-290
4- Kuehl, D.W., Veith, G.D. and Leonard E.N. Water Chlorination. Environmental Impact and Health Effects (Ed. by Jolley R.L. Gorchev, H. and Hamilton, D.H. Jr) Vol. 2, pp 176-192. Ann Arbor Science Publishers MI.
5- Rivera, J., Ventura, F., Guardiola, J., Perramón, J. and Salvatella, N. J. Int. Water Supply AQUA 5 (1982) 469-474.
6- Realini, P.A. J. Chrom. Sci. 19 (1981) 124.
7- Kung-Chao, G. and Suatoni, J.C. J. Chrom. Sci. 20 (1982) 436
8- Ogan, K. and Katz, E. Anal. Chem. 53 (1981) 160.
9- Blo, G., Donti, F. Betti, A. and Bighi, C. J. Chrom. 257 (1983) 69-79.
10- Chau, A.S.Y. and Coburn, J.A. J.A.O.A.C. 57, 2 (1974) 389
11- Coutts, R.T., Hargesheimer, E.E. and Pasutto, F.M. J. Chrom. 179 (1979) 291-299
12- Renberg, L. Chemosphere 10, 7 (1981) 767-773
13- Fogelqvist, E., Josefsson, B. and Roos, C. J. of HRC & CC 3 (1980) 568
14- Thrun, K.E. and Oberholtzer, J.E. Advances in the Identification & Analysis of Organic Pollutants in Water. Chap. 16, pp. 253-267. Ann Arbor Science Publishers Inc. (Ed. by Keith L.H.) Ann Arbor MI.
15- Voss, R.H., Wearing, J.T. and Wong, A. Ibid. Chap. 53, pp. 1059-1097
16- Giger, W. and Schaffner. Ibid. Chap. 8, pp. 141-154
17- Lamparski, L. and Nestrick, T. J. Chrom. 156 (1978) 143-151
18- Chriswell, C.D., Chang, R.C. and Fritz, J.S. Anal. Chem. 47, 8 (1975) 1325-1329
19- Renberg, L. and Lindstrom, K. J. Chrom. 214 (1981) 327-334
20- Krijgsman, W. and Van de Kamp, C. J. Chrom. 131 (1977) 412-416.
21- Kawahara, F.K. Anal. Chem. 40, 6 (1968) 1009-1010
22- Burttschell, R.H., Rosen, A.A., Middleton, F.M. and Ettinger, M.B. J.A.W.W.A. 51 (1959) 205-214.

THE RELATIONSHIP BETWEEN THE CONCENTRATION OF ORGANIC MATTER IN NATURAL
WATERS AND THE PRODUCTION OF LIPOPHILIC VOLATILE ORGANOHALOGEN COMPOUNDS
DURING THEIR CHLORINATION

M. Picer, V. Hocenski and N. Picer
Center for Marine Research Zagreb, "Rudjer Bošković" Institute,
Zagreb, Yugoslavia

Summary

Water chlorination leads to the generation of various halogenated degradation products of water organic matter. In natural waters humic matter usually presents the largest part of the total organic matter, but in water polluted with pulp and in paper waste waters also lignosulphonates are present in significant concentrations. Water samples collected from the Sava river and its ground waters polluted with pulp and paper mill effluents are laboratory chlorinated. Relatively good positive correlation coefficients were obtained in the comparison of the amounts of humic and lignin matter present in the water samples and the production of halomethanes and various other lipophilic volatile organic matter during water chlorination.

1. INTRODUCTION

Pollution of water with pulp and paper mill effluents is characteristic because of heavy additions of various organic matters, especially lignosulphonates. During bleaching processes significant amounts of chlorine are also used and many chlorinated compounds are found in these effluents (1). However, it is also known that during the chlorination of relatively clean ground waters many organohalogenide compounds are produced (2). Isolation and identification of a large number of organic compounds presents a very difficult task requiring use of expensive and complex facilities and skillful personnel. The aim of this work is to investigate the relationship between the concentrations of organic matter in water samples and the production of lipophilic volatile organohalogenides during their chlorination by using the relatively simple ECD fingerprint method (3).

2. METHODOLOGY

Water samples were laboratory chlorinated adding sodium hypochlorite solution to 1 mgl^{-1} active chlorine in the water sample. Concentrations of lignin matter were estimated using the following methods: Method A - measuring water fluorescence at 285/410 excitation/emission line (4); Method B - measuring water fluorescence on two excitation/emission lines, 313/395 and

313/415 (5); Method C - it is a visible spectrophotometric method and concentrations of lignin matter are estimated by using the standard solution of the tannic acid (6); Method D - it is the standard visible spectrophotometric method, so-called "Tannin-Lignin" method (7). For the estimation of humic matter two spectrofluorometric methods were used: Method A - the measuring of fluorescence of water samples at 313/395 nm and 313/415 excitation/ emission lines (the same as Method B applied for lignosulphonates); Method B - was described by Kramer (8) - fluorescence is measured at 365/460 excitation/emission lines. For the estimation of total organic matter in water the UV spectrophotometric method was used, as described by Reid et al. (9).

The amounts of volatile lipophilic organohalides obtained after water chlorination were estimated by the extraction of water samples with solvents (the solvent mixture: hexane/disopropylether for halomethanes fraction and petroleum ether b.p. 40-50 ^{o}C for the other two fractions) before and after water chlorination. Halomethanes fraction is injected directly into the gas chromatograph but for the other two fractions evaporation of the extract down to 5 or 1 ml is performed. By using the EC detector the peaks on the chromatograms are compared with the $CHCl_3$ standard (halomethane fraction), the trichlorobenzene standard (TCB fraction) and the DDT standard (DDT fraction). The amounts of ECD matter produced during chlorination of water samples are obtained by subtracting the amounts of the ECD matter in the water before chlorination from the amounts of the ECD matter in water samples after chlorination.

3. RESULTS AND DISCUSSION

Figure 1 and 2 present the EC chromatograms of the Sava river samples for all the three fractions obtained before and after chlorination process. Water samples were collected before the inflow of pulp and paper mill wastes (Sava I) and after the inflow of effluents (Sava II).

Estimated concentrations of the ECD matter before and after chlorination of water samples and of produced ECD matter for all the three fractions are presented in Table 1. Besides the Sava river samples also ground water samples obtained from a piezometer located about 20 m away from the Sava river bank were investigated and the results are presented in Table 1. The amounts of ECD fingerprint matter for all the samples and fractions are significantly higher after chlorination than before chlorination. It is interesting to note that with higher temperature ranges concentrations of ECD matter get lower and that the amounts of organohalogenides produced during chlorination get lower too. The halomethane fraction produced during the chlorination of the Sava river water collected at a station situated before the site of the inflow of paper mill effluents is not significantly higher in comparison with ground water samples. But it seems that during the chlorination process significantly higher amounts of organohalogenides are produced with higher retention times in river samples even before the site of the inflow of paper mill waste waters. The Sava river water samples collected further down the location of inflow of waste waters produced during chlorination much higher amounts of ECD matter for all the three

Table 1. The amounts of ECD fingerprint matter before and after laboratory chlorination of water samples

	ECD Matter (ngl^{-1})						Produced ECD matter (ngl^{-1})		
	Before chlorination (Fractions)			After chlorination (Fractions)			(Fractions)		
	HM	TCB	DDT	HM	TCB	DDT	HM	TCB	DDT
Sava I	4.3×10^3	125	48	39.2×10^3	476	191	34.9×10^3	351	143
Sava II	520×10^3	1890	1496	862×10^3	8599	2711	342×10^3	6709	1215
Ground Water	2.7×10^3	5	12	21.7×10^3	182	24	19.0×10^3	177	12

fractions in comparison with river water samples collected before the mentioned site and especially in comparison with its ground waters. For the HM fraction the amount was about 10 times higher but for the TCB and DDT fractions the amounts were about 100 times higher in river water samples than in ground water samples.

Table 2 presents linear correlation coefficients obtained after comparing concentrations of lignin, humic and total organic matter (the UV method) in various river and ground water samples with the halomethane fraction of ECD organic matter produced during their chlorination. With respect to the relationship between concentrations of lignin matter and the produced HM fraction the highest correlation coefficient has been obtained measuring lignin matter using Method A. From the correlation coefficients presented it can be concluded that other methods are less reliable for the estimation of lignin matter in water samples investigated, at least not for the part of these complex materials that during chlorination produced halomethanes. Correlation coefficients obtained after comparing the concentrations of humic matter with halomethanes produced during water samples chlorination are practically equally owing to the earlier mentioned correlation coefficients. The correlation coefficient found using Method B is statistically significant. Statistically is not significant the correlation coefficient obtained by measuring concentrations of total organic matter using the UV method.

Table 2. Correlation between produced halomethane fraction and concentrations of organic matter in chlorinated water samples

Correlations between		No. of pairs	Linear correl. coeff.
Produced HM fraction	Conc.of lignin matter (Method A)	10	0.929
"	Lignin matter (Meth.B)	11	0.660
"	Lignin matter (Meth.C)	11	0.734
"	Lignin matter (Meth.D)	11	0.655
"	Humic matter (Meth.A)	9	0.760
"	Humic matter (Meth.B)	11	0.881
"	UV Total organic matter	11	0.629

4. CONCLUSIONS

From the preliminary results obtained during our investigation of organohalogen materials that have formed during water chlorination we can make the following conclusions:
1. There is a positive correlation between the concentrations of organic matter in the water samples investigated and the formation of organohalogen compound estimated by the ECD fingerprint method.
2. The concentrations of lipophilic volatile organohalogen materials having retention times higher than halomethanes and formed during water samples chlorination are about 100 to even 10,000 times lower in comparison with halomethanes. However, because of their lipophility, higher molecular weights and more complex molecular structure they could be environmentally still important even though in so low concentrations. For that reason we believe that it is necessary to identify these chemicals by using more complex methods, such as for example the GC-MS and GC-IR methods.

5. ACKNOWLEDGEMENT

The authors express their gratitude to the Self-management Community of Interest for the Scientific Research of S.R. Croatia for the financial support.

REFERENCES

1. BJØRSETH.A., LUNDE, G. and GJØS, N. (1977). Acta Chemica Scandinavica. B 31, 797.
2. PICER, M. (1981). Pomorski zbornik 19, 663 (in Croatian).
3. PICER, M. (1981). The use of ECD and FID fingerprint techniques for the evaluation of river water purification contaminated with organic pollutants. Proc. of the Second European Symposium "Analysis of Organic Micropollutants in Water", Killarney, Eds. A. Bjorseth and G. Angeletti, 48-50.
4. WILANDER, A., KVARNAST, H. and Lindell T. (1974). Water res. 8, 1037.
5. ALMGREN, T., JOSEFFSON, B. and NYQUIST, G. (1975). Anal. Chim. Acta 78,411.
6. BARNES, C. A. et al. (1963). Tappi 46, 347.
7. APHA-AWWA-WPCF, Standard methods for the examination of water and waste waters, 14th edition 1975, pp 607.
8. KRAMER, C.J.M. (1979). Degradation of sunlight of dissolved fluorescing substances in the upper layers of the Eastern Atlantic Ocean, Neth. J. Sea Res. 14, 325.
9. REID, J.M., CRESSER, M.S. and McLEOD, D.A. (1980). Observations on the estimation of total organic carbon from UV absorbance for unpolluted stream. Water Res. 14, 525.

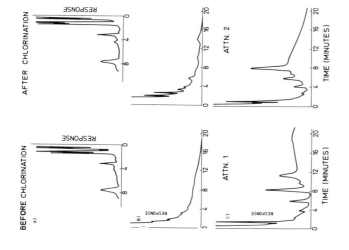

Fig. 1 - The EC chromatograms obtained before and after chlorination of the Sava river water samples (Sava I): a) HM fraction (10% SP 1000 on Chromosorb W, 328°K isothermal); b) TCB fraction (4% SF 96+8% QF1 on Gas ChromQ, 373°K isothermal); c) DDT fraction (the same column as b), 459°K isothermal.

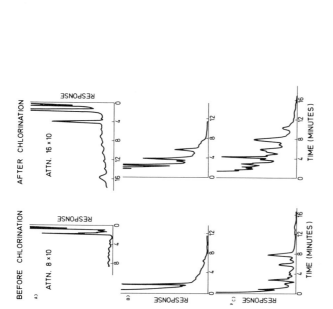

Fig. 2 - The EC chromatograms obtained before and after chlorination of the Sava river water samples (Sava II): a) HM fraction (10% SP 1000 on Chromosorb W, 328 °K isothermal); b) TCB fraction (4% SF 96 + 8% QF1 on Gas ChromQ, 373 °K isothermal); c) DDT fraction (the same column as b), 459 °K isothermal

SESSION IV - FUTURE ENVIRONMENTAL PROBLEMS

Chairman: R. SCHWARZENBACH

Standard setting principles

Integration of the environment research action programme into the framework programme for community scientific and technical activities 1984-1987

COST 64b bis project - Now and the future

STANDARD SETTING PRINCIPLES

R. F. Packham
Water Research Centre,
Henley Road, Medmenham,
Marlow, Buckinghamshire,
ENGLAND

Summary

The general approach to the setting of limits for chemical contaminants of drinking water is considered including the selection of constituents and the toxicological and other information required. In general there is a deficiency of relevant toxicological data but even where good information is available its extrapolation from high to low concentrations and from animal species to man involves considerable uncertainties. The magnitude of these uncertainties has been brought out clearly in the Guidelines for Drinking Water Quality (GDWQ) of the World Health Organization. The use of 'Guideline Values' is a sensible alternative to Maximum Acceptable Concentrations and similar limits used in other standards. GDWQ includes Guideline Values for 27 water constituents of concern in relation to health and of these eighteen are organic compounds.

1. INTRODUCTION

Drinking water standards have been developed to define a quality of water that is safe and acceptable to the consumer. It would be inappropriate to standardise the composition of drinking water in a comprehensive way; the primary purpose of standards is to provide guidance on health related aspects of water quality that are outside the normal competence of those responsible for providing public water supplies. Most drinking water standards therefore set limits for organisms or chemical substances that are dangerous, potentially hazardous or obnoxious to consumers. Obnoxious materials include those that though harmless to health, may have an adverse effect on the use of the water due to taste, odour, colour, turbidity, corrosivity or other undesirable properties.

Drinking water standards of the United States Public Health Service appeared as early as 1914 (1) while those of the World Health Organization (WHO) were first published in 1958 with later editions in 1961, 1962, 1970 and 1971 (2-6). Many countries have national standards and a substantial number of these are based on the WHO International Standards for Drinking Water. The WHO International and European Standards have now been combined and completely revised as WHO Guidelines for Drinking Water Quality (7) to be published in 1983. The European Community has published a Directive on the

Quality of Water for Human Consumption (8) which will be fully implemented in the United Kingdom in 1985.

All drinking water standards include basic criteria for bacteriological quality and there can be no question that these are always of over-riding importance. Concern over chemical components of drinking water has developed in recent years as evidenced by the growth in the number of toxic constituents for which limits have been set (See Table I). The toxic substances included in early drinking water standards were all inorganic and the recent increase in the number of substances of concern listed has been almost entirely due to organic chemicals, in particular organochlorine compounds.

Table I. Limits for drinking water constituents of concern in relation to health

Organisation	Reference	Year	Number of Limits Set		
			Inorganic	Organic	Total
World Health Organization	(2)	1958	5	0	5
	(3)	1961	6	0	6
	(4)	1963	9	0	9
	(5)	1970	8	1	9
	(6)	1971	8	1	9
	(7)	1983	9	18	27
European Community	(8)	1980	9	2	11
US Public Health Service	(9)	1925	1	0	1
	(10)	1942	3	0	3
	(11)	1946	4	0	4
	(12)	1962	8	0	8
US Environmental Protection Agency	(13)	1975	9	6	15
Health & Welfare Canada	(14)	1978	15	17	32

The precise reason for the inclusion of some substances that always appear in drinking water standards is uncertain, but it may stem from anxiety rather than from firm evidence of harmful effects at levels likely to occur in water. The setting of limits has undoubtedly been stimulated by the development of analytical techniques which allow the detection of substances of possible concern to be revealed at very low concentrations. The list of limits is likely to continue to grow for several reasons:

(i) Industry continues to develop new chemicals or new uses of existing chemicals at a rapid rate. The use of industrial chemicals may result in the contamination of a water source.

(ii) There are important gaps in the capability of existing techniques for the analysis of water. These gaps will eventually be filled and the presence of new substances of possible concern will be detected.

(iii) Health effects research may indicate a need to control the concentration of certain water constituents.

The large number of limits for chemical substances does present problems in relation to the feasibility and cost of monitoring particularly where there is a legal requirement to monitor for compliance. Although there is no doubt that all water supplies should be regularly monitored for bacteriological contamination the same is not true for all chemical contaminants many of which can be identified clearly with specific circumstances which do not arise in all supplies i.e. lead from lead pipes and tanks, cyanide and chromium from industrial effluents, high pesticide levels with their use in vector control, arsenic with certain groundwaters. It is doubtful whether routine monitoring of a supply for such substances can be justified unless there is evidence that the supply is at risk.

2. SELECTION OF CONSTITUENTS

In compiling standards or guidelines for the quality of drinking water it is desirable to include any chemical constituents of widespread distribution which are known or suspected to have an effect on health in or near to the concentration range relevant to water. Although almost any toxic chemical could in theory find its way into water, the inclusion of largely irrelevant substances can cause confusion. Thus toxicity, frequency of occurrence and concentration are the most important criteria for the selection of substances for inclusion in water quality standards or guidelines. From a practical point of view it is also important that a suitable analytical method is available for the substances of interest. There have been several examples of limits having been set at levels for which there was no proven analytical technique available at all.

It is also important to establish that technology is available to meet any limits specified and to have some notion of the costs involved. The final decision as to whether or not to set a limit for any particular substance must include a judgement on the extent to which the benefit to health of a limit is justified by the cost either of remedial action or even the loss of an otherwise acceptable resource.

3. SETTING THE LIMITS

Judged relative to any other kind of food the chemical content of drinking water is minute and there are very few substances for which water represents a significant proportion of the total dietary intake. Because of their low concentration the chemical components of drinking water are unlikely to cause effects on health in the short term. Any such effects that exist are more likely to result from intake over a long period of time. In setting limits for such substances acute toxicity data is usually almost irrelevant but comprehensive information on chronic toxicity and carcinogenicity is highly desirable. In considering such information is is necessary to identify any population groups particularly at risk such as infants, pregnant women or other consumers with specific physiological defects.

Information is also required on the normal intake of the substance from different routes of exposure i.e. air, food and water. On the basis of this it is necesary to make a judgement as to the acceptable daily intake of the substance and its apportionment between the various routes of exposure and in particular to water. It is necessary also to apply a safety factor, the size of which will be determined by the strength of the toxicological evidence and the nature and magnitude of any health risk. Table II shows the criteria adopted by the US National Academy of Sciences (15) for setting safety factors. It will be noticed that a safety factor as low as 10 is only used where there is "good chronic or acute human data plus chronic or acute data in other species".

Table II. "Safety" or "Uncertainty" factors adopted by the US National Academy of Sciences

Criteria	Factor
Good chronic or acute human data + chronic or acute data in other species	10
Good chronic or acute toxicity data in one or more species	100
Limited or incomplete acute or chronic	1000

In practice the information available on the health effects and the routes of human exposure of water constituents is invariably inadequate. There is an almost total lack of hard evidence for health effects of chemicals at the concentration levels found in water and for several reasons this is not surprising.

Health effects associated with submicrogram per litre concentrations of organic compounds in drinking water, if they exist at all, are likely to be very small and to be manifested on the incidence of diseases such as cancer that are already widespread in the population. In most epidemiological studies such small increases in disease incidence due to drinking water are very difficult to perceive against a normal background incidence due to other more important factors. The considerable induction period (20-30 years) for cancer also creates enormous problems in trying to link cause and effect particularly as exposure information relating to 30 years ago for water constituents of interest today is almost non-existent.

The toxicological information available for water constituents is also defective in many respects and normally relates to levels of exposure that are orders of magnitude greater than those experienced from drinking water. Such data also invariably relates to tests with laboratory animals rather than man himself. The necessity to extrapolate from data resulting from high exposures with laboratory animals to very low exposures with man leads to considerable difficulty in assigning a limit to some water constituents.

As an illustration of the difficulty of advising on limits for water constituents on the basis of available toxicological data, the example of trichloroethylene will be considered. This is a compound which has been found to be a widespread contaminant of groundwater and which has given rise to some concern. Surveys of trichloroethylene levels in groundwater are being conducted in many parts of the world and there is clearly a need for guidance as to what would be considered a safe level.

The best published information currently available on the carcinogenicity of trichloroethylene following oral exposure is from a study undertaken by the National Cancer Institute (16). The purpose of this study was to assess the carcinogenicity of some widely used chlorinated organic compounds; it must be emphasised that the study was not designed to derive limits for the respective compounds in drinking water. The work included an investigation of the effect of trichloroethylene on the incidence of liver tumours in mice and rats. Male and female groups were exposed at two dosage levels and the proportion of animals developing liver tumours compared with the proportion in a control group. The results, summarised in Table III, typify the kind of toxicological information that may be available for a water constituent and on which some decision regarding a safe level of exposure may have to be made.

Table III. Trichloroethylene (technical) - oral carcinogenicity studies

Species	Dosage mg/kg/day*	Animals with liver tumours
MOUSE B6C3F1		
Males	2339	31/48
	1169	26/50
	0	1/20
Females	1739	11/47
	869	4/50
	0	0/20
RAT (OSBOURNE MENDEL)		
Males and Females	1097	No differences observed
	549	
	0	

* Time weighted average

With only two levels of exposure the information is obviously crude and the results clearly show differences in effect between the two species and between male and female mice. The particular figure of interest in relation to the setting of limits is the lowest level of exposure that has been demonstrated to lead to an increased incidence of tumours. In this data this is the exposure of 869 mg/kg/day in female mice and it is this figure on which any guidelines as to safe levels in drinking water for humans will be based.

Table IV. Comparison of experimental exposure in mammalian tests and human exposure from groundwater

		Trichloro-ethylene	Tetrachloro-ethylene
(a)	Lowest dose in mouse study mg/kg/day	869	368
(b)	Maxiumum concentration in groundwater µg/l	100	10
(c)	Exposure from groundwater, 60 kg/man 2 litre per day, mg/kg/day	0.004	0.0004
(d)	Ration experimental exposure: maximum exposure in practice (a/c)	217 250	920 000

In Table IV it is shown that this level of exposure is more than 200 000 times greater than that experienced through the consumption of a contaminated groundwater. For tetrachloroethylene, another groundwater contaminant of current interest, the available chronic toxicity data was obtained at a level of exposure nearly a million times greater than that experienced with contaminated groundwater. The validity of toxicity extrapolations from high to such very low exposures seems very questionable. A further problem with this data is that the technical trichloroethylene used in the mouse and rat studies almost certainly contained epichlorohydrin, a known mutagen. Thus it is not clear whether the measured incidence of liver tumours in mice reflected the concentration of trichloroethylene, of epichlorohydria or both compounds.

This example illustrates several problems in the use of available toxicological information to derive limits for water constituents. At the present time only tentative guideline values have been established for trichloroethylene and tetrachloroethylene. For some of the water constituents of interest, the toxicological data available is much better than that illustrated. It is clear however that the development of a limit on the basis of even reasonably good information will depend to a large extent on professional judgement. It is inevitable that such limits become highly controversial when they have significant cost or operational implications. The precautionary nature of some limits is emphasised where these are set, not on the basis of direct toxicological evidence at all, but on the minimum concentration detectable by analysis (e.g. chromium (11), cadmium (6)). Although such limits may seem strange they are preferable to those set below the limit of detection. There were several examples of this e.g. phenols (10) pesticides (8).

4. TYPES OF LIMIT

The difficulties and uncertainties involved in setting limits for drinking water constituents are especially great with carcinogenic substances and the best available techniques for extrapolating toxicological data for these substances can give a limit which could be out by an order of magnitude either way (7). The specification of limits in such widely used absolute terms as 'Maximum Permissible Concentration' and 'Maximum Acceptable Level' is therefore inappropriate and this has been brought out clearly in WHO Guidelines for Drinking Water Quality in which the limits have been set in terms of Guideline Values. The nature of these is defined as follows:

(a) A Guideline Value represents a concentration or number which ensures an aesthetically pleasing water and does not result in any significant risk to health of the consumer.

(b) The quality of water defined by the Guidelines for Drinking Water Quality is such that it is suitable for human consumption and for all usual domestic purposes including personal hygiene. However water of a higher quality may be required for some special purposes such as renal dialysis.

(c) A Guideline Value is to be used as a signal:

  (i) to investigate the cause when values are exceeded with a view to taking remedial action,

  (ii) to consult with authorities responsible for public health advice.

(d) Although the Guideline Values describe a quality of water acceptable for lifelong consumption, the establishment of these guidelines should not be regarded as implying that the quality of drinking-water may be degraded to the recommended level. Indeed, a continuous effort should be made to maintain drinking-water quality at the highest level of purity.

(e) The Guideline Values specified have been derived to safeguard health on the basis of life-long consumption. Short-term exposures to higher levels of chemical constituents such as might occur following an accidental spill, may be tolerated but need to be assessed on a case-by-case basis, taking into account, for example, the acute toxicity of the substance involved.

(f) Short-term excursions above the Guideline Values do not, necessarily, imply that the water is unsuitable for consumption. The amount by which, and the duration for which, any guideline value can be exceeded without affecting public health depends on the specific substance involved.

It is recommended that when a Guidelines Value is exceeded, the surveillance agency (usually the authority responsible for public health) be consulted for advice on suitableaction based upon considerations such as the intake of the substance from sources other than drinking-water (for chemical constituents), the likelihood of adverse effects, the practicality of remedial measures and other similar factors.

(g) In developing national drinking water standards based on the WHO Guidelines it will be necessary to take account of a variety of local, geographical, socio-economic, dietary and industrial conditions. This may lead to national standards that differ appreciably from the Guideline Values.

(h) In the case of radioactive substances, the term Guideline Value is used in the sense of "Reference Level" as defined by the International Commission on Radiological Protection (ICRP).

5. WHO GUIDELINES FOR DRINKING WATER QUALITY

The drinking water standards of the World Health Organization have previously been available as separate 'European' and 'International' editions but these have now been combined and completely revised as 'WHO Guidelines for Drinking Water Quality' (GDWQ). The term 'guideline' has been used instead of standards to emphasise the advisory nature of the document. The intention is that GDWQ should provide a basis for national organisations to set thir own limits taking account of local circumstances. To facilitate this GDWQ will be published in three volumes the first of which will include the Guideline Values together with brief information on their basis and their application. The second volume will set out in detail the criteria on the basis of which the Guideline Values were established while Volume 3 will deal with the application of GDWQ in developing countries.

Some important aspects of GDWQ in particular the Guideline Value concept have already been dealt with. Considerable emphasis is given in GDWQ to the overwhelming importance of microbiological aspects of water quality and the uncertainties surrounding the Guideline Values for many chemical constituents. Guideline Values are set only for well-defined chemical species. None are set for group parameters such as trihalomethanes, polyaromatic hydrocarbons or pesticides although individual members of these categories are included.

Table V. Guideline Values of chemical constituents of possible health significance

INORGANIC CONSTITUENTS	mg/l
Arsenic	0.05
Cadmium	0.005
Chromium	0.05
Cyanide	0.1
Fluoride	1.5
Lead	0.05
Mercury	0.001
Nitrate (as N)	10.0
Selenium	0.01

ORGANIC CONSTITUENTS	µg/l
Benzene	10.0
Chlorinated Alkanes and Alkenes	
Carbon tetrachloride	3.0 *
1,2-Dichloroethane	10.0
1,1-Dichloroethylene	0.3
Tetrachloroethylene	10.0 *
Trichloroethylene	30.0 *
Chlorophenols	
Pentachlorophenol	10.0
2,4,6-Trichlorophenol	10.0 (Odour Threshold Concn 0.1 µg/l)
Polynuclear Aromatic Hydrocarbons	
Benzo (a) pyrene	0.01
Trihalomethanes	
Chloroform	30.0
Pesticides	
Aldrin/Dieldrin	0.03
Chlordane	0.3
2,4 D	100.0
DDT	1.0
Heptachlor and Heptachlor Epoxide	0.1
Hexachlorobenzene	0.01
Lindane	3.0
Methoxyclor	30.0

* = Tentative Guideline Value. When available carcinogenicity data could not support a Guideline Value, but the compounds were judged to be of importance in drinking water and guidance was considered essential, a tentative guideline was set on available health-related data.

Table V sets out the Guideline Values for chemical substances of possible health significance and it will be seen that eighteen organic compounds are included of which sixteen are chloro-organics. The need to review and revise at regular intervals the Guideline Values for chemical substances is emphasised.

6. CONCLUSIONS

Despite the absence of clear cut evidence of health effects associated with the chemical contaminants of drinking water, the number of those for which limits have been set in drinking water standards is increasing rapidly. Limits are helpful in the management of drinking water quality but an understanding of their basis is essential if they are to be used properly. There are considerable uncertainties attached to the limits set for many toxic constituents and their over rigid application could lead to enormous expenditure with little benefit to public health.

The type of limit defined in the new WHO Guidelines for Drinking Water Quality enables a more pragmatic approach to the use of drinking water standards than has been possible hitherto. There is however a desperate shortage of relevant information on the long term toxic effects of trace levels of many environmental contaminants and the resources needed to provide this information would be considerable. This would seem to be a field ripe for international collaborative research.

7. ACKNOWLEDGEMENT

The author was actively involved in the WHO Task Group formed to develop the WHO Guidelines for Drinking Water Quality. Material for this paper has been drawn from Task Group papers but the opinions expressed are his own. The permission of the Water Research Centre to publish this paper is acknowledged.

REFERENCES

1. UNITED STATES TREASURY DEPARTMENT. Bacteriological Standards for Drinking Water. Public Health Report, 1914, 29, 2959-2966.

2. WORLD HEALTH ORGANIZATION. International Standards for Drinking Water. WHO, Geneva, 1958.

3. WORLD HEALTH ORGANIZATION. European Standards for Drinking Water. WHO, Geneva, 1961.

4. WORLD HEALTH ORGANIZATION. International Standards for Drinking Water. WHO, Geneva, 1963.

5. WORLD HEALTH ORGANIZATION. European Standards for Drinking Water. WHO, Geneva, 1970.

6. WORLD HEALTH ORGANIZATION. International Standards for Drinking Water, WHO, Geneva, 1971.

7. WORLD HEALTH ORGANIZATION. Guidelines for Drinking Water Quality, WHO, Geneva, (in press).

8. COUNCIL OF THE EUROPEAN COMMUNITIES. Council Directive of 15 July 1980 relating to the quality of water intended for human consumption (80/778/EEC). Official Journal of the European Communities No. L229/11, August 30 1980.

9. US PUBLIC HEALTH SERVICE. Report of Advisory Committee on Official Water Standards. Public Health Repert, 1925, 40, pp 693-721.

10. UNITED STATES PUBLIC HEALTH SERVICE. Public Health Service Drinking Water Standards, 1942, U.S. Gov. Printing Office, Washington D.C., 1943.

11. US DEPARTMENT OF HEALTH, EDUCATION AND WELFARE. Public Health Service Drinking Water Standards 1946. Public Health Report, 1946, **61**, pp 371-384 (Reissued March 1956).

12. US DEPARTMENT OF HEALTH EDUCATION AND WELFARE. Public Health Service Drinking Water Standards, 1962. US Gov. Printing Office, Washington D.C. 1962.

13. US ENVIRONMENTAL PROTECTION AGENCY. National Interim Primary Drinking Water Regulations, EDA-570/9-76-003. US Gov. Printing office, Washington D.C. 1976.

14. MINISTRY OF NATIONAL HEALTH & WELFARE. Guidelines for Canadian Drinking Water Quality, 1978. Canadian Gov. Publishing Centre, Quebec, 1978.

15. US NATIONAL ACADEMY OF SCIENCES. Drinking Water and Health Vol. 3, 1980. National Academy Press, Washington D.C.

16. WEISBURGER, ELIZABETH K. Carcinogenicity studies on halogenated hydrocarbons. Environmental Health Directives 1977, 21, pp 7-16.

INTEGRATION OF THE ENVIRONMENT RESEARCH ACTION PROGRAMME INTO THE FRAMEWORK
PROGRAMME FOR COMMUNITY SCIENTIFIC AND TECHNICAL ACTIVITIES 1984-1987

A. KLOSE, G. ANGELETTI and C. WHITE
Directorate-General for Science, Research and Development
Commission of the European Communities, Brussels

Summary

A general overview of the first Framework Programme of the European Communities for a common strategy in the field of science and technology is given. Emphasis is put on the introduction of the Environment Research Action Programme into the Framework Programme as a mean to support the implementation of the environmental policy of the European Communities.
An outline of the Third R & D Programme in the field of Environment and of its revision, to be effective from 1st January 1984, is presented. Particular stress is laid upon the follow-up programme of the Concerted Action " Analysis of Organic Micropollutants in Water ", the scientific content of which has been substantially revised and the title changed to " Organic Micropollutants in the Aquatic Environment ".

1. INTRODUCTION

The aim of this paper is to give a general description of the first Framework Programme for Community Scientific and Technical Activities and illustrate how environmental research, implemented by means of the Research Action Programme (RAP) dealing with the Environment, is integrated into it as a tool in order to improve living and working conditions.
Let us start with the Framework Programme and then later examine the integration of environmental research into it.
As an answer to requests from the Council of Ministers, the Commission of the European Communities prepared, in December 1982 and May 1983 (1, 2,) proposals for a Council Decision on the first Framework programme from 1984 to 1987.

2. FRAMEWORK PROGRAMME

This Framework Programme is designed to become, for both the Member States and the Community Institutions :
- a concertation tool and one for choosing the scientific and technical objectives to be adopted at community level as a function of the major socio-economic goals of the EEC;
- a programming tool making it possible to guide, as much in terms of orientation as in ambition, the preparation of action programmes and activities to be developed in order to achieve the selected objectives (choosing between national, international and community level action) ;
- a financial forecasting tool capable of facilitating decisions as to financial allocations for programmes, and the preparation of annual budgets.

To establish a suitable basis for the Community action strategy which should be followed during the 1990's, the Commission is proposing that a strategy of adaptation be implemented during the years 1984 to 1987, with the aim of reorienting, developing and supplementing the range of current Community actions.

Implementation of this strategy of adaptation is clearly needed, in the Commission's view, in order to tackle the situation in which the Community finds itself in the international context :
- the similarity of the major problems with which the Member States are faced (the economic crisis, the need to strengthen competitiveness);
- excessive duplication of the research carried out at various levels and with differing resources by the different Member States. Wasteful repetition of this kind has the effect of limiting Europe's competitive capacity;
- the often unacceptable level of dependence upon non Member States; (for example vegetable proteins, raw materials, American embargo upon Euro-Siberian gas pipeline);
- the progressive deterioration in the technological balance both North/North and North/South;
- the gaps of the inadequacies as between scientific "supply", that is the products of research, and socio-economic "demand".

In the light of these observations and on the basis of the socio-economic goals which have been adopted (see table I), the Commission has set out a range of major scientific and technical objectives to be pursued at Community level, and has identified the priorities which should be allocated to them in terms of financial implications. Within this overall framework the Commission has specified, goal by goal, the specific objectives which should be attained and the relative weighting to be given to them. This forms a Community programming guide which should serve as a basis for the preparation of action programmes during the period 1984-1987.

The adoption of this Framework Programme should thus make it possible to check the extent to which the agreed ambitions of the Community institutions are being progressively realised.

The strategic approach which is suggested, the overall balance which is being aimed at, the financial implications expressed in absolute values and the relative weighting proposed by the Commission in the Framework Programme are listed in table I.

The Commission believes that these figures are not at all overambitious in respect of the needs which have to be satisfied and the lost ground which has to be made up so far as competitiveness with the Community's major trading partners is concerned.

This proposal is a commitment by Europe of a new intensity, comparable to what which the Member States decided upon and followed through during the 1960's. That was the period when J.J. Servan Schreiber wrote his " Défi Américain " and Europe tried to bridge the " technological gap " with U.S.A. and Japan. In this period (1968) Community Research and Development (R & D) activities corresponded to 2,5 % of the total public expenditure by Member States in this field whilst in 1982 (when it was 590 Mio ECUS) Community Research, Development and Demonstration (R, D & D) corresponded to only 2.2 % of that expenditure.

In the Community everything depends on decisions made by the Member States.

So this strategy presupposes common agreement and an explicit choice of significant objectives which are of joint interest and which correspond to the needs of the Member States.

Table I

Scientific and technical objectives and the amounts considered necessary to achieve them

	1984 - 1987	
	Mio ECUs [1]	%
1. Promoting agricultural competitiveness	130	3,5
- developing agricultural productivity and improving products : agriculture	115	
fisheries	15	
2. Promoting industrial competitiveness	1.060	28,2
- removing and reducing impediments	30	
- new techniques and products for the conventional industries	350	
- new technologies	680	
3. Improving the management of raw materials	80	2,1
4. Improving the management of energy resources	1.770	47,2
- developing nuclear fission energy	460	
- controlled thermonuclear fusion	480	
- developing renewable energy sources	310	
- rational use of energy	520	
5. Reinforcing development aid	150	4,0
6. Improving living and working conditions	385	10,3
- improving safety and protecting health	190	
- protecting the environment	195	
7. Improving the efficacy of the Community's scientific and technical potential	85	2,3[2]
- Horizontal activities	90	2,4
	3.750	100,0

1. : in ECUs at 1982 constant values
2. : corresponds to 5 % by the end of the period.

It is not necessary to tell you that the world and especially Europe, is undergoing a period of crisis and is facing serious and difficult social and economic problems, notably unemployment. This leads to financial difficulties for the governments and their immediate reaction is to cut down R + D budgets.

It is the Commission's opinion that since R, D & D is <u>one of the most productive of investments in the long and medium term</u>, the financial outlay to be accorded to it needs not only to be significantly increased but also guaranteed against pressures of circumstances that might bear upon the redistribution of Community resources.

The adoption of the Framework Programme - expressing as it does the joint wish to attain major priority goals, the scope of effort to be undertaken and the relative weighting to be accorded to specific objectives - should make it possible both to facilitate individual programme decisions and to maintain the overall balance which is considered desirable.

The Council of the European Communities recently resolved (3) that a common scientific and technical strategy be developed along the lines given above, that it should be expressed in the form of Framework Programmes to be prepared by the Commission and put before the Council and the Parliament, that within these Framework Programmes the Commission should prepare specific Action Programmes, and that it was necessary to increase expenditure upon R, D & D. (The financial figures shown in table I were accepted as an indication of the order of magnitude of the increase in expenditure required). At the same time the Council approved the scientific and technical objectives which are contained within the first Framework Programme and agreed that this be reexamined, by 1985 at the latest, on the basis of experience gained during the first phase of implementation.

3. INTEGRATION OF ENVIRONMENTAL RESEARCH INTO THE FRAMEWORK PROGRAMME

In the introduction we discussed the Framework Programme as a Programming tool for the preparation of Research Action Programmes.

A research action programme consists of a number of research and development activities which are related to each other in that each activity contributes to a common overall programme aim. The common aim of the activities which make up the research action programme may be of a scientific or a policy nature.

The Third R & D Programme in the field of the Environment (Environmental Protection and Climatology) - Indirect and Concerted Actions - 1981 to 1985 (Scientific and technical content in table II), has been proposed to be revised by the Commission. A decision of the Council of Ministers is expected for December 1983. This revision will be part of the Research Action Programme (RAP) " Environment ", which, besides the environmental protection and climatology, contains two other sub-programmes : industrial risk and remote sensing.

It has been shown before that one of the scientific and technical objectives of the Framework Programme, the protection of the environment, is related to the Community goal for the improvement of living and working conditions.

Apart from the aim of introducing the environmental dimension into other policies, the Environment RAP will contribute to various other objectives identified in the Framework Programme and it will be relevant not only to the two objectives " protecting the environment " and " improving safety and protecting health ", but especially to :

Table II
Scientific and technical content of the third R & D Environment Programme 1981-1985.

Sub-Programme I : Environment protection

	Indirect action (contracts and coordination)	Concerted action
Research area 1 : Sources, pathways and effects of pollutants		
1.1. Heavy metals	x	–
1.2. Organic micro-pollutants and new chemical products	x	(1) Analysis of organic micro-pollutants in water (COST 64b bis) (until 31 December 1983)
1.3. Asbestos and other fibres	x	–
1.4. Air quality	x	(2) Physico-chemical behaviour of atmospheric pollutants (COST 61a bis)(until 31 December 1983)
1.5. Surface and underground freshwater quality	x	–
1.6. Thermal pollution	x	–
1.7. Marine environment quality	x	(3) Benthic coastal ecology (COST 47)
1.8. Noise pollution	x	–
Research area 2 : Reduction and prevention of pollution and nuisances		
2.1. Sewage sludge	–	(4) Treatment and use of sewage sludge (COST 68 bis) (until 31 December 1983)
2.2. Pollution abatement technologies	x	–
2.3. Clean technologies	x	–
2.4. Ecological effects of solid waste disposal	x	–
2.5. Oil pollution cleaning techniques	x	–
2.6. Impact of new technologies	x	–
Research area 3 : Protection, conservation and management of natural environments		
3.1. Ecosystems studies	x	–
3.2. Biogeochemical cycles	x	–
3.3. Ecosystems conservation	x	–
3.4. Bird protection	x	x
3.5. Reclamation of damaged ecosystems	x	–

Table II - cont.

	Indirect action (contracts and coordination)	Concerted action
Research area 4 : Environment information management		
4.1. Data bank on environmental chemicals	x	-
4.2. Evaluation, storage and exploration of data	x	-
4.3. Ecological cartography	x	-
Research area 5 : Complex interactive systems : man-environment interactions	x	-

Sub-Programme II : Climatology (indirect action)

Research area 1 : Understanding climate

1.1. Reconstruction of past climates.
 Exploration and analysis of :
 (a) natural records;
 (b) observational and other historical records.

1.2. Climate modelling and prediction.
 Investigations to improve models which are capable of simulating climate, especially by including the slowly-varying components of the climatic system, and of assessing climate predictability on time and space scales that are of interest to the Community.

Research area 2 : Man-climate interactions

2.1. Climate variability and European resources :

 (a) impact on agricultural and water resources;
 (b) climatic hazards evaluation;
 (c) impact on energy requirements, use and production.

2.2. Man's influence on climate :

 (a) chemical pollution of the atmosphere, with special emphasis on carbon dioxide accumulation;
 (b) release of energy.

- Developing agricultural productivity and improving the quality of products.
- Improving and developing new techniques and new products for the traditional industries.
- Rational use of energy.
- Improving the management of raw materials.
- Reinforcing development aid.

Table III gives an idea of the structure and funding estimated to be necessary for the implementation of the Environment RAP. This implementation is done in three ways : direct action (Joint Research Centre), shared cost contracts research and concerted action with the participation of interested Non-Member Countries.

Table III

Environment Research Action Programme
Structure and funding, Mio ECU

Subprogramme	Revised Programme (2)			Existing Programme		
	Contract Research 1981-85	Concerted Actions 1981-85	JRC(1) 1984-87	Contract Research 1981-85	Concerted Actions 1981-83	JRC 1980-83
Environmental protection	43.3	3.2	48.2	33.0	1.0	37.5
Climatology	8.0	-	-	8.0	-	-
Remote Sensing	-	-	29.0	-	-	20.5
Industrial Risk	-	-	21.5	-	-	-

(1) 1983 ECU's   (2) Amounts proposed by the Commission

Community research activities in the field of the environment are intended to provide scientific support for the implementation of the environmental policy of the European Communities.

The revision of the Third R & D Environment Programme takes into account the items which have been recommended as priorities in the Third Environment Action Programme 1982-1986 (Council Resolution of 7 February 1983, 4) in particular :
- environmental chemicals
- atmospheric pollution
- fresh water and marine pollution
- pollution of the soil
- transfrontier pollution
- waste management
- development of clean technologies.

The scientific content of the contract research remains essentially unchanged, with emphasis on ecological effects of air pollutants (" acid rain ") and management, recycling and disposal of toxic and dangerous waste.

Three Concerted Actions "Physico-Chemical behaviour of atmospheric pollutants", "Analysis of Organic Micropollutants in Water", "Treatment and use of sewage sludge", ending on 31 December 1983, are proposed to be extended in a revised form. Three new concerted actions are proposed dealing with :
- Effects of Air pollutants on Terrestrial and Aquatic Ecosystems.
- Indoor Air Quality.
- Bird Protection.

Of special interest for participants in this symposium is of course the follow-up programme (to be effective from 1st January 1984) of the Concerted Action " Analysis of Organic Micropollutants in Water ". A detailed description of the revision of this particular project will be the subject of the next lecture (A. Bjørseth and H. Ott).

Nevertheless, it can be mentioned that, taking into account the fact that instrument development has already reached a phase permitting detailed and reliable analysis of environmental samples, more emphasis has to be given in future to the practical application of analytical methods to the real environmental problems, extending the research to new topics such as :
- the physico-chemical behaviour of organic micropollutants in the aquatic environment,
- the transformation reactions in such environment,
- the behaviour and transformation of organic micropollutants in water treatment processes.

Consequently, the title of the Project has been changed to "Organic Micropollutants in the Aquatic Environment".

In conclusion, the aim of this paper was to demonstrate the Community's intentions towards future activities in the field of environmental research by means of the Environment Research Action Programme. We have tried to illustrate particularly the ways and means by which these activities, the subject of this conference, will continue to be carried out in the coming years.

REFERENCES

1. Commission of the European Communities (1982).
   Proposals for a European Scientific and Technical Strategy -
   Framework Programme 1984-1987, 21 December 1982, COM (82) 865.

2. Commission of the European Communities (1983).
   Proposal for a Council Decision on the Framework Programme for
   Community Scientific and technical activities 1984-1987.
   17 May 1983, COM (83) 260.

3. Council of the European Communities (1983).
   Council Resolution (dated 25 July 1983) on Framework Programmes for
   Community Research, Development and Demonstration activities and
   first Framework Programme 1984 to 1987.
   O.J. n° C 208 of 4 August 1983.

4. Council of the European Communities (1983).
   Council Resolution of 7 February 1983 on the third Environment
   Action Programme 1982-1986.
   O.J. n° C 46 of 17 February 1983.

COST 64b bis PROJECT - NOW AND THE FUTURE

A. BJØRSETH and H. OTT
Central Institute for Industrial Research
Oslo 3, Norway
and
Commission of the European Communities
DG XII/G-1, Brussels, Belgium

SUMMARY

This paper summarizes the achievements of the present COST 64b project and outlines the technical framework for a follow-up project. In particular, the selection of topics for the new project is discussed. The main activity of the new project will concentrate on behaviour and transformations of organic pollutants in the aquatic environment and in water treatment processes. There will also be work directed towards basic analytical techniques. However, in this area, the main emphasis will be put on specific analytical methods for selected compounds and for groups of compounds. The work will be implemented through working parties and activity centers with defined responsibilities in the mentioned areas.

1. INTRODUCTION

This paper has a dual purpose. First it summarizes the achievements of the COST 64b project. Secondly gives some idea how it is intended to carry this work further.

A follow-up project naturally has to be based on the results of the COST 64b project and to take new research requirements into account. The present symposium was organized partly to point to the most recent achievements of the COST 64b project. It will also be the basis for a final report which will give accountancy to the governments of the participating countries and to the scientific community of the achievements.

The main objectives of the original program were :

- development and assessment of equipment and methods for the identification and quantitative determination of organic micropollutants, present in all types of water, excluding toxicological investigation

- elaboration and collection of data on the characteristics of such pollutants (e.g. chromatographic data, mass spectra), necessary as reference for the identification of unknown pollutants

- collection of information on polluting substances which have actually been identified in various waters (inventory of pollutants).

This paper will briefly describe the technical approach of the COST 64b project in order to meet these objectives, using this as a basis for outlining the future work in the area of organic micropollutants in water.

.../...

2. WHAT HAS BEEN ACHIEVED ?

The program was initiated in 1972 and has made a significant contribution to the rapid growth of competence in this research area among the member countries as well as the participating non-member countries. There are several achievements worth mentioning, particularly concerning development of separation techniques.

When the program started, gas chromatography was already a rather developed technique. As regards the separation by gas chromatography, two main lines of separation were followed. The separating power of packed columns was considerably increased. Furthermore, capillary column technology, which was introduced in the early years of COST 64b and with a strong input from some of the cooperating laboratories, was developed to a very high standard. During this period, the coupling of capillary gas chromatographs with mass spectrometers and the performance of mass spectrometers was considerably improved. Also computers simplified the identification of the compounds analyzed and are now available in any high standard laboratory. The laboratories collaborating in COST 64b made substantial contributions to these developments, a contribution which is recognized within the scientific community. After the presentations made earlier in this symposium, there should be no need to go into detail here.

We also know that there are certain disadvantages in gas chromatography. Only fairly volatile compounds can be analyzed by this technique although the definition of volatility is extended every year. Furthermore, the possibility to form volatile derivatives of highly polar compounds is limited. For many polar and thermolabil compounds, gas chromatography is not a suitable choice.

However, at the beginning of the project COST 64b the technique of high pressure liquid chromatography was introduced. Also in this area the contributions of COST 64b to available knowledge are remarkable. The separation techniques are now at a very high standard and the separative power of newly developed columns is remarkable.

The remaining problem of high pressure liquid chromatography, the lack of a universal detector, has not yet been overcome. The possibilities for constructing such detectors have limits which seem to be difficult to overcome. However, the combination of high performance liquid chromatography and gas chromatography is a powerful combination, as discussed previously at this symposium.

In addition to the analytical techniques in the strict sense, substantial work has been done on sampling and sample preparation. For the introduction to this topic we refer to the review papers given at this symposium and at the two preceding symposia (1,2). A substantial effort has also been made on disseminating knowledge of substances which have been identified in various waters and to compile these into a data base. Furthermore, collections of mass spectra both in hard copies and in computerized form have been established, and considerable efforts have been made in the development of software for handling such spectra.

In the late seventies the scope of project COST 64b was enlarged to incorporate also specific analytical methods for a number of selected compounds and groups of compounds. These groups include petroleum hydrocarbons, organic halogen, phenols and aromatic compounds. In our opinion, this symposium has given a fair account of these achievements.

.../...

3. THE CURRENT SITUATION

After more than 10 years of coordinated research, it is time to review and evaluate the results, particularly in view of the extension of the project.

Separation techniques are not yet fully developed and further improvements are still possible. However, it is not likely that the coordinated effort in this area can contribute to the same extent to further developments as it has done in the past. Manufacturers of analytical instrumentation have supplied equipment for most requirements, which are sophisticated and of high quality. Research is carried out further in industry to a much greater extent than previously. For the environmental chemist the time has therefore come to go more into the analysis of individual compounds and groups of compounds which have to be analysed. The reason for this is that legislators gradually increase the requirements for monitoring of waterpollutions. However, monitoring is only a part of the task. It is also important to develop and apply general concepts on the behaviour and fate of pollutant and to use these concepts as an aid in procuring relevant analytical data. These arguments are in favour of a thorough revision of project COST 64b (or the design of an entirely new project). This revision goes hand in hand with a revision of the environment research program of the Commission of the European Communities, effective as of 1984. The new program is designed for an initial period of two years. This may be called an experimental period, and it is intended to come up with a proposal for a final project in 1985. We intend to use the forthcoming two years as a trial period for a long term program for the late eighties and the beginning of the nineties.

4. FUTURE WORK

A new project has been proposed with the tentative title "Organic micropollutants in the aquatic environment". There are four main lines of the new program. The order of priority is not necessarily the order in which it is presented here.

1) Basic Analytical Techniques

This topic is to a certain degree an extension of the ongoing COST 64b. It should be limited to an evaluation of techniques rather than to design of new basic techniques. This also means that we will give priority to the practical application of the various analytical methods. These techniques include gas chromatography, high pressure liquid chromatography and mass spectrometry but also new techniques as, for instance, MSMS. It is clear that also the activities on sampling and sample treatment will have to go on because there is in this area a strong need for further research.

The main emphasis in this area, however, will be put on specific analytical methods for selected compounds and for groups of compounds. In order to avoid too much dispersion, a number of compounds to which the main interest should be focussed, has to be identified. First, there are the compounds which are likely to be regulated by the European Communities in the context of the Council Directive of discharge of toxic and dangerous compounds into surface water, recently communicated by the Commission to the Council. From this list, we will select some groups of important compounds which are suitable for such a coordination of European efforts.

In addition to this list we propose focussing on the following groups of chemical pollutants :

.../...

Chlorinated paraffins. These compounds constitute a difficult analytical problem. What we find in substantial concentrations in the environment and in many organisms is a mixture of short and long chained compounds which are chlorinated to various degrees.

Tensides. We will select a number of tensides or individual surface active compounds in order to elaborate good analytical methods for their detection and quantitative determination. These compounds include both cation-active, anion-active and neutral surfactants.

Optical brightners. Attemps will be made to elaborate good and reliable methods for the detection of optical brightners. These compounds represent serious analytical problems since they are difficult to isolate and purify from water samples.

Metalorganic compounds. We will also work with the problem of characterization of metalorganic compounds. Several of these are a major concern from a health point of view and there is a need for specific analytical techniques for their determination.

The ongoing activities on the collection and treatment of analytical data will of course continue, but we shall give more careful thought as to how to organize this initiative.

The other three topics are all designed to apply analytical techniques to environmental problems and hopefully to use the expertise acquired to contribute to a solution of these environmental problems.

2) Behaviour of Organic Micropollutants in the Aquatic Environment.

This topic concerns investigations of mechanisms for the distribution and transport of pollutants in the aquatic environment to assess the bioavailability and the potential for bioaccumulation or biomagnification. These studies should also include an examination of structure/activity relationships. We will select a number of indicator compounds of various characteristics and trace their way from the source to the sinks. Furthermore, we will try to relate the behaviour of these compounds to their chemical structures.

3) Transformation Reactions in the Aquatic Environment

The fate of chemicals in the environment is strongly influenced by possible transformation reactions. There are simple cases like hydrolysis of reactive compounds, there are also more complex cases of chemical interaction with environmental constituents which may happen in part in the atmosphere, but the degradation products are brought back to the aquatic environment. Under this heading explorations should also be made as to what extent photochemical reactions occur in the aquatic environment. Last, but not least, there is the complex problem of biological transformation as a consequence of the metabolic activities of aquatic organisms ranging from bacteria to higher animals.

4) Behaviour and Transformation of Organic Micropollutant in Water Treatment Processes

Under this heading we will particularly study the fate of organic micropollutants during water treatment processes. Special attention will be paid to the development of an integrated approach to cover sampling and analytical methodologies in order to better investigate the fate of pollutants in relevant processes like infiltration, waste water and drinking water treatment.

In this case, however, the overlap with other COST projects has to be discussed before we specify further work.

.../...

5. CONCLUSION

The Council of Ministers of the European Communities has been presented with a formal proposal along the lines described above, and the negotiations with non-member states will be initiated soon. We are confident that the proposal will be endorsed as it has already been thoroughly discussed by the Community-COST Concertation Committee.

The new project will require an adaptation of the structures, and new groups of researchers will have to be sought.

We do hope to continue the fruitful cooperation that has been established in COST 64b, together with scientists from other disciplines, in order to cope with the extended scope of the new project.

REFERENCES

1. Proceedings of the First European Symposium on "Analysis of Organic Micropollutants in Water", held in Berlin, 11-13 December 1979.
   (OMP/29/82 - XII/ENV/17/82)

2. Proceedings of the Second European Symposium on "Analysis of Organic Micropollutants in Water", held in Killarney (Ireland), 17-19 November 1981, published by D. Reidel Publishing Company, edited by A. Bjørseth and G. Angeletti.
   (EUR 7623)

LIST OF PARTICIPANTS

AHEL, M.
 Institute for Water Resources
 and Water Pollution Control
 (EAWAG)
 CH - 8600 DUEBENDORF

ANGELETTI, G.
 Commission of the European
 Communities
 Directorate-General "Science,
 Research and Development"
 200, rue de la Loi
 B - 1049 BRUSSELS

BECHER, G.
 Central Institute for
 Industrial Research
 Forskningsv. 1
 P.B. 350 - Blindern
 N - OSLO 3

BENESTAD, C.
 Central Institute for
 Industrial Research
 Forskningsv. 1
 P.B. 350 - Blindern
 N - OSLO 3

BJØRSETH, A.
 Central Institute for
 Industrial Research
 Forskningsv. 1
 P.B. 350 - Blindern
 N - OSLO 3

BØLER, J.
 Central Institute for
 Industrial Research
 Forskningsv. 1
 P.B. 350 - Blindern
 N - OSLO 3

BOTTA, D.
 Politecnico - Istituto di
 Chimica Industriale
 Piazza Leonardo da Vinci, 32
 I - 20133 MILANO

BRENER, L.
 Société Degrémont
 183, ave du 18 juin 1940
 F - 92508 RUEIL MALMAISON

BRINKMANN, F.
 Rijksinstituut voor de Drink-
 watervoorziening
 Voorburg, Nieuwe Havenstraat 6
 NL - 2260 AD LEIDSCHENDAM

BRUCHET, A.
 Laboratoire Central
 Société Lyonnaise des Eaux
 38, rue du Président Wilson
 F - 78230 LE PECQ

BUECHERT, A.
 National Food Institute
 19 Moerkhoej Bygade
 DK - 2860 SOEBORG

CABRIDENC, R.
 I.R.C.H.A.
 B.P. no 1
 F - 91710 VERT-LE-PETIT

CARLBERG, G.
 Central Institute for
 Industrial Research
 Forskningsv. 1
 P.B. 350 - Blindern
 N - OSLO 3

DAASVATN, K.
 University of Oslo
 Dept. of Chemistry
 P.B. 1033 - Blindern
 N - OSLO 3

DRANGSHOLT, H.
 Central Institute for
 Industrial Research
 P.B. 350 - Blindern
 N - OSLO 3

DREVENKAR, V.
　Institute for Medical Research
　and Occupational Health
　Mose Pijade 158
　YU - 41000 ZAGREB

EADES, J.
　An Foras Talunais
　The Agricultural Institute
　Oak Park Research Centre
　IRL - CARLOW

EISEN, L.
　Commission of the European
　Communities
　Directorate-General "Information
　Market and Innovation"
　P.O.B. 1907
　L - 2920 LUXEMBOURG

FERRAND, R.
　Cerchar
　B.P. no 2
　F - 60550 VERNEUIL

FIELDING, M.
　Water Research Centre
　Henley Road
　UK - MEDMENHAM, Marlow
　　　Bucks SL7 2HD

FLANAGAN, P.
　An Foras Forbartha
　St. Martin's House
　Waterloo Road
　IRL - DUBLIN 4

FOLKE, J.
　Water Quality Institute
　11 Agern Alle
　DK - 2970 HORSHOLM

GAMES, D.E.
　University College
　P.O. Box 78
　UK - CARDIFF CF 1 IXL

GIGER, W.
　Institute for Water Resources
　and Water Pollution Control
　(EAWAG)
　CH - 8600 DUEBENDORF

GJØS, N.
　Central Institute for
　Industrial Research
　Forskningsv. 1
　P.B. 350 - Blindern
　N - OSLO 3

GONZALES-NICOLAS, J.
　Centro de Estudios Hidrograficos
　Paseo Bajo Virgen del Puerto, 3
　E - MADRID 5

GROB, K.
　Institute for Water Resources
　and Water Pollution Control
　(EAWAG)
　CH - 8600 DUEBENDORF

GROLL, P.
　Nuclear Research Center Karlsruhe
　Institut for Hot Chemistry
　Postfach 3640
　D - 7500 KARLSRUHE

HANSEN, N.
　Levnedsmiddelkontrollen I/S
　1, Dyregaardsvej
　DK - 2740 SKOVLUNDE

HENNEQUIN, C.
　I.R.C.H.A.
　B.P. no 1
　F - 91710 VERT-LE-PETIT

HENSCHEL, P.
　Umweltbundesamt
　Bismarckplatz 1
　D - 1000 BERLIN 33

HUNT, D.
　Department of Chemistry
　University of Virginia
　USA - VIRGINIA,
　　　Charlotteville 22901

JOSEFSSON, B.
　Department of Analytical
　and Marine Chemistry
　Chalmers University of Technology
　and University of Göteborg
　S - 412 96 GOETEBORG

KARRENBROCK, F.
　E.S.W.E. - Institut für Wasser-
　forschung und Wassertechnologie
　Söhnleinstr. 158
　D - 6200 WIESBADEN

KLOSE, A.
  Commission of the European
  Communities
  Directorate-General "Science,
  Research and Development"
  200, rue de la Loi
  B - 1049 BRUSSELS

KLUNGSØYR, J.
  Institute of Marine
  Research
  P.O. Box 1870
  N - 5011 BERGEN

KNOEPPEL, H.
  Commission of the European
  Communities
  Joint Research Centre
  I - 21020 ISPRA (Varese)

KRAAK, J.
  Lab. voor Analytische Scheikunde
  Universiteit van Amsterdam
  Nieuwe Achtergracht 166
  NL - 1018 WV AMSTERDAM

KRINGSTAD, A.
  Central Institute for
  Industrial Research
  Forskningsv. 1
  P.B. 350 - Blindern
  N - OSLO 3

KRISTENSEN, K.
  Commission of the European
  Communities
  Directorate-General "Science,
  Research and Development"
  200, rue de la Loi
  B - 1049 BRUSSELS

KUEHN, W.
  DVGW-Forschungsstelle am Engler-
  Bunte-Institute - Wasserchemie
  Richard-Willstätter-Allee 5
  D - 7500 KARLSRUHE

KVESETH, K.
  Central Institute for
  Industrial Research
  Forskningsv. 1
  P.B. 350 - Blindern
  N - OSLO 3

LA NOCE, T.
  Istituto di Ricerca sulle Acque
  C.N.R.
  Via Reno 1
  I - 00198 ROMA

LANDMARK, V.
  Central Institute for
  Industrial Research
  Forskningsv. 1
  P.B. 350 - Blindern
  N - OSLO 3

LARSSON, M.
  Department of Analytical
  and Marine Chemistry
  Chalmers University of Technology
  and University of Göteborg
  S - 412 96 GOETEBORG

LEVSEN, K.
  Institut für Physikalische
  Chemie
  Wegelerstr. 12
  D - 5300 BONN 1

LIBERATORI, A.
  Istituto di Ricerca sulle Acque
  C.N.R.
  Via Reno 1
  I - 00198 ROMA

LICHTENTHALER, R.
  Central Institute for
  Industrial Research
  Forskningsv. 1
  P.B. 350 - Blindern
  N - OSLO 3

LUND, U.
  V.K.I.
  Water Quality Institute
  11 Agern Alle
  DK - 2970 HORSHOLM

MAILAHN, W.
  Institut für Wasser-, Boden-
  und Lufthygiene des Bundes-
  gesundheitsamtes
  Correnplatz 1
  D - 1000 BERLIN 33

MANTICA, E.
　Politecnico - Istituto di
　Chimica Industriale
　Piazza Leonardo da Vinci, 32
　I - 20133 MILANO

MARTINSEN, K.
　Central Institute for
　Industrial Research
　Forskningsv. 1
　P.B. 350 - Blindern
　N - OSLO 3

MASSOT, R.
　S.E.A. - C.E.N.G.
　B.P. 85 X
　F - 38041 - GRENOBLE Cedex

MEIJERS, A.P.
　W.R.K.
　P.O.B. 10
　NL - 3430 AA NIEUWEGEN

MIMICOS, N.
　Nuclear Research
　Center "Democritos"
　Chemistry Division
　Aghia Paraskevi Attikis
　GR - ATHENS

MONARCA, S.
　Institute of Hygiene
　Fac. Pharmacy
　Via del Giochetto
　I - 06100 PERUGIA

NEUMAYR, V.
　Institut für Wasser-, Boden-
　und Lufthygiene des Bundes-
　gesundheitsamtes - Aussenstelle
　Langen Voltastr. 10
　D - 6070 LANGEN

O'DONNELL, C.
　An Foras Forbartha
　St. Martin's House
　Waterloo Road
　IRL - DUBLIN 4

OSVIK, A.
　Central Institute for
　Industrial Research
　Forskningsv. 1
　P.B. 350 - Blindern
　N - OSLO 3

PACKHAM, R.F.
　Water Research Centre
　Henley Road
　UK - MEDMENHAM,
　　　Marlow Bucks SL7 2HD

PICER, M.
　Center for Marine Research
　Institute "Rudjer Boskovic"
　Bijenicka, 54
　YU - ZAGREB

PIET, G.
　National Institute for
　Water Supply
　P.O. Box 150
　NL - 2260 AD LEIDSCHENDAM

QUAGHEBEUR, D.
　Instituut voor Hygiëne en
　Epidemiologie
　Ministry of Public Health
　Juliette Wytsmanstraat 14
　B - 1050 BRUSSELS

RAMDAHL, T.
　Central Institute for
　Industrial Research
　Forskningsv. 1
　P.B. 350 - Blindern
　N - OSLO 3

RENBERG, L.
　National Swedish Environment
　Protection Board
　Special Analytical Laboratory
　Wallenberg Laboratory
　S - 10691 STOCKHOLM

RIVERA ARANDA, J.
　Institut Quimica
　Bio-Organica (C.S.I.C.)
　c/ Jorge Girona Salgado S/M
　E - BARCELONA 34

SANDRA, P.
　University of Ghent
　Laboratory of Organic Chemistry
　Krijgslaan 271 S4
　B - 9000 GHENT

SCHMITZ, W.
　Landesanstalt für Umweltschutz
　Hebelstr. 2
　D - 7500 KARLSRUHE

SCHNITZLER, M.
 DVGW-Forschungstelle am Engler-
 Bunte-Institute - Wasserchemie
 Richard-Willstätter-Allee 5
 D - 7500 KARLSRUHE 1

SCHWARZENBACH, R.P.
 Institute for Water Resources and
 Water Pollution Control (EAWAG)
 CH - 8600 DUEBENDORF

SKIDA, A.
 I.R.C.H.A.
 B.P. no 1
 F - 91710 VERT-LE-PETIT

SLETTEN, T.
 Central Institute for
 Industrial Research
 Forskningsv. 1
 P.B. 350 - Blindern
 N - OSLO 3

SPORSTØL, S.
 Central Institute for
 Industrial Research
 Forskningsv. 1
 P.B. 350 - Blindern
 N - OSLO 3

STIEGLITZ, L.
 Nuclear Research Center
 I.H.C.H.
 Postfach 3640
 D - 7500 KARLSRUHE

STUMM, W.
 Institute for Water Resources and
 Water Pollution Control (EAWAG)
 CH - 8600 DUEBENDORF

SUNDSTROEM, G.
 National Swedish Environment
 Protection Board
 Special Analytical Laboratory
 Wallenberg Laboratory
 S - 10691 STOCKHOLM

TERMONIA, M.
 I.R.C. - I.S.O.
 Ministry of Agriculture
 Museumlaan 5
 B - 1980 TERVUREN

TOIVANEN, E.
 Helsinki City Waterworks
 Kuninkaantàmmentie 11
 SF - 00430 HELSINKI 43

TVETEN, G.
 Central Institute for
 Industrial Research
 Forskningsv. 1
 P.B. 350 - Blindern
 N - OSLO 3

URDAL, K.
 Central Institute for
 Industrial Research
 Forskningsv. 1
 P.B. 350 - Blindern
 N - OSLO 3

VEENENDAAL, G.
 The Netherlands Waterworks'
 Testing and Research Institute
 Sir W. Churchilllaan 273
 NL - 2280 AB RIJSWIJK

VINCENT, R.
 Thames Water
 177 Roseberg Avenue
 UK - LONDON EC1R 4TP

WAGGOTT, A.
 Water Research Centre
 Elder Way
 UK - STEVENAGE, Herts. SG1 1TH

WATTS, C.
 Water Research Centre
 Henley Road
 UK - MEDMENHAM,
 Marlow Bucks SL7 2HD

WEGMAN, R.
 National Institute of
 Public Health
 P.O. Box 1
 NL - 3720 BA BILTHOVEN

WESTRHEIM, K.
 Institute of Marine
 Research
 P.O. Box 1870
 N - 5011 BERGEN

WILHELMSEN, S.
　Institute of Marine
　Research
　P.O. Box 1870
　N - 5011 BERGEN

XIE, T.M.
　Department of Analytical and
　Marine Chemistry
　Chalmers University of Technology
　and University of Göteborg
　S - 412 96 GOETEBORG

INDEX OF AUTHORS

AHEL, M., 91, 280
ANGELETTI, G., 320
BIRKLUND, J., 242
BJØRSETH, A., 328
BOERBOOM, A.J.H., 132
BOTTA, D., 261
BRINKMANN, F.J.J., 3
BRUCHET, A., 27
BUECHERT, A., 173
CARLBERG, G.E., 276
CASTELLANI PIRRI, L., 261
COGNET, L., 27
CONNOR, K.J., 153
CRATHORNE, B., 120
DAEHLING, P., 132
DAVID, F., 169
DOURTE, P., 162
DREVENKAR, V., 289
FARR, J.A., 234
FIELDING, M., 120
FOLKE, J., 242
FROEBE, Z., 289
GAMES, D.E., 68
GIGER, W., 91, 280
GLENYS FOSTER, M., 68
GROB, K., 43
GROLL, P., 77
HABERER, K., 179
HARVEY, T.M., 53
HOCENSKI, V., 301
HUNT, D.F., 53
KARRENBROCK, F., 179
KISTEMAKER, P.G., 132
KJAER SØRENSEN, A., 234, 242
KLOSE, A., 320
KRAAK, J.C., 110
KRINGSTAD, A., 276
KUEHN, W., 191
KUNINGAS, I., 255
LAINE, S., 255
LEVSEN, K., 132

LICHTENTHALER, R.G., 225
LUND, U., 234, 242
MALLEVIALLE, J., 27
MANTICA, E., 261
McDOWALL, M.A., 68
McLUCKEY, S.A., 132
MERESZ, O., 68
MOLNAR-KUBICA, E., 280
MONSEUR, X., 162
NEUMAYR, V., 5
O'DONNEL, C., 36
OTT, H., 328
PACKHAM, R.F., 309
PICER, M., 301
PICER, N., 301
RENBERG, L., 214
RIVERA, J., 294
ROELLGEN, F.W., 132
ROTH, W., 141
SANDRA, P., 84, 165, 169
SCHAFFNER, C., 91, 280
SCHMITZ, W., 15
SCHNEIDER, E., 132
SCHNITZLER, M., 191
SHABANOWITZ, J., 53
SPORSTØL, S., 147
STEEL, C.P., 120
STENGL, B., 289
STIEGLITZ, L, 141
TERMONIA, M., 162
TKALCEVIC, B., 289
TOIVANEN, E., 255
URDAL, K., 147
VEENENDAAL, G., 205
VENTURA, F., 294
VERZELE, M., 169
WAGGOTT, A., 153
WALRAVENS, J., 162
WATTS, C.D., 120
WHITE, C., 320